RF and Microwave Modeling and Measurement Techniques for Field Effect Transistors

RF and Microwave Modeling and Measurement Techniques for Compound Field Effect Transistors

Jianjun Gao

SciTech Publishing, Inc.

Published by SciTech Publishing, Inc.
911 Paverstone Drive, Suite B
Raleigh, NC 27615
(919) 847-2434, fax (919) 847-2568
scitechpublishing.com

Editor: Dudley R. Kay
Production Manager: Robert Lawless
Typesetting: J. K. Eckert & Company, Inc.
Cover Design: Aaron Lawhon
Printer: IBT Global

Printed in the United States of America

10 9 8 7 6 5 4 3 2 1

ISBN: 9781891121890

Library of Congress Cataloging-in-Publication Data

Gao, Jianjun, 1968-
RF and microwave modeling and measurement techniques for field effect transistors / Jianjun Gao.
p. cm.
Includes bibliographical references and index.
ISBN 978-1-891121-89-0 (hardcover : alk. paper)
1. Field-effect transistors--Testing. 2. Compound semiconductors--Testing. 3. Field-effect transistors--Mathematical models. 4. Compound semiconductors--Mathematical models. 5. Microwave measurements. 6. Radio measurements. I. Title.
TK7871.95.G36 2009
621.3815'284--dc22

2009017804

To my wife Dongyan and son Hanqi for the lost hours and to my parents for their unquestioning support.

Contents

Preface

Radio and microwave compound field-effect transistors (FETs), such as metal semiconductor field effect transistors (MESFETs), and heterojunction field effect transistors (HFETs), which include high electron mobility transistors (HEMTs) and pseudomorphic high electron mobility transistors (PHEMTs), extend the advantages of silicon counterparts to significantly higher frequencies. This advanced performance of FETs is attractive for high-frequency circuit design in view of a system-on-a-chip realization, where digital, mixed-signal baseband, and radio frequency (RF) transceiver blocks would be integrated on a single chip.

Accurate microwave and RF measurement techniques are the basis of characterization of the microwave and RF devices along with the corresponding model parameter extraction. Existing books on microwave and RF devices traditionally lack a thorough treatment of the high-frequency measurement techniques. The primary objective of the present book is to bridge the gap between device modeling and state-of-the-art microwave measurement technique.

This book combines both measurement technique and its application in an example of compound semiconductor FETs. The book shows an approach on how to do the measurement and based on the measurement data, to start the small signal, nonlinear modeling, and parameter extraction for the devices. The book includes all detailed information for FETs, which seldom appears in other books like this, so this should be helpful for new researchers to make their way in their career. Even for those without a good microwave background, the contents of this book can be easily understood. The presentation of this book assumes only a basic course in electronic circuits as a prerequisite. Instead of using electromagnetic fields as most of the microwave engineering books do, the subject is introduced via circuit concepts.

This book is intended to serve as a reference book for practicing engineers and technicians working in the areas of RF, microwave and solid-state device, optoelectronic integrated circuit design. The book should also be useful as a text-book for RF and microwave courses designed for senior undergraduate and first-year graduate students. Especially in student design projects, we foresee that this book will be a valuable handbook as well as a reference, both on basic modeling issues and on specific FET models encountered in circuit simulators. For a senior course, chapters 4-7 are complementary to active microwave courses.

It is hoped that this book will offer the type of practical information necessary for use in today's complex and rapidly changing RF and microwave circuit design.

Chapter 1

Introduction

1.1 Overview of III–V Compound Semiconductor Devices

There are a wide variety of solid-state device technologies available for implementing microwave and radio frequency (RF) integrated circuits (ICs). The common used semiconductors are as follows [1–11]:

1. Silicon-based bipolar junction transistors (BJTs)
2. Silicon-based mental-oxide semiconductor field effect transistors (MOSFETs)
3. Silicon-based laterally diffused metal-oxide field-effect transistors (LDMOSFETs)
4. Silicon germanium heterojunction bipolar transistors (SiGe HBTs)
5. Gallium arsenide based metal semiconductor field-effect transistors (GaAs MESFETs)
6. Gallium arsenide high electron mobility transistors (GaAs HEMTs)
7. Gallium arsenide heterojunction bipolar transistors (GaAs HBTs)
8. Indium phosphide high electron mobility transistors (InP HEMTs)
9. Indium phosphide heterojunction bipolar transistors (InP HBTs)

Semiconductor material systems can be categorized into silicon-based and III–V-compound-semiconductor-based devices. Silicon-based semiconductor devices, with their low-cost, high-volume production, have improved frequency response significantly as the channel length is made smaller and up to 45 nm. In contrast, compound semiconductor-based devices take advantages of their intrinsic material properties and offer superior device performance in high-frequency applications such as monolithic microwave integrated circuits (MMICs). The III–V semiconductor industries have also increased their production yield and integration scale in response to the increasing demand of RF circuits in terrestrial and mobile wireless communications.

Alternatively, in terms of the operation mechanism, microwave and RF semiconductors can be categorized into field-effect transistors and bipolar transistors.

Table 1.1 Comparison of FET and Bipolar Transistor

Parameters	FET/HEMT	BJT/HBT
Physical dimension limitation	Gate length	Base and collector thickness
Turn-on characteristics	Gate threshold voltage	Base-emitter voltage
Output current density	Medium	High
Input impedance controller	Gate voltage	Base current
Noise-source	Gate-induced noise. Channel current noise. Low frequency noise. Gate leakage current noise.	Shot noise. Low-frequency noise.
Processing complexity	Medium	High

Table 1.1 shows the comparison of some device parameters for both FET and bipolar transistor devices [1]. First, the physical dimension limitation sets the ultimate device speed performance. Fundamentally, a shorter gate length in an FET can reduce the carrier transport time and a narrower base as well as thinner collector in a bipolar device can decrease the carrier transit time. Device turn-on characteristics of an HBT are controlled by a material band gap or the turn-on voltage in the base-emitter junction. In contrast, the pinchoff voltage of an FET depends on the doping and thickness of active channel. Typically, the noise sources in FET devices include gate-induced noise, channel current noise, low-frequency noise, gate leakage current noise, and thermal noise sources generated from the extrinsic resistances. For bipolar transistors, the noise figures are determined by shot noise, which is related to the operating currents.

Depending on applications, it is preferable that device have some of the following features [1]:

1. Maximum power gain bandwidth
2. Minimum noise figure
3. Maximum power-added efficiency (PAE)

4. Low thermal resistance
5. High temperature of operation and reliability
6. High linearity
7. Low leakage current under cutoff operation
8. Low-frequency noise
9. Multi functionality, low single-power supply
10. Semi-insulating substrate, mature technology, low cost

Different RF/microwave circuits require different transistor parameters. For example, power amplifiers (PAs) use transistors with higher power densities; low-noise amplifiers (LNAs) employ transistors with low-noise characteristics. A qualitative performance summary of each device technology is listed in Table 1.2 [9]. In most designs, the minimum noise figure, maximum power gain, and stability factor voltage standing wave ratio (VSWR) usually do not occur at the same input/output impedance in the Smith chart. Therefore, the selection of bias point and input/output impedance is determined by the spec requirements of each component.

Table 1.2 A Comparison Chart for Different Device Technologies in Wireless Communication RF Transceiver Applications

Parameters	GaAs MESFET	GaAs HBT	GaAs HEMT	Si RF CMOS	SiGe HBT	InP HBT
Device speed	Good	Good	Good	Fair	Good	Excellent
Chip density	Low	High	Low	Low	High	High
Transconductance	Medium	High	High	Low	High	High
Device matching	Poor	Good	Poor	Poor	Good	Good
PAE	Medium	High	High	Medium	Medium	High
Linearity	High	High	High	Low	Medium	High
Low-frequency noise	Poor	Good	Poor	Poor	Good	Good
Breakdown voltage	High	High	High	Medium	Medium	High
Integration level	MSI, LSI	MSI,LSI	MSI,LSI	VLSI	LSI,VLSI	MSI,LSI

1.2 RF/Microwave Device and Circuit CAD

The application of modern computer-aided design (CAD) tools offers an improved approach. As the sophistication and accuracy of these tools improve, significant reductions in design cycle time can be realized. The goal is to develop CAD tools with sufficient accuracy that can achieve first pass design. The CAD tools need to be improved until the simulated and measured RF performance of the component being designed are in good agreement. This will permit the design to be completed, simulated, and fully tested by an engineer working at a computer workstation before fabrication is implemented. To achieve this goal, improved accuracy CAD tools are required.

There are two kinds of commercial RF and microwave CAD software: physical-based and equivalent circuit based CAD software. The physical-based CAD software as a starting point of analysis, considers fundamental equations of transport in semiconductors. The equivalent circuit-based CAD software addresses the issue of what needs to be known about the device in addition to its equivalent circuit to predict the noise performance. State-of-the-art CAD methods for active microwave circuits rely heavily on models of real devices. The model permits the RF performance of a device or integrated circuit to be determined as a function of process and device design information or of bias and RF operating conditions. Table 1.3 summarizes the semiconductor device models in the commercial RF/microwave CAD software, in most cases, BJT, MOSFET, and MESFET models are available. This book focuses on the III–V compound FET device modeling and measurement technique.

Table 1.3 Semiconductor Device Models in Commercial RF/Microwave CAD Software

Software	BJT	MOSFET	MESFET	HEMT	HBT
ADS	☑	☑	☑	☑	☑
Cadence	☑	☑	☑	☑	☑
MDS	☑	☑	☑	☑	
EESOF	☑	☑	☑	☑	☑
PSPICE	☑	☑	☑		
HSPICE	☑	☑	☑		

1.3 Organization of This Book

We will spend the rest of this book trying to convey the microwave and RF modeling and measurement techniques for FET devices. This book focuses on how to measure the microwave performance and to build the linear, nonlinear, and noise models for FET devices.

In Chapter 2, the concept of two-port networks is discussed. The reader is introduced to the characterization of two-port networks and its representation in terms of a set of parameters (impedance, admittance, hybrid, transmission, scattering, and chain-scattering parameters) that can be cast into a matrix format. The relationship between two-port and three-port networks is then illustrated.

Chapter 3 presents basic concepts of the commonly used microwave and RF measurement techniques, which include S parameters and noise and power measurements. The corresponding calibration methods for each measurement are also summarized. The signal and noise de-embedding methods for microwave components and circuits in on-wafer and coaxial measurement systems are discussed in more detail.

In Chapter 4, we introduce the physical structure and operation concept of FET devices. The small signal modeling and parameter extraction method are described, especially determination methods for pad capacitances, feedline inductances, extrinsic resistances, and intrinsic elements. The scaling rules for intrinsic elements are given also.

The nonlinear models for FETs, which include the physics-based nonlinear model, table-based nonlinear model, and empirical equivalent-circuit-based model are introduced in Chapter 5, as well as compact modeling techniques.

Chapter 6 first deals with the noise modeling and parameter extraction methods for FETs, and then turns to the determination of noise parameters, including tuner-based and noise-figure-based methods.

Artificial neural network (ANN) is very useful for neural-based microwave computer-aided design, and for analytically unified dc, small signal, and nonlinear device modeling. In Chapter 7, the microwave nonlinear device modeling technique based on a combination of the conventional equivalent circuit model and ANN is presented.

Chapter 2

Representation of Microwave Two-Port Network

The microwave signal and noise matrix analysis techniques are the basis of representation of the microwave network, and are the important tools of the radio frequency (RF) and microwave semiconductor modeling and parameter extraction. RF and microwave device, circuit and components can be classified as one-, two-, three-, and N-port networks. A majority of circuits under analysis are two-port networks. Therefore we focus in this chapter primarily on two-port characterization and study its representation in terms of a set of parameters that can be cast into a matrix format. The definition of a two-port network is that a network that has only two access ports: one for input or excitation and one for output or response. This chapter introduces the important linear parameters (including signal and noise parameters) that are currently used to characterize two-port networks. These parameters enable manipulation and optimization.

2.1 Signal Parameters

There are several ways to characterize the two-port network. The most commonly used parameters are the impedance Z, admittance Y, hybrid H, transmission $ABCD$, scattering S, and chain scattering T parameters. These parameters are used to describe linear networks fully and are interchangeable. Conversion between them is often used as an aid to circuit design when, for example, conversion enables easy deconvolution of certain parts of an equivalent circuit. This is because the terminating impedance's and driving sources vary. Further, if components are added in parallel the admittance parameters can be directly added;

similarly, if they are added in series impedance parameters can be used. Matrix manipulation also enables easy conversion among for example, common base (gate), common emitter (source), and common collector (drain) configurations. The impedance Z, admittance Y, hybrid H, and transmission $ABCD$ normally are called the low-frequency signal parameters, and are based on the voltages and currents at each port. The scattering S and chain scattering T parameters normally are called the high-frequency signal parameters, and are based on traveling waves applied to a network. Each of them can be used to characterize linear networks fully, and all show a generic form. This chapter concentrates on two-port networks, though all the rules described can be extended to N-port devices. A two-port network based on the Z-, Y-, H-, and $ABCD$ parameters is shown in Figure 2.1. It can be seen that the two-port network has four port variables: V_1, V_2, I_1, and I_2. We can use two of variables as excitation variables and the other two as response variables.

2.1.1 Impedance Parameters

The open-circuit impedance parameters (i.e., Z parameters) characterization of two-port networks are based on the exciting the network by the voltage V_1 at input port and the voltage V_2 at output port. In this case I_1 and I_2 are the independent variables, and V_1 and V_2 are the dependent variables. The network operation can be described by the following two equations:

$$V_1 = Z_{11} \cdot I_1 + Z_{12} \cdot I_2 \tag{2.1}$$

$$V_2 = Z_{21} \cdot I_1 + Z_{22} \cdot I_2 \tag{2.2}$$

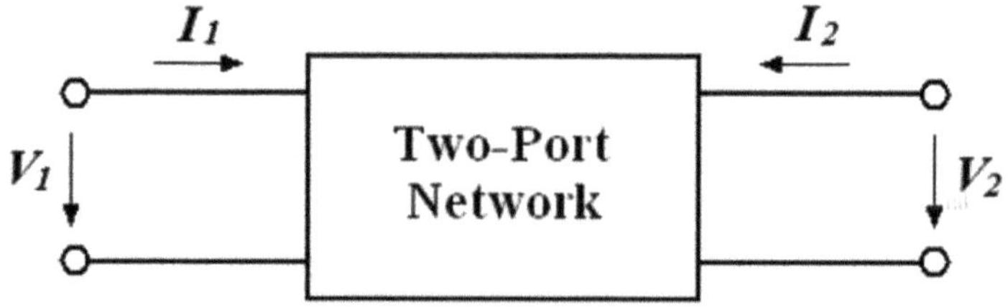

Figure 2.1 A block diagram of a two-port network

Using the matrix representation, we can write:

$$\begin{bmatrix} V_1 \\ V_2 \end{bmatrix} = \begin{bmatrix} Z_{11} & Z_{12} \\ Z_{21} & Z_{22} \end{bmatrix} \begin{bmatrix} I_1 \\ I_2 \end{bmatrix} \tag{2.3}$$

or

$$[V] = [Z][I] \tag{2.4}$$

where the $[Z]$ is called the impedance matrix of a two-port network. Since voltages V_1 and V_2 are in volts, and I_1 and I_2 are in amperes, the four parameters Z_{11}, Z_{12}, Z_{21}, and Z_{22} must be in ohms. Therefore, these are called impedance parameters, and their values completely characterize the linear two-port network. The definition equations and physical meaning of Z parameters are summarized in Table 2.1, and Figure 2.2 shows the corresponding equivalent circuit model of Z parameters. It is noted that the units of all Z parameters are in ohms (Ω).

The open circuits are not very easy to implement at a higher frequency range because of fringing capacitances; therefore, these parameters were measured only at low-frequency range. When measuring an active device or circuit, a bias network (or Bias-Tee) is required. This should still present an open-circuit at the signal frequency but of course should be a short circuit to the bias voltage. This would usually consist of a large inductor with a low series resistance.

2.1.2 Admittance Parameters

Now let us look at the short-circuit Y parameters where the voltages are the independent variables. These are called the short-circuit admit-

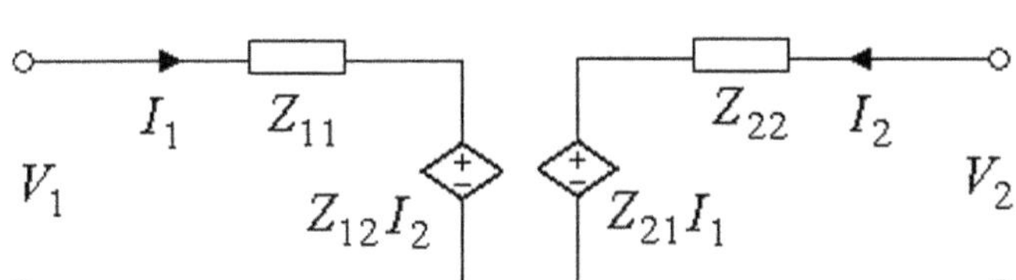

Figure 2.2 Equivalent circuit model for Z parameters

Table 2.1 Definition of Z Parameters

Parameters	Definition Equation	Physical Meaning
Z_{11}	$\left.\frac{V_1}{I_1}\right\|_{I_2=0}$	Input impedance with output port open-circuited
Z_{12}	$\left.\frac{V_1}{I_2}\right\|_{I_1=0}$	Reverse transmission impedance with input port open-circuited
Z_{21}	$\left.\frac{V_2}{I_1}\right\|_{I_2=0}$	Forward transmission impedance with output port open-circuited
Z_{22}	$\left.\frac{V_2}{I_2}\right\|_{I_1=0}$	Output impedance with input port open-circuited

tance parameters and describe the input, output, forward, and reverse admittances with the opposite port terminated in a short circuit. The short-circuit admittance parameters (i.e., Y parameters) characterization of two-port networks based on the exciting the network by the current I_1 at input port and I_2 at output port. In this case voltage V_1 and voltage V_2 are the independent variables, and I_1 and I_2 are the dependent variables. The network operation can be described by the following two equations:

$$I_1 = Y_{11} \cdot V_1 + Y_{12} \cdot V_2 \tag{2.5}$$

$$I_2 = Y_{21} \cdot V_1 + Y_{22} \cdot V_2 \tag{2.6}$$

Using the matrix representation, we can write:

$$\begin{bmatrix} I_1 \\ I_2 \end{bmatrix} = \begin{bmatrix} Y_{11} & Y_{12} \\ Y_{21} & Y_{22} \end{bmatrix} \begin{bmatrix} V_1 \\ V_2 \end{bmatrix} \tag{2.7}$$

or

$$[I] = [Y][I] \tag{2.8}$$

where the [Y] is called the admittance matrix of a two-port network. Since currents I_1 and I_2 are in amperes, and V_1 and V_2 are in volts, the four parameters Y_{11}, Y_{12}, Y_{21}, andY_{22} must be in siemens. Therefore, these are called admittance parameters, and their values completely characterize the linear two-port network. The definition equations and physical meaning of Y parameters are summarized in Table 2.2, and Figure 2.3 shows the corresponding equivalent circuit model of Y parameters. It is noted that the units of all Y parameters are in siemens ($S = 1/\Omega$).

Table 2.2 Definition of Y Parameters

Parameters	Definition Equation	Physical Meaning
Y_{11}	$\left.\frac{I_1}{V_1}\right\|_{V_2=0}$	Input admittance with output port short-circuited
Y_{12}	$\left.\frac{I_1}{V_2}\right\|_{V_1=0}$	Reverse transmission admittance with input port short-circuited
Y_{21}	$\left.\frac{I_2}{V_1}\right\|_{V_2=0}$	Forward transmission admittance with output port short-circuited
Y_{22}	$\left.\frac{I_2}{V_2}\right\|_{V_1=0}$	Output admittance with input port short-circuited

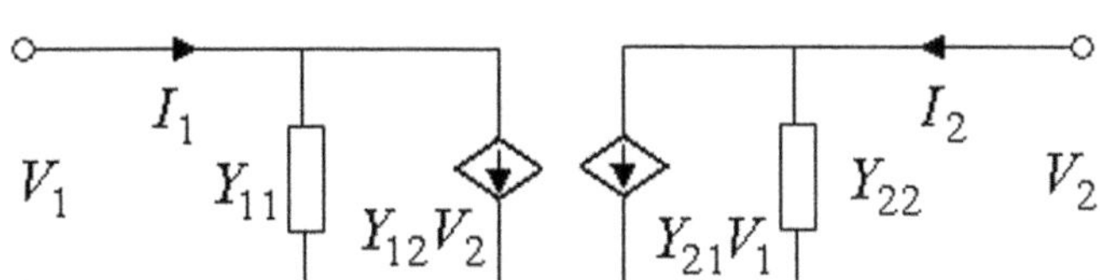

Figure 2.3 Equivalent circuit model for Y parameters

2.1.3 Hybrid Parameters

The hybrid parameters characterization of two-port networks based on the exciting the network by the current I_1 at input port and the voltage V_2 at output port. In this case I_1 and V_2 are the independent variables, and V_1 and I_2 are the dependent variables. The network operation can be described by the two equations:

$$V_1 = H_{11} \cdot I_1 + H_{12} \cdot V_2 \tag{2.9}$$

$$I_2 = H_{21} \cdot I_1 + H_{22} \cdot V_2 \tag{2.10}$$

Using the matrix representation, we can write:

$$\begin{bmatrix} V_1 \\ I_2 \end{bmatrix} = \begin{bmatrix} H_{11} & H_{12} \\ H_{21} & H_{22} \end{bmatrix} \begin{bmatrix} I_1 \\ V_2 \end{bmatrix} \tag{2.11}$$

or

$$\begin{bmatrix} V_1 \\ I_2 \end{bmatrix} = [H] \begin{bmatrix} I_1 \\ V_2 \end{bmatrix} \tag{2.12}$$

where the $[H]$ is called the hybrid matrix of a two-port network. Since the currents I_1 and I_2 are in amperes, and V_1 and V_2 are in volts, parameter H_{11} must be in ohms, H_{12} and H_{21} must be dimensionless, and H_{22} must be in siemens. The parameters H_{11} and H_{21} represent the input impedance and forward current gain, respectively, when the output port is shorted. Similarly, the parameters H_{12} and H_{22} represent the reverse voltage gain and output admittance, when the input port is opened. Because of this mix, therefore these are called hybrid parameters, and their values completely characterize the linear two-port network. The definition equations and physical meaning of H parameters are summarized in Table 2.3, and Figure 2.4 shows the corresponding equivalent circuit model of H parameters.

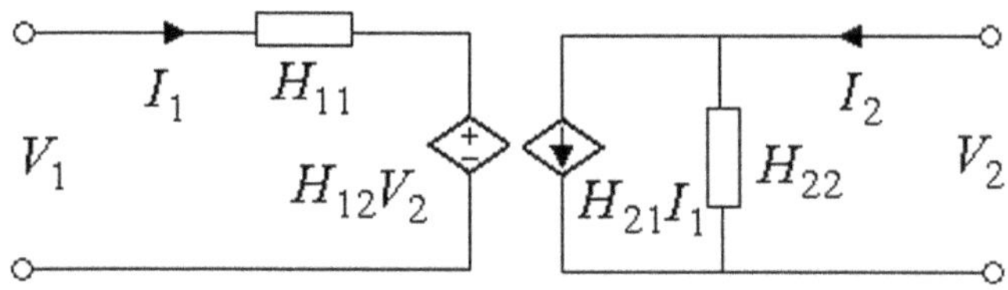

Figure 2.4 Equivalent circuit model for *H* parameters

Table 2.3 Definition of *H* Parameters

Parameters	Definition Equation	Physical Meaning
H_{11}	$\left.\frac{V_1}{I_1}\right\|_{V_2=0}$	Input impedance with output port short-circuited
H_{12}	$\left.\frac{V_1}{V_2}\right\|_{I_1=0}$	Reverse voltage gain with input port open-circuited
H_{21}	$\left.\frac{I_2}{I_1}\right\|_{V_2=0}$	Forward current gain with output port short-circuited
H_{22}	$\left.\frac{I_2}{V_2}\right\|_{I_1=0}$	Output admittance with input port open-circuited

2.1.4 Transmission Parameters

The transmission parameters characterization of two-port networks are on the exciting the network by the voltage and current at input port. In this case V_1 and I_1 are the independent variables, and V_2 and I_2 are the dependent variables. The network operation can be described by the following two equations:

$$V_1 = A \cdot V_2 - B \cdot I_2 \tag{2.13}$$

$$I_1 = C \cdot V_2 - D \cdot I_2 \tag{2.14}$$

Using the matrix representation, we can write:

$$\begin{bmatrix} V_1 \\ I_1 \end{bmatrix} = \begin{bmatrix} A & B \\ C & D \end{bmatrix} \begin{bmatrix} V_2 \\ -I_2 \end{bmatrix} \tag{2.15}$$

or

$$\begin{bmatrix} V_1 \\ I_1 \end{bmatrix} = [A] \begin{bmatrix} V_2 \\ -I_2 \end{bmatrix} \tag{2.16}$$

where $[A]$ is called the chain matrix of a two-port network. Since currents I_1 and I_2 are in amperes, and V_1 and V_2 are in volts, parameter A and D must be dimensionless, B must be in ohms, and C must be in siemens. These are called *ABCD* parameters. The chain matrix is very convenient for the calculation of the cascade connections. Note that they related the input voltage to the output voltage and the input current to the negative of the output current. This means that they are just multiplied for cascade connections as the output parameters become the input parameters for the next stage. The definition equations and physical meaning of *ABCD* parameters are summarized in Table 2.4, and Figure 2.5 shows the corresponding equivalent circuit model of *ABCD* parameters.

2.1.5 Conversion of Impedance, Admittance, Chain, and Hybrid Parameters

One type of the network parameters can be converted into another via the respective defining equation. For convenience, Table 2.5 summarizes the formulas for the previously defined four parameter sets.

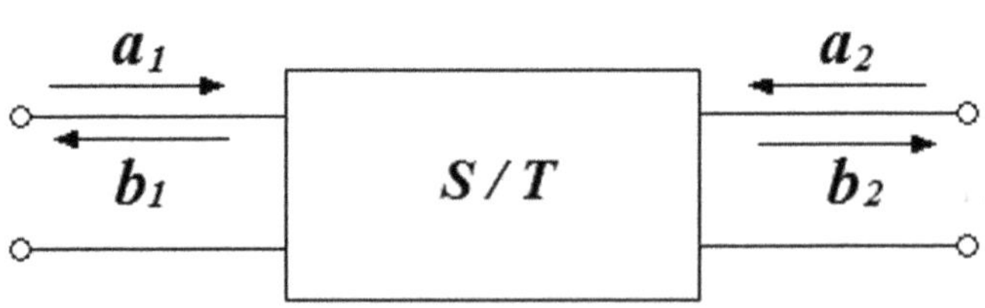

Figure 2.5 A two-port network with incident and reflected waves at each port

Table 2.4 Definition of *ABCD* Parameters

Parameters	Definition Equation	Physical Meaning
A	$\left.\frac{V_1}{V_2}\right\|_{I_2=0}$	Reverse voltage gain with output port open-circuited
B	$\left.\frac{V_1}{-I_2}\right\|_{V_2=0}$	Negative reverse transmission impedance with input port short-circuited
C	$\left.\frac{I_1}{V_2}\right\|_{I_2=0}$	Reverse transmission admittance with output port open-circuited
D	$\left.\frac{I_1}{-I_2}\right\|_{V_2=0}$	Negative reverse current gain with input port short-circuited

2.2 *S*- and *T*-parameters

The *Z*, *Y*, *H*, and *ABCD* parameters cannot be accurately measured at higher frequencies because the required short- and open-circuit tests are difficult to achieve over a broad range of frequencies. The scattering (*S*) parameters are currently the easiest parameters to measure at frequencies above a few tens of MHz as they are measured with 50 or 75 Ω network analyzers. The network analyzer is the basic measurement tool required for most RF and microwave circuit design, and the modern instrument offers rapid measurement and high accuracy through a set of basic calibrations. The high-frequency *S*- and *T*-parameters are used to characterize high RF/microwave two-port networks. These parameters are based on the concept of traveling wave and provide a complete characterization of any two-port network under analysis or test at high RF/microwave frequencies.

Figure 2.5 shows a network along with the incident and reflected waves at its two ports, where $a_i(i = 1{,}2)$ are the incident normalized power waves, and $b_i(i = 1{,}2)$ are the reflected normalized power waves. Hence, a_1 is an incident wave while b_1 is reflected wave at input port. Similarly, a_2 and b_2 represent incident and reflected waves at the out-

Table 2.5 Conversion between Different Network Representations

Parameter	Z	Y
Z	$\begin{bmatrix} Z_{11} & Z_{12} \\ Z_{21} & Z_{22} \end{bmatrix}$	$\begin{bmatrix} \frac{Z_{22}}{\Delta Z} & -\frac{Z_{12}}{\Delta Z} \\ -\frac{Z_{21}}{\Delta Z} & \frac{Z_{11}}{\Delta Z} \end{bmatrix}$
Y	$\begin{bmatrix} \frac{Y_{22}}{\Delta Y} & -\frac{Y_{12}}{\Delta Y} \\ -\frac{Y_{21}}{\Delta Y} & \frac{Y_{11}}{\Delta Y} \end{bmatrix}$	$\begin{bmatrix} Y_{11} & Y_{12} \\ Y_{21} & Y_{22} \end{bmatrix}$
H	$\begin{bmatrix} \frac{\Delta H}{H_{22}} & \frac{H_{12}}{H_{22}} \\ -\frac{H_{21}}{H_{22}} & \frac{1}{H_{22}} \end{bmatrix}$	$\begin{bmatrix} \frac{1}{H_{11}} & -\frac{H_{12}}{H_{11}} \\ \frac{H_{21}}{H_{11}} & \frac{\Delta H}{H_{11}} \end{bmatrix}$
$ABCD$	$\begin{bmatrix} \frac{A}{C} & \frac{\Delta A}{C} \\ \frac{1}{C} & \frac{D}{C} \end{bmatrix}$	$\begin{bmatrix} \frac{D}{B} & -\frac{\Delta A}{B} \\ -\frac{1}{B} & \frac{A}{B} \end{bmatrix}$

put port. It can be seen that the two-port network has four port variables: a_1, a_2, b_1, and b_2. Normally, the two incident waves are used as excitation variables, and other two are used as response variables.

The incident and reflected normalized power waves can be expressed as follows:

$$a_n = \frac{1}{2\sqrt{Z_o}}(V_n + Z_o I_n) \quad (n = 1, 2) \tag{2.17}$$

Table 2.5 (continued)

Parameter	H	$ABCD$
Z	$\begin{bmatrix} \frac{\Delta Z}{Z_{22}} & \frac{Z_{12}}{Z_{22}} \\ -\frac{Z_{21}}{Z_{22}} & \frac{1}{Z_{22}} \end{bmatrix}$	$\begin{bmatrix} \frac{Z_{11}}{Z_{21}} & \frac{\Delta Z}{Z_{21}} \\ \frac{1}{Z_{21}} & \frac{Z_{22}}{Z_{21}} \end{bmatrix}$
Y	$\begin{bmatrix} \frac{1}{Y_{11}} & -\frac{Y_{12}}{Y_{11}} \\ \frac{Y_{21}}{Y_{11}} & \frac{\Delta Y}{Y_{11}} \end{bmatrix}$	$\begin{bmatrix} -\frac{Y_{22}}{Y_{21}} & -\frac{1}{Y_{21}} \\ -\frac{\Delta Y}{Y_{21}} & -\frac{Y_{11}}{Y_{21}} \end{bmatrix}$
H	$\begin{bmatrix} H_{11} & H_{12} \\ H_{21} & H_{22} \end{bmatrix}$	$\begin{bmatrix} -\frac{\Delta H}{H_{21}} & -\frac{H_{11}}{H_{21}} \\ -\frac{H_{22}}{H_{21}} & -\frac{1}{H_{21}} \end{bmatrix}$
$ABCD$	$\begin{bmatrix} \frac{B}{D} & \frac{\Delta A}{D} \\ -\frac{1}{D} & \frac{C}{D} \end{bmatrix}$	$\begin{bmatrix} A & B \\ C & D \end{bmatrix}$

$$b_n = \frac{1}{2\sqrt{Z_o}}(V_n - Z_o I_n) \qquad (n = 1, 2) \tag{2.18}$$

where the index n refers either to port number 1 or 2, and impedance Z_o is the characteristic impedance of the connecting lines on the input and output side of the network.

Inverting (2.17) and (2.18) leads to the following voltage and current expressions:

$$V_n = \sqrt{Z_o}(a_n + b_n) \quad (n = 1, 2) \tag{2.19}$$

$$I_n = \frac{1}{\sqrt{Z_o}}(a_n - b_n) \quad (n = 1, 2) \tag{2.20}$$

with

$$a_n = \frac{V_n^+}{\sqrt{Z_o}} = \sqrt{Z_o} I_n^+ \quad (n = 1, 2) \tag{2.21}$$

$$b_n = \frac{V_n^-}{\sqrt{Z_o}} = -\sqrt{Z_o} I_n^- \quad (n = 1, 2) \tag{2.22}$$

2.2.1 *S* Parameters

The *S* parameters characterization of two-port networks based on the exciting the network by the incident waves at input port and output port. In this case a_1 and a_2 are the independent variables, and b_1 and b_2 are the dependent variables. The network operation can be described by the following two equations:

$$b_1 = S_{11} \cdot a_1 + S_{12} \cdot a_2 \tag{2.23}$$

$$b_2 = S_{21} \cdot a_1 + S_{22} \cdot a_2 \tag{2.24}$$

Using the matrix representation, we can write:

$$\begin{bmatrix} b_1 \\ b_2 \end{bmatrix} = \begin{bmatrix} S_{11} & S_{12} \\ S_{21} & S_{22} \end{bmatrix} \begin{bmatrix} a_1 \\ a_2 \end{bmatrix} \tag{2.25}$$

or

$$[b] = [S][a] \tag{2.26}$$

where [S] is called the scattering matrix of two-port network, and $S_{ij}(i,j = 1,2)$ are known as the scattering parameters of the two-port network. Since the units of the incident and reflected waves are same, scattering parameters must be dimensionless. As already mentioned, the S-parameters can be determined only under conditions of perfect matching on the input or output side. For instance, to record S_{11} and S_{21} we have to ensure that on the output side the line impedance Z_o is matched for $a_2 = 0$ to be enforced.

Now we will discuss the relationship between the S-parameters and the voltages at input and output ports. Figure 2.6 shows the block diagram for calculation of S_{11} and S_{21}. When the output port is terminated in a matched load ($Z_L = Z_o$), the input reflection coefficient can be expressed as follows:

$$\Gamma_{in} = \frac{V_1^-}{V_1^+} = \left.\frac{b_1}{a_1}\right|_{a_2=0} = S_{11} \tag{2.27}$$

It can be found that the physical meaning of S_{11} is the input reflection coefficient. S_{11} also can be expressed in terms of the input port impedance and characteristic impedance:

$$S_{11} = \frac{Z_{in} - Z_o}{Z_{in} + Z_o} \tag{2.28}$$

where Z_{in} is the input impedance:

$$Z_{in} = \frac{V_1}{I_1} = \frac{V_S - Z_o I_1}{I_1} \tag{2.29}$$

Substituting (2.29) into (2.28), we have

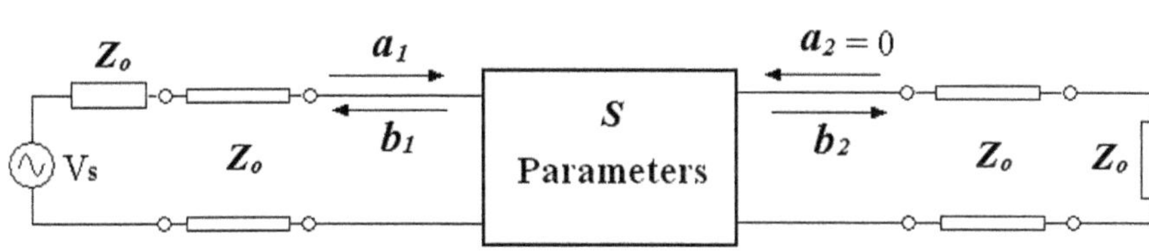

Figure 2.6 Block diagram for calculation of S_{11} and S_{21}

$$S_{11} = \frac{V_1 - Z_o I_1}{V_1 + Z_o I_1} = \frac{V_1 - (V_S - V_1)}{V_S} = \frac{2V_1}{V_S} - 1 \tag{2.30}$$

From (2.30), it can be found that S_{11} can be directly determined from input port node voltage and source voltage.

Moreover, with output port properly terminated, S_{21} can be expressed as follows:

$$S_{21} = \left.\frac{b_2}{a_1}\right|_{a_2=0} = \left.\frac{V_2^- / \sqrt{Z_o}}{(V_1 + Z_o I_1)/2\sqrt{Z_o}}\right|_{I_2^+ = V_2^+ = 0} \tag{2.31}$$

Since $a_2 = 0$, we can set to zero the positive traveling voltage and current waves at output port. Replacing V_1 by the source voltage V_s minus the voltage drop over the source impedance Z_o ($V_s - Z_o I_1$), we have:

$$S_{21} = \frac{2V_2}{V_S} \tag{2.32}$$

Here, it can be observed that the voltage recorded at output port is directly related to the source voltage and thus specifies the forward voltage gain of the network.

Figure 2.7 shows the block diagram for calculation of S_{12} and S_{22}. When the input port is terminated in a matched load ($Z_s = Z_o$), the output reflection coefficient can be expressed as follows:

$$\Gamma_{out} = \frac{V_2^-}{V_2^+} = \left.\frac{b_2}{a_2}\right|_{a_1=0} = S_{22} \tag{2.33}$$

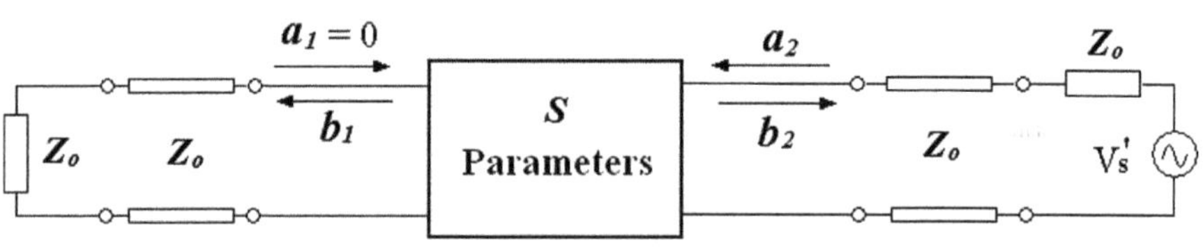

Figure 2.7 Block diagram for calculation of S_{12} and S_{22}

It can be found that the physical meaning of S_{22} is the output reflection coefficient. S_{22} also can be expressed in terms of the output port impedance and characteristic impedance:

$$S_{22} = \frac{Z_{out} - Z_o}{Z_{out} + Z_o} \tag{2.34}$$

where Z_{out} is the output impedance:

$$Z_{out} = \frac{V_2}{I_2} = \frac{V_S' - Z_o I_2}{I_2} \tag{2.35}$$

Substituting (2.35) into (2.34), we have:

$$S_{22} = \frac{V_2 - Z_o I_2}{V_2 + Z_o I_2} = \frac{V_2 - (V_S' - V_2)}{V_S'} = \frac{2V_2}{V_S'} - 1 \tag{2.36}$$

From (2.36), it can be found that S_{11} can be directly determined from output port node voltage and source voltage.

Moreover, with output port properly terminated, S_{12} can be expressed as follows:

$$S_{12} = \left.\frac{b_1}{a_2}\right|_{a_1=0} = \left.\frac{V_1^- / \sqrt{Z_o}}{(V_2 + Z_o I_2)/2\sqrt{Z_o}}\right|_{I_1^+ = V_1^+ = 0} \tag{2.37}$$

Since $a_1 = 0$, we can set to zero the positive traveling voltage and current waves at input port. Replacing V_2 by the source voltage $V_s{}'$ minus the voltage drop over the source impedance Z_o ($V_s{}' - Z_o I_2$), we have

$$S_{12} = \frac{2V_1}{V_S'} \tag{2.38}$$

Here, it can be observed that the voltage recorded at output port is directly related to the source voltage and thus specifies the forward voltage gain of the network.

The definition equations and physical meaning of S parameters are summarized in Table 2.6.

Table 2.6 Definition of S Parameters

Parameters	Definition Equation	Physical Meaning
S_{11}	$\left.\frac{b_1}{a_1}\right\|_{a_2=0} = \frac{2V_1}{V_S} - 1$	Input reflection coefficient when output port is terminated in a matched load
S_{21}	$\left.\frac{b_2}{a_1}\right\|_{a_2=0} = \frac{2V_2}{V_S}$	Forward transmission coefficient when output port is terminated in a matched load
S_{12}	$\left.\frac{b_1}{a_2}\right\|_{a_1=0} = \frac{2V_1}{V_S'}$	Reverse transmission coefficient when input port is terminated in a matched load
S_{22}	$\left.\frac{b_2}{a_2}\right\|_{a_1=0} = \frac{2V_2}{V_S'}$	Output reflection coefficient when input port is terminated in a matched load

When the characteristic impedances at each port are complex, the corresponding incident and reflected normalized power waves can be expressed as follows:

$$a_n = \frac{1}{\sqrt{2(Z_{on} + Z_{on}^*)}}(V_n + Z_{on} I_n) \qquad (n = 1,2) \tag{2.39}$$

$$b_n = \frac{1}{\sqrt{2(Z_{on} + Z_{on}^*)}}(V_n - Z_{on}^* I_n) \qquad (n = 1,2) \tag{2.40}$$

Table 2.7 summarizes conversion between S parameters and Z, Y, H, and $ABCD$ parameters [1].

2.2.2 *T* parameters

The chain scattering parameters are also known as the scattering transfer parameters or T-parameters and are useful for network in cas-

Table 2.7 Conversion between S parameters and Z, Y, H, and $ABCD$

$$S \leftrightarrow Z$$

$$S_{11} = \frac{(Z_{11} - Z_{o1}^*)(Z_{22} + Z_{o2}) - Z_{12}Z_{21}}{(Z_{11} + Z_{o1})(Z_{22} + Z_{o2}) - Z_{12}Z_{21}}$$

$$Z_{11} = \frac{(S_{11}Z_{o1} + Z_{o1}^*)(1 - S_{22}) + S_{12}S_{21}Z_{o1}}{(1\text{-}S_{11})(1\text{-}S_{22}) - S_{12}S_{21}}$$

$$S_{12} = \frac{2Z_{12}\sqrt{R_{o1}R_{o2}}}{(Z_{11} + Z_{o1})(Z_{22} + Z_{o2}) - Z_{12}Z_{21}} \qquad Z_{12} = \frac{2S_{12}\sqrt{R_{o1}R_{o2}}}{(1\text{-}S_{11})(1\text{-}S_{22}) - S_{12}S_{21}}$$

$$S_{21} = \frac{2Z_{21}\sqrt{R_{o1}R_{o2}}}{(Z_{11} + Z_{o1})(Z_{22} + Z_{o2}) - Z_{12}Z_{21}} \qquad Z_{21} = \frac{2S_{21}\sqrt{R_{o1}R_{o2}}}{(1\text{-}S_{11})(1\text{-}S_{22}) - S_{12}S_{21}}$$

$$S_{22} = \frac{(Z_{11} + Z_{o1})(Z_{22} - Z_{o2}^*) - Z_{12}Z_{21}}{(Z_{11} + Z_{o1})(Z_{22} + Z_{o2}) - Z_{12}Z_{21}}$$

$$Z_{22} = \frac{(S_{22}Z_{o2} + Z_{o2}^*)(1 - S_{11}) + S_{12}S_{21}Z_{o2}}{(1\text{-}S_{11})(1\text{-}S_{22}) - S_{12}S_{21}}$$

$$S \leftrightarrow Y$$

$$S_{11} = \frac{(1\text{-}Y_{11}Z_{o1}^*)(1 + Y_{22}Z_{o2}) + Y_{12}Y_{21}Z_{o1}^*Z_{o2}}{(1 + Y_{11}Z_{o1})(1 + Y_{22}Z_{o2}) - Y_{12}Y_{21}Z_{o1}Z_{o2}}$$

$$Y_{11} = \frac{(1\text{-}S_{11})(Z_{o2}^* + S_{22}Z_{o2}) + S_{12}S_{21}Z_{o2}}{(Z_{o1}^* + S_{11}Z_{o1})(Z_{o2}^* + S_{22}Z_{o2}) - S_{12}S_{21}Z_{o1}Z_{o2}}$$

$$S_{12} = \frac{\text{-}2Y_{12}\sqrt{R_{o1}R_{o2}}}{(1 + Y_{11}Z_{o1})(1 + Y_{22}Z_{o2}) - Y_{12}Y_{21}Z_{o1}Z_{o2}}$$

Table 2.7 Conversion between S parameters and Z, Y, H, and $ABCD$ (Continued)

$$Y_{12} = \frac{-2S_{12}\sqrt{R_{o1}R_{o2}}}{(Z_{o1}^* + S_{11}Z_{o1})(Z_{o2}^* + S_{22}Z_{o2}) - S_{12}S_{21}Z_{o1}Z_{o2}}$$

$$S_{21} = \frac{-2Y_{21}\sqrt{R_{o1}R_{o2}}}{(1+Y_{11}Z_{o1})(1+Y_{22}Z_{o2}) - Y_{12}Y_{21}Z_{o1}Z_{o2}}$$

$$Y_{21} = \frac{-2S_{21}\sqrt{R_{o1}R_{o2}}}{(Z_{o1}^* + S_{11}Z_{o1})(Z_{o2}^* + S_{22}Z_{o2}) - S_{12}S_{21}Z_{o1}Z_{o2}}$$

$$S_{22} = \frac{(1+Y_{11}Z_{o1})(1-Y_{22}Z_{o2}^*) + Y_{12}Y_{21}Z_{o1}Z_{o2}^*}{(1+Y_{11}Z_{o1})(1+Y_{22}Z_{o2}) - Y_{12}Y_{21}Z_{o1}Z_{o2}}$$

$$Y_{22} = \frac{(1-S_{22})(Z_{o1}^* + S_{11}Z_{o1}) + S_{12}S_{21}Z_{o1}}{(Z_{o1}^* + S_{11}Z_{o1})(Z_{o2}^* + S_{22}Z_{o2}) - S_{12}S_{21}Z_{o1}Z_{o2}}$$

$$S \leftrightarrow H$$

$$S_{11} = \frac{(H_{11} - Z_{o1}^*)(1+H_{22}Z_{o2}) - H_{12}H_{21}Z_{o2}}{(H_{11} + Z_{o1})(1+H_{22}Z_{o2}) - H_{12}H_{21}Z_{o2}}$$

$$H_{11} = \frac{(Z_{o1}^* + S_{11}Z_{o1})(Z_{o2}^* + S_{22}Z_{o2}) - S_{12}S_{21}Z_{o1}Z_{o2}}{(1-S_{11})(Z_{o2}^* + S_{22}Z_{o2}) + S_{12}S_{21}Z_{o2}}$$

$$S_{12} = \frac{2H_{12}\sqrt{R_{o1}R_{o2}}}{(H_{11} + Z_{o1})(1+H_{22}Z_{o2}) - H_{12}H_{21}Z_{o2}}$$

$$H_{12} = \frac{2S_{12}\sqrt{R_{o1}R_{o2}}}{(1-S_{11})(Z_{o2}^* + S_{22}Z_{o2}) + S_{12}S_{21}Z_{o2}}$$

Table 2.7 Conversion between S parameters and Z, Y, H, and $ABCD$ (Continued)

$$S_{21} = \frac{-2H_{21}\sqrt{R_{o1}R_{o2}}}{(H_{11}+Z_{o1})(1+H_{22}Z_{o2})-H_{12}H_{21}Z_{o2}}$$

$$H_{21} = \frac{-2S_{21}\sqrt{R_{o1}R_{o2}}}{(1-S_{11})(Z_{o2}^{*}+S_{22}Z_{o2})+S_{12}S_{21}Z_{o2}}$$

$$S_{22} = \frac{(H_{11}+Z_{o1})(1-H_{22}Z_{o2}^{*})+H_{12}H_{21}Z_{o2}^{*}}{(H_{11}+Z_{o1})(1+H_{22}Z_{o2})-H_{12}H_{21}Z_{o2}}$$

$$H_{22} = \frac{(1\text{-}S_{11})(1\text{-}S_{22})-S_{12}S_{21}}{(1-S_{11})(Z_{o2}^{*}+S_{22}Z_{o2})+S_{12}S_{21}Z_{o2}}$$

$$S \leftrightarrow A$$

$$S_{11} = \frac{AZ_{o2}+B-CZ_{01}^{*}Z_{02}-DZ_{o1}^{*}}{AZ_{o2}+B+CZ_{o1}Z_{o2}+DZ_{o1}}$$

$$A = \frac{(Z_{o1}^{*}+S_{11}Z_{o1})(1\text{-}S_{22})+S_{12}S_{21}Z_{o1}}{2S_{21}\sqrt{R_{o1}R_{o2}}}$$

$$S_{12} = \frac{2(AD-BC)\sqrt{R_{o1}R_{o2}}}{AZ_{o2}+B+CZ_{o1}Z_{o2}+DZ_{o1}}$$

$$B = \frac{(Z_{o1}^{*}+S_{11}Z_{o1})(Z_{o2}^{*}+S_{22}Z_{o2})\text{-}S_{12}S_{21}Z_{o1}Z_{o2}}{2S_{21}\sqrt{R_{o1}R_{o2}}}$$

$$S_{21} = \frac{2\sqrt{R_{o1}R_{o2}}}{AZ_{o2}+B+CZ_{o1}Z_{o2}+DZ_{o1}}$$

Table 2.7 Conversion between S parameters and Z, Y, H, and $ABCD$ (Continued)

$$C = \frac{(1\text{-}S_{11})(1\text{-}S_{22}) - S_{12}S_{21}}{2S_{21}\sqrt{R_{o1}R_{o2}}}$$

$$S_{22} = \frac{-AZ_{o2}^{*} + B - CZ_{01}Z_{02}^{*} + DZ_{o1}}{AZ_{o2} + B + CZ_{o1}Z_{o2} + DZ_{o1}}$$

$$D = \frac{(1\text{-}S_{11})(Z_{o2}^{*} + S_{22}Z_{o2}) + S_{12}S_{21}Z_{o2}}{2S_{21}\sqrt{R_{o1}R_{o2}}}$$

cade. These are defined on the basis of waves a_1 and b_1 at input port as dependent variables and waves a_2 and b_2 at output port as independent variables. The network operation can be described by the following two equations:

$$a_1 = T_{11} \cdot b_2 + T_{12} \cdot a_2 \tag{2.41}$$

$$b_1 = T_{21} \cdot b_2 + T_{22} \cdot a_2 \tag{2.42}$$

Using the matrix representation, we can write:

$$\begin{bmatrix} a_1 \\ b_1 \end{bmatrix} = \begin{bmatrix} T_{11} & T_{12} \\ T_{21} & T_{22} \end{bmatrix} \begin{bmatrix} b_2 \\ a_2 \end{bmatrix} \tag{2.43}$$

or

$$[b] = [T][a] \tag{2.44}$$

where the $[T]$ is called the chain scattering matrix of a two-port network, and $T_{ij}(i, j = 1, 2)$ are known as the chain scattering parameters of the two-port network. Since the units of the incident and reflected waves are the same, chain scattering parameters must be dimensionless.

The relationship between S- and T-parameters can be derived using the previous basic definition as follows:

$$T_{11} = \left.\frac{a_1}{b_2}\right|_{a_2=0} = \frac{a_1}{S_{21}a_1} = \frac{1}{S_{21}} \tag{2.45}$$

$$T_{12} = \left.\frac{a_1}{a_2}\right|_{b_2=0} = -\frac{S_{22}}{S_{21}} \tag{2.46}$$

$$T_{21} = \left.\frac{b_1}{b_2}\right|_{a_2=0} = \frac{S_{11}}{S_{21}} \tag{2.47}$$

$$T_{22} = \left.\frac{b_1}{a_2}\right|_{b_2=0} = -\frac{S_{11}S_{22} - S_{12}S_{21}}{S_{21}} = -\frac{\Delta S}{S_{21}} \tag{2.48}$$

The reverse relationship expressing *S*-parameters in terms of *T*-parameters can also be derived with the following result:

$$S_{11} = \frac{T_{21}}{T_{11}} \tag{2.49}$$

$$S_{12} = \frac{T_{11}T_{22} - T_{21}T_{12}}{T_{11}} = \frac{\Delta T}{T_{11}} \tag{2.50}$$

$$S_{21} = \frac{1}{T_{11}} \tag{2.51}$$

$$S_{22} = \frac{T_{12}}{T_{11}} \tag{2.52}$$

Table 2.8 summarizes conversion between *T*-parameters and *Z*, *Y*, *H*, and *ABCD* parameters [1].

Table 2.8 Conversion between T- parameters and Z, Y, H, and $ABCD$

$$T \leftrightarrow Z$$

$$T_{11} = \frac{(Z_{11} + Z_{o1})(Z_{22} + Z_{o2}) - Z_{12}Z_{21}}{2Z_{21}\sqrt{R_{o1}R_{o2}}}$$

$$Z_{11} = \frac{Z_{o1}^{*}(T_{11} + T_{12}) + Z_{o1}(T_{21} + T_{22})}{T_{11} + T_{12} - T_{21} - T_{22}}$$

$$T_{12} = \frac{(Z_{11} + Z_{o1})(Z_{o2}^{*} - Z_{22}) + Z_{12}Z_{21}}{2Z_{21}\sqrt{R_{o1}R_{o2}}}$$

$$Z_{12} = \frac{2\sqrt{R_{o1}R_{o2}}(T_{11}T_{22} - T_{12}T_{21})}{T_{11} + T_{12} - T_{21} - T_{22}}$$

$$T_{21} = \frac{(Z_{11} - Z_{o1}^{*})(Z_{o2} + Z_{22}) - Z_{12}Z_{21}}{2Z_{21}\sqrt{R_{o1}R_{o2}}} \qquad Z_{21} = \frac{2\sqrt{R_{o1}R_{o2}}}{T_{11} + T_{12} - T_{21} - T_{22}}$$

$$T_{22} = \frac{(Z_{o1}^{*} - Z_{11})(Z_{22} - Z_{o2}^{*}) + Z_{12}Z_{21}}{2Z_{21}\sqrt{R_{o1}R_{o2}}}$$

$$Z_{22} = \frac{Z_{o2}^{*}(T_{11} - T_{21}) - Z_{o2}(T_{12} - T_{22})}{T_{11} + T_{12} - T_{21} - T_{22}}$$

$$T \leftrightarrow Y$$

$$T_{11} = \frac{(-1\text{-}Y_{11}Z_{o1})(1 + Y_{22}Z_{o2}) + Y_{12}Y_{21}Z_{o1}Z_{o2}}{2Y_{21}\sqrt{R_{o1}R_{o2}}}$$

$$Y_{11} = \frac{Z_{o2}^{*}(T_{11} - T_{21}) - Z_{o2}(T_{12} - T_{22})}{T_{11}Z_{o1}^{*}Z_{o2}^{*} - T_{12}Z_{o1}^{*}Z_{o2} + T_{21}Z_{o1}Z_{o2}^{*} - T_{22}Z_{o1}Z_{o2}}$$

Table 2.8 Conversion between T- parameters and *Z*, *Y*, *H*, and *ABCD* (Continued)

$$T_{12} = \frac{(1+Y_{11}Z_{o1})(1-Y_{22}Z_{o2}^*)+Y_{12}Y_{21}Z_{o1}Z_{o2}^*}{2Y_{21}\sqrt{R_{o1}R_{o2}}}$$

$$Y_{12} = \frac{-2\sqrt{R_{o1}R_{o2}}(T_{11}T_{22}-T_{12}T_{21})}{T_{11}Z_{o1}^*Z_{o2}^* - T_{12}Z_{o1}^*Z_{o2} + T_{21}Z_{o1}Z_{o2}^* - T_{22}Z_{o1}Z_{o2}}$$

$$T_{21} = \frac{(Y_{11}Z_{o1}^*\text{-}1)(1+Y_{22}Z_{o2})-Y_{12}Y_{21}Z_{o1}^*Z_{o2}}{2Y_{21}\sqrt{R_{o1}R_{o2}}}$$

$$Y_{21} = \frac{-2\sqrt{R_{o1}R_{o2}}}{T_{11}Z_{o1}^*Z_{o2}^* - T_{12}Z_{o1}^*Z_{o2} + T_{21}Z_{o1}Z_{o2}^* - T_{22}Z_{o1}Z_{o2}}$$

$$T_{22} = \frac{(1\text{-}Y_{11}Z_{o1}^*)(1-Y_{22}Z_{o2}^*)-Y_{12}Y_{21}Z_{o1}^*Z_{o2}}{2Y_{21}\sqrt{R_{o1}R_{o2}}}$$

$$Y_{22} = \frac{Z_{o1}^*(T_{11}+T_{12})+Z_{o1}(T_{21}+T_{22})}{T_{11}Z_{o1}^*Z_{o2}^* - T_{12}Z_{o1}^*Z_{o2} + T_{21}Z_{o1}Z_{o2}^* - T_{22}Z_{o1}Z_{o2}}$$

$$T \leftrightarrow H$$

$$T_{11} = \frac{(\text{-}H_{11}-Z_{o1})(1+H_{22}Z_{o2})+H_{12}H_{21}Z_{o2}}{2H_{21}\sqrt{R_{o1}R_{o2}}}$$

$$H_{11} = \frac{Z_{o2}^*(T_{11}Z_{o1}^*+T_{21}Z_{o1})-Z_{o2}(T_{12}Z_{o1}^*+T_{22}Z_{o1})}{Z_{o2}^*(T_{11}-T_{21})-Z_{o2}(T_{12}+T_{22})}$$

$$T_{12} = \frac{(H_{11}+Z_{o1})(1-H_{22}Z_{o2}^*)+H_{12}H_{21}Z_{o2}^*}{2H_{21}\sqrt{R_{o1}R_{o2}}}$$

Table 2.8 Conversion between T- parameters and Z, Y, H, and $ABCD$ (Continued)

$$H_{12} = \frac{2\sqrt{R_{o1}R_{o2}}(T_{11}T_{22} - T_{12}T_{21})}{Z^*_{o2}(T_{11} - T_{21}) - Z_{o2}(T_{12} + T_{22})}$$

$$T_{21} = \frac{(Z^*_{o1}\text{-}H_{11})(1 + H_{22}Z_{o2}) + H_{12}H_{21}Z_{o2}}{2H_{21}\sqrt{R_{o1}R_{o2}}}$$

$$H_{21} = \frac{-2\sqrt{R_{o1}R_{o2}}}{Z^*_{o2}(T_{11} - T_{21}) - Z_{o2}(T_{12} + T_{22})}$$

$$T_{22} = \frac{(H_{11} - Z^*_{o1})(1 - H_{22}Z^*_{o2}) + H_{12}H_{21}Z^*_{o2}}{2H_{21}\sqrt{R_{o1}R_{o2}}}$$

$$H_{22} = \frac{T_{11} + T_{12} - T_{21} - T_{22}}{Z^*_{o2}(T_{11} - T_{21}) - Z_{o2}(T_{12} + T_{22})}$$

$$T \leftrightarrow A$$

$$T_{11} = \frac{AZ_{o2} + B + CZ_{01}Z_{02} + DZ_{o1}}{2\sqrt{R_{o1}R_{o2}}}$$

$$A = \frac{Z^*_{o1}(T_{11} + T_{12}) + Z_{o1}(T_{21} + T_{22})}{2\sqrt{R_{o1}R_{o2}}}$$

$$T_{12} = \frac{AZ^*_{o2} - B + CZ_{01}Z^*_{o2} - DZ_{o1}}{2\sqrt{R_{o1}R_{o2}}}$$

$$B = \frac{Z^*_{o2}(T_{11}Z^*_{01} + T_{21}Z_{01})\text{-}Z_{o2}(T_{12}Z^*_{01} + T_{22}Z_{01})}{2\sqrt{R_{o1}R_{o2}}}$$

Table 2.8 Conversion between T- parameters and *Z*, *Y*, *H*, and *ABCD* (Continued)

$$T_{21} = \frac{AZ_{o2} + B - CZ_{01}^{*}Z_{02} - DZ_{o1}^{*}}{2\sqrt{R_{o1}R_{o2}}} \qquad C = \frac{T_{11} + T_{12} - T_{21} - T_{22}}{2\sqrt{R_{o1}R_{o2}}}$$

$$T_{22} = \frac{AZ_{o2}^{*} - B - CZ_{01}^{*}Z_{02}^{*} + DZ_{o1}^{*}}{2\sqrt{R_{o1}R_{o2}}}$$

$$D = \frac{Z_{o2}^{*}(T_{11} - T_{21})\text{-}Z_{o2}(T_{12} - T_{22})}{2\sqrt{R_{o1}R_{o2}}}$$

2.3 Representation of Noisy Two-Port Network

The circuit theory of linear noisy networks shows that any noisy two-port can be replaced by a noise equivalent circuit, which consists of original two-port (now assumed to be noiseless) and two additional correlated noise sources. Spectral representation of noise plays a very important role in the analysis of narrow band circuits with center frequency f_o. By assuming the power noise spectral density is constant around f_o, two noise sources can be described completely by their noise power spectral densities ($S_{n_1}(f)$ and $S_{n_2}(f)$) and correlated spectral density ($S_{n_1 n_2}(f)$).

The cross-correlation term of two correlation noise sources can be expressed as:

$$S_{n_1 n_2}(f) = C_{12}\sqrt{S_{n_1}(f) \cdot S_{n_2}(f)} \tag{2.53}$$

where C_{12} is the so-called cross-correlation coefficient.

Arranging these noise spectral densities in matrix form leads to the so-called noise correlation matrix as follows:

$$C = \frac{1}{4kT\Delta f}\begin{bmatrix} S_{n_1}(f) & S_{n_1 n_2}(f) \\ S_{n_2 n_1}(f) & S_{n_2}(f) \end{bmatrix} \tag{2.54}$$

From (2.54), it can be found that the noise correlation matrices are Hermitian matrices:

$$C = C^{+} = \begin{bmatrix} C_1 & C_{12} \\ C_{21} & C_2 \end{bmatrix} \tag{2.55}$$

where the plus sign (+) is use to denote the Hermitian conjugation. The properties of the noise correlation matrices are as follows:

$$\mathrm{Im}(C_{11}) = \mathrm{Im}(C_{22}) = 0 \tag{2.56}$$

$$C_{11} \geq 0 \;, \; C_{22} \geq 0 \tag{2.57}$$

$$C_{12} = C_{21}^{*} \tag{2.58}$$

$$\Delta C = C_{11}C_{22} - |C_{12}|^2 \geq 0 \tag{2.59}$$

Therefore, the properties of the noisy linear two-port networks are fully described by four real numbers: C_{11}, C_{22}, real part of C_{12}, and imaginary part of C_{12}.

In general, six different forms of noise equivalent circuits exist depending upon the type of the two additional noise sources and their arrangement relative to the noiseless two-port. Each is called a noise representation. The two additional noise sources do not actually exist in the position marked as already mentioned; they are merely concentrated equivalent representation of the effect of all noise sources inside the two-port network.

For common applications only three of these representations are required. In this section, we discuss five different forms of noise equivalent circuits and their corresponding correlation matrices: impedance, admittance, chain, *S*-parameter, and *T*-parameter noise representations.

2.3.1 Impedance Noise Correlation Matrix

The impedance noise representation of a noisy two-port network can be described by the following two equations [2–3]:

$$V_1 = Z_{11} \cdot I_1 + Z_{12} \cdot I_2 + V_{N1} \tag{2.60}$$

$$V_2 = Z_{21} \cdot I_1 + Z_{22} \cdot I_2 + V_{N2} \tag{2.61}$$

Using the matrix representation, we can write:

$$\begin{bmatrix} V_1 \\ V_2 \end{bmatrix} = \begin{bmatrix} Z_{11} & Z_{12} \\ Z_{21} & Z_{22} \end{bmatrix} \begin{bmatrix} I_1 \\ I_2 \end{bmatrix} + \begin{bmatrix} V_{N1} \\ V_{N2} \end{bmatrix} \tag{2.62}$$

Equation (2.62) expresses the fact that a noisy two-port develops noise voltages V_{N1} and V_{N2} across both of its ports if they are simultaneously open-circuited:

$$V_{N1} = V_1 \Big|_{I_1=0, I_2=0} \tag{2.63}$$

$$V_{N2} = V_2 \Big|_{I_1=0, I_2=0} \tag{2.64}$$

The corresponding equivalent circuit model of the impedance noise representation of a noisy two-port network is shown in Figure 2.8.

The impedance noise representation normalization with respect to the thermal noise available power $kT\Delta f$ is used:

$$C_Z = \frac{1}{4kT\Delta f} \begin{bmatrix} \langle V_{N1} \cdot V_{N1}^* \rangle & \langle V_{N1} \cdot V_{N2}^* \rangle \\ \langle V_{N1}^* \cdot V_{N2} \rangle & \langle V_{N2} \cdot V_{N2}^* \rangle \end{bmatrix} \tag{2.65}$$

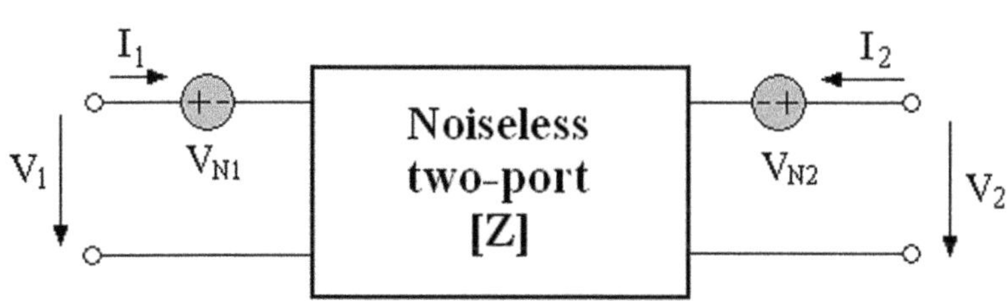

Figure 2.8 Impedance noise representation of a noisy two-port network

where k is the Boltzmann constant, T is the absolute temperature, Δf is the bandwidth, and * the conjugate of complex. The noise parameters according to the impedance noise matrix representation of the two-port network are as follows:

$$R_1 = \frac{\langle V_{N1} \cdot V_{N1}^* \rangle}{4kT\Delta f} \tag{2.66}$$

$$R_2 = \frac{\langle V_{N2} \cdot V_{N2}^* \rangle}{4kT\Delta f} \tag{2.67}$$

$$\rho_\upsilon = \frac{\langle V_{N1}^* \cdot V_{N2} \rangle}{\sqrt{\langle V_{N1} \cdot V_{N1}^* \rangle \langle V_{N2} \cdot V_{N2}^* \rangle}} \tag{2.68}$$

where R_1 and R_2 are the equivalent noise resistances, and ρ_υ is the correlation coefficient:

$$\rho_\upsilon = |\rho_\upsilon| e^{j\phi_\upsilon} \tag{2.69}$$

Equation (2.65) can be rewritten as follows:

$$C_Z = \begin{bmatrix} R_1 & |\rho_\upsilon| e^{-j\phi_\upsilon} \sqrt{R_1 R_2} \\ |\rho_\upsilon| e^{j\phi_\upsilon} \sqrt{R_1 R_2} & R_2 \end{bmatrix} \tag{2.70}$$

2.3.2 Admittance Noise Correlation Matrix

The admittance noise representation of a noisy a two-port network can be described by the following two equations [2–3]:

$$I_1 = Y_{11} \cdot V_1 + Y_{12} \cdot V_2 + I_{N1} \tag{2.71}$$

$$I_2 = Y_{21} \cdot V_1 + Y_{22} \cdot V_2 + I_{N2} \tag{2.72}$$

Using the matrix representation, we can write:

$$\begin{bmatrix} I_1 \\ I_2 \end{bmatrix} = \begin{bmatrix} Y_{11} & Y_{12} \\ Y_{21} & Y_{22} \end{bmatrix} \begin{bmatrix} V_1 \\ V_2 \end{bmatrix} + \begin{bmatrix} I_{N1} \\ I_{N2} \end{bmatrix} \tag{2.73}$$

Equation (2.73) expresses the fact that a noisy two-port develops noise currents I_{N1} and I_{N2} across both of its ports if they are simultaneously short-circuited:

$$I_{N1} = I_1 \big|_{V_1=0, V_2=0} \tag{2.74}$$

$$I_{N2} = I_2 \big|_{V_1=0, V_2=0} \tag{2.75}$$

The corresponding equivalent circuit model of the admittance noise representation of a noisy two-port network is shown in Figure 2.9.

The admittance noise representation normalization with respect to the thermal noise available power $kT\Delta f$ is used:

$$C_Y = \frac{1}{4kT\Delta f} \begin{bmatrix} \langle I_{N1} \cdot I_{N1}^* \rangle & \langle I_{N1} \cdot I_{N2}^* \rangle \\ \langle I_{N1}^* \cdot I_{N2} \rangle & \langle I_{N2} \cdot I_{N2}^* \rangle \end{bmatrix} \tag{2.76}$$

The noise parameters according to the impedance noise matrix representation of the two-port network are as follows:

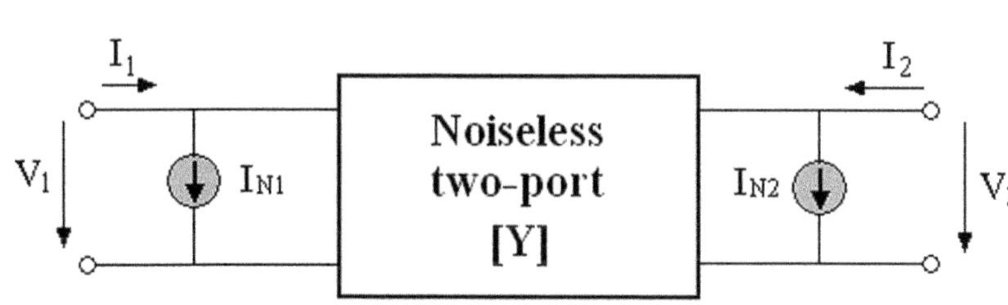

Figure 2.9 Admittance noise representation of noisy two-port network

$$G_1 = \frac{\langle I_{N1} \cdot I_{N1}^* \rangle}{4kT\Delta f} \tag{2.77}$$

$$G_2 = \frac{\langle I_{N2} \cdot I_{N2}^* \rangle}{4kT\Delta f} \tag{2.78}$$

$$\rho_I = \frac{\langle I_{N1}^* \cdot I_{N2} \rangle}{\sqrt{\langle I_{N1} \cdot I_{N1}^* \rangle \langle I_{N2} \cdot I_{N2}^* \rangle}} \tag{2.79}$$

where G_1 and G_2 are the equivalent noise conductances, and ρ_I is the correlation coefficient:

$$\rho_I = |\rho_I| e^{j\phi_I} \tag{2.80}$$

Equation (2.76) can be rewritten as follows:

$$C_Y = \begin{bmatrix} G_1 & |\rho_I| e^{-j\phi_I} \sqrt{G_1 G_2} \\ |\rho_I| e^{j\phi_I} \sqrt{G_1 G_2} & G_2 \end{bmatrix} \tag{2.81}$$

2.3.3 Chain Noise Correlation Matrix

The chain noise representation of a noisy two-port network can be described by the following two equations [2–3]:

$$V_1 = A \cdot V_2 - B \cdot I_2 + V_N \tag{2.82}$$

$$I_1 = C \cdot V_2 - D \cdot I_2 + I_N \tag{2.83}$$

Using the matrix representation, we can write:

$$\begin{bmatrix} V_1 \\ I_1 \end{bmatrix} = \begin{bmatrix} A & B \\ C & D \end{bmatrix} \begin{bmatrix} V_2 \\ -I_2 \end{bmatrix} + \begin{bmatrix} V_N \\ I_N \end{bmatrix} \tag{2.84}$$

The corresponding equivalent circuit model of the chain noise representation of a noisy two-port network is shown in Figure 2.10.

It is noted that two input noise voltage V_N and current I_N are used to represent the noise property of the noisy two-port, and can be determined as follows:

$$V_N = (V_1 - AV_2)\big|_{I_2=0} \tag{2.85}$$

$$I_N = (I_1 + DI_2)\big|_{V_2=0} \tag{2.86}$$

The chain noise representation normalization with respect to the thermal noise available power $kT\Delta f$ is used:

$$C_A = \frac{1}{4kT\Delta f} \begin{bmatrix} \langle V_N \cdot V_N^* \rangle & \langle V_N \cdot I_N^* \rangle \\ \langle V_N^* \cdot I_N \rangle & \langle I_N \cdot I_N^* \rangle \end{bmatrix} \tag{2.87}$$

The noise parameters according to the chain noise matrix representation of the two-port network are as follows:

$$R_N = \frac{\langle V_N \cdot V_N^* \rangle}{4kT\Delta f} \tag{2.88}$$

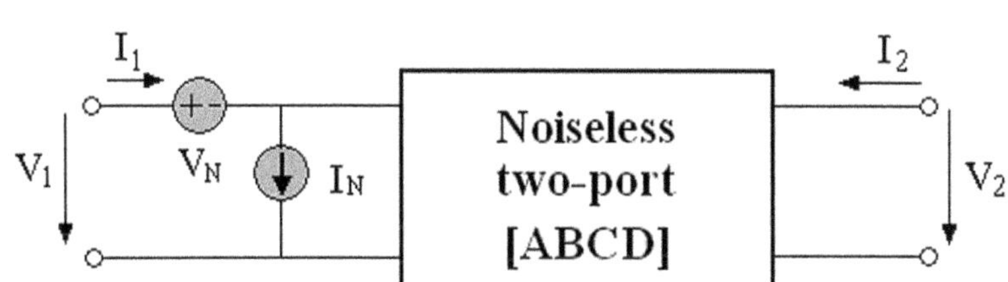

Figure 2.10 Chain noise representation of noisy two-port network

$$g_N = \frac{\langle I_N \cdot I_N^* \rangle}{4kT\Delta f} \tag{2.89}$$

$$\rho = \frac{\langle V_N^* \cdot I_N \rangle}{\sqrt{\langle V_N \cdot V_N^* \rangle \langle I_N \cdot I_N^* \rangle}} \tag{2.90}$$

where R_N and g_N are the equivalent noise resistance and conductance, and ρ is the correlation coefficient:

$$\rho = |\rho| e^{j\phi} \tag{2.91}$$

Equation (2.87) can be rewritten as follows:

$$C_A = \begin{bmatrix} R_N & |\rho| e^{-j\phi} \sqrt{R_N G_N} \\ |\rho| e^{j\phi} \sqrt{R_N G_N} & g_N \end{bmatrix} \tag{2.92}$$

2.3.4 S-parameter Noise Correlation Matrix

Assuming B_{N1} is the reflected noise power wave at input port, and B_{N2} is the reflected noise power wave at output port, the noise wave complex amplitudes at the ports are related by a linear matrix equation:

$$b_1 = S_{11} \cdot a_1 + S_{12} \cdot a_2 + b_{N1} \tag{2.93}$$

$$b_2 = S_{21} \cdot a_1 + S_{22} \cdot a_2 + b_{N2} \tag{2.94}$$

Using the matrix representation, we can write:

$$\begin{bmatrix} b_1 \\ b_2 \end{bmatrix} = \begin{bmatrix} S_{11} & S_{12} \\ S_{21} & S_{22} \end{bmatrix} \begin{bmatrix} a_1 \\ a_2 \end{bmatrix} + \begin{bmatrix} b_{N1} \\ b_{N2} \end{bmatrix} \tag{2.95}$$

The noise reflected waves b_{N1} and b_{N2} can be determined when input and output ports are simultaneously terminated in the matched load ($Z_S = Z_L = Z_o$):

$$b_{N1} = b_1\big|_{a_1=0, a_2=0} \tag{2.96}$$

$$b_{N2} = b_2\big|_{a_1=0, a_2=0} \tag{2.97}$$

The corresponding equivalent circuit model of the S-parameter noise representation of a noisy two-port network is shown in Figure 2.11.

The S-parameter noise representation normalization with respect to the thermal noise available power $kT\Delta f$ is used:

$$C_S = \frac{1}{kT\Delta f}\begin{bmatrix} \langle b_{N1}\cdot b_{N1}^*\rangle & \langle b_{N1}\cdot b_{N2}^*\rangle \\ \langle b_{N1}^*\cdot b_{N2}\rangle & \langle b_{N2}\cdot b_{N2}^*\rangle \end{bmatrix} \tag{2.98}$$

The noise parameters according to the S-parameter noise matrix representation of the two-port network are

$$T_{B1} = \frac{\langle b_{N1}\cdot b_{N1}^*\rangle}{k\Delta f} \tag{2.99}$$

$$T_{B2} = \frac{\langle b_{N2}\cdot b_{N2}^*\rangle}{k\Delta f} \tag{2.100}$$

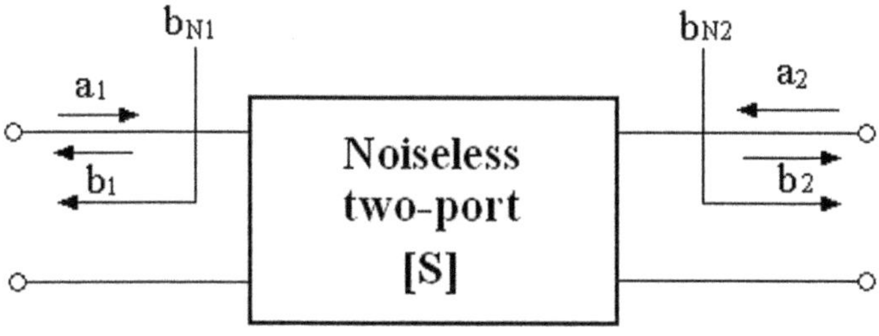

Figure 2.11 S-parameter noise representation of noisy two-port network

$$\rho_B = \frac{\langle b_{N1}^* \cdot b_{N2} \rangle}{\sqrt{\langle b_{N1} \cdot b_{N1}^* \rangle \langle b_{N2} \cdot b_{N2}^* \rangle}} \tag{2.101}$$

where T_{B1} and T_{B2} are the equivalent noise temperatures, and ρ_B is the correlation coefficient:

$$\rho_B = |\rho_B| e^{j\phi_B} \tag{2.102}$$

Equation (2.98) can be rewritten as follows:

$$C_S = \frac{1}{T} \begin{bmatrix} T_{B1} & |\rho_B| e^{-j\phi_B} \sqrt{T_{B1} T_{B2}} \\ |\rho_B| e^{j\phi_B} \sqrt{T_{B1} T_{B2}} & T_{B2} \end{bmatrix} \tag{2.103}$$

2.3.5 *T*-parameter Noise Correlation Matrix

Figure 2.12 shows the corresponding equivalent circuit model of the *T*-parameter noise representation of a noisy two-port network. Assuming a_N and b_N and are the incident and reflected normalized power waves at input port, the noise wave complex amplitudes at the ports are related by a linear matrix equation:

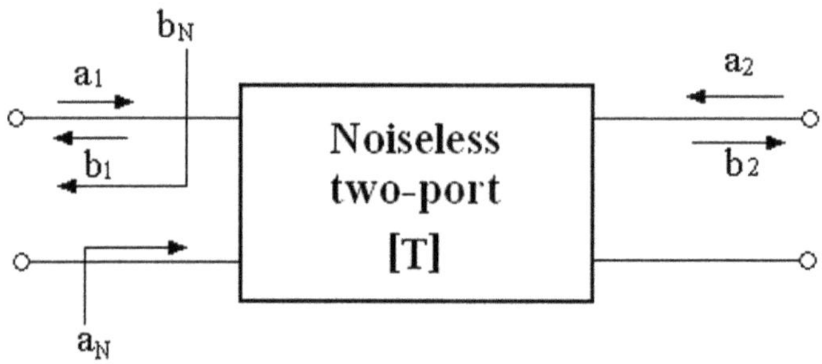

Figure 2.12 *T*-parameter noise representation of noisy two-port network

$$\begin{bmatrix} a_1 \\ b_1 \end{bmatrix} = \begin{bmatrix} T_{11} & T_{12} \\ T_{21} & T_{22} \end{bmatrix} \begin{bmatrix} b_2 \\ a_2 \end{bmatrix} + \begin{bmatrix} a_N \\ b_N \end{bmatrix} \tag{2.104}$$

a_N and b_N can be determined as follows:

$$a_N = (a_1 - T_{11} b_2)\big|_{a_2=0} \tag{2.105}$$

$$b_N = (b_1 + T_{21} b_2)\big|_{a_2=0} \tag{2.106}$$

The T-parameter noise representation normalization with respect to the thermal noise available power $kT\Delta f$ is used:

$$C_T = \frac{1}{kT\Delta f} \begin{bmatrix} \langle a_N \cdot a_N^* \rangle & \langle a_N \cdot b_N^* \rangle \\ \langle a_N^* \cdot b_N \rangle & \langle b_N \cdot b_N^* \rangle \end{bmatrix} \tag{2.107}$$

The noise parameters according to the T-parameter noise matrix representation of the two-port network are

$$T_A = \frac{\langle a_N \cdot a_N^* \rangle}{k\Delta f} \tag{2.108}$$

$$T_B = \frac{\langle b_N \cdot b_N^* \rangle}{k\Delta f} \tag{2.109}$$

$$\rho_{AB} = \frac{\langle a_N^* \cdot b_N \rangle}{\sqrt{\langle a_N \cdot a_N^* \rangle \langle b_N \cdot b_N^* \rangle}} \tag{2.110}$$

where T_A and T_2 are the equivalent noise temperatures, and ρ_{AB} is the correlation coefficient:

$$\rho_{AB} = |\rho_{AB}| e^{j\phi_{AB}} \tag{2.111}$$

Equation (2.107) can be rewritten as follows:

$$C_T = \frac{1}{T}\begin{bmatrix} T_A & |\rho_{AB}|e^{-j\phi_{AB}}\sqrt{T_A T_B} \\ |\rho_{AB}|e^{j\phi_{AB}}\sqrt{T_A T_B} & T_B \end{bmatrix} \tag{2.112}$$

2.3.6 Relationship between Correlation Noise Matrices

One type of the correlation noise matrices can be converted into another via the respective defining equation. For convenience, the formulas for the previously defined five parameter sets are summarizes as follows.

(1) $C_Z \rightarrow C_Y$

Substituting (2.62) into (2.73), we have

$$\begin{bmatrix} I_{N1} \\ I_{N2} \end{bmatrix} = -\begin{bmatrix} Y_{11} & Y_{12} \\ Y_{21} & Y_{22} \end{bmatrix}\begin{bmatrix} V_{N1} \\ V_{N2} \end{bmatrix} \tag{2.113}$$

Substituting (2.113) into (2.76), we have

$$C_{Y11} = |Y_{11}|^2 C_{Z11} + |Y_{12}|^2 C_{Z22} + Y_{11}Y_{12}^* C_{Z12} + Y_{12}Y_{11}^* C_{Z21} \tag{2.114}$$

$$C_{Y22} = |Y_{21}|^2 C_{Z11} + |Y_{22}|^2 C_{Z22} + Y_{21}Y_{22}^* C_{Z12} + Y_{22}Y_{21}^* C_{Z21} \tag{2.115}$$

$$C_{Y12} = C_{Z11}Y_{11}Y_{21}^* + C_{Z22}Y_{12}Y_{22}^* + Y_{11}Y_{22}^* C_{Z12} + Y_{12}Y_{21}^* C_{Z21} \tag{2.116}$$

$$C_{Y21} = C_{Z11}Y_{21}Y_{11}^* + C_{Z22}Y_{12}^*Y_{22} + Y_{12}^*Y_{21} C_{Z12} + Y_{11}^*Y_{22} C_{Z21} \tag{2.117}$$

(2) $C_Y \rightarrow C_Z$

Substituting (2.73) into (2.62), we have

$$\begin{bmatrix} V_{N1} \\ V_{N2} \end{bmatrix} = -\begin{bmatrix} Z_{11} & Z_{12} \\ Z_{21} & Z_{22} \end{bmatrix} \begin{bmatrix} I_{N1} \\ I_{N2} \end{bmatrix} \tag{2.118}$$

Substituting (2.118) into (2.65), we have

$$C_{Z11} = C_{Y11}|Z_{11}|^2 + C_{Y22}|Z_{12}|^2 + Z_{11}Z_{12}^*C_{Y12} + Z_{12}Z_{11}^*C_{Y21} \tag{2.119}$$

$$C_{Z22} = C_{Y11}|Z_{21}|^2 + C_{Y22}|Z_{22}|^2 + Z_{21}Z_{22}^*C_{Y12} + Z_{22}Z_{21}^*C_{Y21} \tag{2.120}$$

$$C_{Z12} = C_{Y11}Z_{11}Z_{21}^* + C_{Y22}Z_{22}^*Z_{12} + Z_{11}Z_{22}^*C_{Y12} + Z_{12}Z_{21}^*C_{Y21} \tag{2.121}$$

$$C_{Z21} = C_{Y11}Z_{11}^*Z_{21} + C_{Y22}Z_{22}Z_{12}^* + Z_{12}^*Z_{21}C_{Y12} + Z_{11}^*Z_{22}C_{Y21} \tag{2.122}$$

(3) $C_A \rightarrow C_Z$

Substituting (2.62) into (2.84), we have

$$V_{N1} = V_N - I_N Z_{11} \tag{2.123}$$

$$V_{N2} = -I_N Z_{21} \tag{2.124}$$

Substituting (2.123) and (2.124) into (2.65), we have

$$C_{Z11} = C_{A11} + C_{A22}|Z_{11}|^2 - Z_{11}^*C_{A12} - Z_{11}C_{A21} \tag{2.125}$$

$$C_{Z12} = C_{A22}Z_{21}^*Z_{11} - C_{A12}Z_{21}^* \tag{2.126}$$

$$C_{Z21} = C_{A22}Z_{11}^{*}Z_{21} - C_{A21}Z_{21} \tag{2.127}$$

$$C_{Z22} = C_{A22}\left|Z_{21}\right|^{2} \tag{2.128}$$

(4) $C_Z \rightarrow C_A$

When (2.124) and (2.123) are combined, we have

$$V_N = V_{N1} - \frac{Z_{11}V_{N2}}{Z_{21}} \tag{2.129}$$

$$I_N = -\frac{V_{N2}}{Z_{21}} \tag{2.130}$$

Substituting (2.129) and (2.130) into (2.87), we have

$$C_{A11} = C_{Z11} + C_{Z22}\frac{\left|Z_{11}\right|^{2}}{\left|Z_{21}\right|^{2}} - \frac{Z_{11}^{*}}{Z_{21}^{*}}C_{Z12} - \frac{Z_{11}}{Z_{21}}C_{Z21} \tag{2.131}$$

$$C_{A12} = C_{Z22}\frac{Z_{11}}{\left|Z_{21}\right|^{2}} - C_{Z12}\frac{1}{Z_{21}^{*}} \tag{2.132}$$

$$C_{A21} = C_{Z22}\frac{Z_{11}^{*}}{\left|Z_{21}\right|^{2}} - C_{Z21}\frac{1}{Z_{21}} \tag{2.133}$$

$$C_{A22} = C_{Z22}\frac{1}{\left|Z_{21}\right|^{2}} \tag{2.134}$$

(5) $C_A \rightarrow C_Y$

When (2.73) and (2.84) are combined, we have

$$I_{N1} = I_N - Y_{11}V_N \tag{2.135}$$

$$I_{N2} = -V_N Y_{21} \tag{2.136}$$

Substituting (2.135) and (2.136) into (2.76), we have

$$C_{Y11} = C_{A22} + C_{A11}|Y_{11}|^2 - Y_{11}^* C_{A21} - Y_{11}C_{A12} \tag{2.137}$$

$$C_{Y12} = C_{A11}Y_{21}^* Y_{11} - C_{A21}Y_{21}^* \tag{2.138}$$

$$C_{Y21} = C_{A11}Y_{11}^* Y_{21} - C_{A12}Y_{21} \tag{2.139}$$

$$C_{Y22} = C_{A11}|Y_{21}|^2 \tag{2.140}$$

(6) $C_Y \to C_A$

Equations (2.135) and (2.136) can be rewritten as follows:

$$I_N = I_{N1} - \frac{Y_{11}}{Y_{21}} I_{N2} \tag{2.141}$$

$$V_N = -\frac{I_{N2}}{Y_{21}} \tag{2.142}$$

Substituting (2.141) and (2.142) into (2.87), we have

$$C_{A11} = \frac{C_{Y22}}{|Y_{21}|^2} \tag{2.143}$$

$$C_{A12} = \frac{Y_{11}^* C_{Y22} - Y_{21}^* C_{Y21}}{|Y_{21}|^2} \tag{2.144}$$

$$C_{A21} = (C_{A12})^* \tag{2.145}$$

$$C_{A22} = C_{Y11} + \frac{|Y_{11}|^2 C_{Y22}}{|Y_{21}|^2} - 2\,\mathrm{Re}(\frac{Y_{11}}{Y_{21}} C_{Y21}) \tag{2.146}$$

(7) $C_S \rightarrow C_T$

When (2.95) and (2.104) are combined, we have

$$a_N = \frac{b_{N2}}{S_{21}} \tag{2.147}$$

$$b_N = b_{N1} - \frac{S_{11}}{S_{21}} b_{N2} \tag{2.148}$$

Substituting (2.147) and (2.148) into (2.107), we have

$$C_{T11} = \frac{C_{S22}}{|S_{21}|^2} \tag{2.149}$$

$$C_{T12} = \frac{C_{S21}}{S_{21}} - \frac{S_{11}^* C_{S22}}{|S_{21}|^2} \tag{2.150}$$

$$C_{T21} = (C_{T12})^* \tag{2.151}$$

$$C_{T22} = C_{S11} + \frac{|S_{11}|^2 C_{S22}}{|S_{21}|^2} - 2\,\mathrm{Re}(\frac{S_{11}}{S_{21}} C_{S21}) \tag{2.152}$$

(8) $C_T \rightarrow C_S$

From (2.147) and (2.148), we have:

$$b_{N1} = b_N + S_{11} a_N \tag{2.153}$$

$$b_{N2} = S_{21}a_N \tag{2.154}$$

Substituting (2.153) and (2.154) into (2.98), we have

$$C_{S11} = |S_{11}|^2 C_{T11} + C_{T22} + S_{11}C_{T12} + S_{11}^* C_{T21} \tag{2.155}$$

$$C_{S12} = S_{11}S_{21}^* C_{T11} + C_{T21}S_{21}^* \tag{2.156}$$

$$C_{S21} = (C_{S12})^* \tag{2.157}$$

$$C_{S22} = |S_{21}|^2 C_{T11} \tag{2.158}$$

A set of matrices covering all possible transformations among impedance, admittance, chain, S-parameter noise and T-parameter noise representations is presented in Table 2.9 and Table 2.10 [6–7], where C and C' denote the correlation matrix of the original and resulting representation, respectively, and T is the transformation matrix. The relationship between the correlation matrix of the original and resulting representation is as follows:

$$C' = TCT^+ \tag{2.159}$$

2.3.7 Conversion between the Noise Parameters

Based on the different noise representations for a two-port network, the corresponding types of noise parameters are different. The most commonly used types of noise parameters are the impedance, admittance, and reflection coefficient.

1. Admittance

When the two-port network is driven by a signal source with the internal admittance, the noise figure is written in the following form:

$$F = F_{\min} + \frac{R_n}{G_s}|Y_s - Y_{opt}|^2 \tag{2.160}$$

Table 2.9 Conversion between Different Network Noise Representations C_Y, C_Z, and C_A

Resulting Matrix C'	Original Matrix C: C_Y	C_Z	C_A
C_Y	$\begin{bmatrix} 1 & 0 \\ 0 & 1 \end{bmatrix}$	$\begin{bmatrix} Y_{11} & Y_{12} \\ Y_{21} & Y_{22} \end{bmatrix}$	$\begin{bmatrix} -Y_{11} & 1 \\ -Y_{21} & 0 \end{bmatrix}$
C_Z	$\begin{bmatrix} Z_{11} & Z_{12} \\ Z_{21} & Z_{22} \end{bmatrix}$	$\begin{bmatrix} 1 & 0 \\ 0 & 1 \end{bmatrix}$	$\begin{bmatrix} 1 & -Z_{11} \\ 0 & -Z_{21} \end{bmatrix}$
C_A	$\begin{bmatrix} 0 & A_{12} \\ 1 & A_{22} \end{bmatrix}$	$\begin{bmatrix} 1 & -A_{11} \\ 0 & -A_{21} \end{bmatrix}$	$\begin{bmatrix} 1 & 0 \\ 0 & 1 \end{bmatrix}$

Table 2.10 Conversion between Different Network Noise Representation C_S and C_T

Resulting Matrix C'	Original Matrix C: C_S	C_T
C_S	$\begin{bmatrix} 1 & 0 \\ 0 & 1 \end{bmatrix}$	$\begin{bmatrix} S_{11} & 1 \\ S_{21} & 0 \end{bmatrix}$
C_T	$\begin{bmatrix} 0 & +T_{11} \\ 1 & -T_{21} \end{bmatrix}$	$\begin{bmatrix} 1 & 0 \\ 0 & 1 \end{bmatrix}$

2. Impedance

When the two-port network is driven by a signal source with the internal impedance, the noise figure is written in the following form:

$$F = F_{\min} + \frac{g_n}{R_s}\left|Z_s - Z_{opt}\right|^2 \tag{2.161}$$

3. Reflection coefficient

By using definition of source reflection coefficient, (2.160) and (2.161) can be rewritten as follows:

$$F = F_{\min} + N\frac{\left|\Gamma_s - \Gamma_{opt}\right|^2}{(1-\left|\Gamma_s\right|^2)(1-\left|\Gamma_{opt}\right|^2)} \tag{2.162}$$

$$F = F_{\min} + \frac{4R_n}{Z_o}\frac{\left|\Gamma_s - \Gamma_{opt}\right|^2}{(1-\left|\Gamma_s\right|^2)\left|1+\Gamma_{opt}\right|^2} \tag{2.163}$$

where Y_s is the source admittance, Z_s is the source impedance, and Γ_s is the source reflection coefficient:

$$Y_s = G_s + jB_s$$

$$Z_s = R_s + jX_s$$

Table 2.11 summarizes the noise parameters of impedance, admittance, and reflection coefficient types for a two-port network, and Table 2.12 gives the conversion between the different noise parameter types.

Table 2.11 Noise Parameters for Two-Port Network

	Impedance	Admittance	Reflection Coefficient
Noise Parameters	Optimum noise figure F_{min}	Optimum noise figure F_{min}	Optimum noise figure F_{min}
	Optimum source resistance R_{opt}	Optimum source conductance G_{opt}	Magnitude of optimum source reflection coefficient $\left|\Gamma_{opt}\right|$
	Optimum source reactance X_{opt}	Optimum source susceptance B_{opt}	Phase of optimum source reflection coefficient $\angle\Gamma_{opt}$
	Equivalent noise admittance g_n	Equivalent noise resistance R_n	Equivalent noise factor N

Table 2.12 Conversion between Impedance, Admittance, and Reflection Coefficient

Impedance ↔ Admittance	
$F_{min} = F_{min}$ $R_n = g_n(R_{opt}^2 + X_{opt}^2)$ $G_{opt} = \dfrac{R_{opt}}{R_{opt}^2 + X_{opt}^2}$ $B_{opt} = -\dfrac{X_{opt}}{R_{opt}^2 + X_{opt}^2}$	$F_{min} = F_{min}$ $g_n = R_n(G_{opt}^2 + B_{opt}^2)$ $R_{opt} = \dfrac{G_{opt}}{G_{opt}^2 + B_{opt}^2}$ $X_{opt} = -\dfrac{B_{opt}}{G_{opt}^2 + B_{opt}^2}$
Reflection coefficient ↔ Admittance	
$F_{min} = F_{min}$ $N = 4R_n G_{opt}$ $\lvert\Gamma_{opt}\rvert = \sqrt{\dfrac{(Y_o - G_{opt})^2 + B_{opt}^2}{(Y_o + G_{opt})^2 + B_{opt}^2}}$ $\Phi_{opt} = \arctan\left\{\dfrac{B_{opt}}{Y_o - G_{opt}}\right\} - \arctan\left\{\dfrac{B_{opt}}{Y_o + G_{opt}}\right\}$	$F_{min} = F_{min}$ $R_n = \dfrac{N}{4G_{opt}}$ $G_{opt} = \mathrm{Re}\left\{Y_o \dfrac{1-\Gamma_{opt}}{1+\Gamma_{opt}}\right\}$ $B_{opt} = \mathrm{Im}\left\{Y_o \dfrac{1-\Gamma_{opt}}{1+\Gamma_{opt}}\right\}$

The most commonly used type of noise parameters is admittance; the corresponding noise parameters are the optimum noise figure F_{min}, optimum source conductance G_{opt}, optimum source admittance B_{opt}, and equivalent noise resistance R_n. The relationship among admittance, chain, S-parameter, and T-parameter noise correlation matrices and admittance noise parameters are discussed in the following section.

1. Relationship between admittance correlation matrix and admittance noise parameters:

$$F_{\min} = 1 + 2[\operatorname{Re}(\frac{C_{Y22}Y_{11}^{*}}{|Y_{21}|^{2}} - \frac{C_{Y21}}{Y_{21}}) + G_{opt}R_{n}] \tag{2.164}$$

$$R_{n} = \frac{C_{Y22}}{|Y_{21}|^{2}} \tag{2.165}$$

$$G_{opt} = \sqrt{\frac{C_{Y11} + \frac{C_{Y22}|Y_{11}|^{2}}{|Y_{21}|^{2}} - 2\operatorname{Re}(\frac{Y_{11}C_{Y21}}{Y_{21}})}{R_{n}} - |B_{opt}|^{2}} \tag{2.166}$$

$$B_{opt} = \frac{1}{R_{n}}\operatorname{Im}(\frac{C_{Y22}Y_{11}^{*}}{|Y_{21}|^{2}} - \frac{C_{Y21}}{Y_{21}}) \tag{2.167}$$

2. Relationship between chain correlation matrix and admittance noise parameters:

$$R_{n} = C_{A11} \tag{2.168}$$

$$G_{opt} = \sqrt{\frac{C_{A22}}{C_{A11}} - (\frac{\operatorname{Im}(C_{A12})}{C_{A11}})^{2}} \tag{2.169}$$

$$B_{opt} = -\frac{\operatorname{Im}(C_{A12})}{C_{A11}} \tag{2.170}$$

$$F_{\min} = 1 + 2[\operatorname{Re}(C_{A12}) + G_{opt}C_{A11}] \tag{2.171}$$

3. Relationship between *T*-parameter correlation matrix and admittance noise parameters:

$$F_{\min} = 1 + \frac{1}{2}(C_{T11} - C_{T22} + N) \tag{2.172}$$

$$\Gamma_{opt} = \frac{2C_{T12}}{C_{T11} + C_{T22} + N} \tag{2.173}$$

$$N = \sqrt{(C_{T11} + C_{T22})^2 - 4|C_{T12}|^2} \tag{2.174}$$

4. Relationship between S-parameter correlation matrix and admittance noise parameters:

$$F_{\min} = 1 + \frac{1}{2}\left[C_{S22}\frac{1-|S_{11}|^2}{|S_{21}|^2} - C_{S11} + 2\,\mathrm{Re}\left\{C_{S12}\frac{S_{11}^*}{S_{21}^*}\right\} + N\right] \tag{2.175}$$

$$\Gamma_{opt} = \frac{2\left(C_{S12}^*\dfrac{1}{S_{21}} - C_{S22}\dfrac{S_{11}^*}{|S_{21}|^2}\right)}{C_{S22}\dfrac{1+|S_{11}|^2}{|S_{21}|^2} + C_{S11} - 2\,\mathrm{Re}\left\{C_{S12}\dfrac{S_{11}^*}{S_{21}^*}\right\} + N} \tag{2.176}$$

$$N = \sqrt{\left[C_{S22}\frac{1+|S_{11}|^2}{|S_{21}|^2} + C_{S11} - 2\,\mathrm{Re}\left\{C_{S12}\frac{S_{11}^*}{S_{21}^*}\right\}\right]^2 - 4\left|C_{S12}^*\frac{1}{S_{21}} - C_{S22}\frac{S_{11}^*}{|S_{21}|^2}\right|^2} \tag{2.177}$$

2.4 Interconnections of Two-Port Network

For application in noise analysis, interconnections of two two-port networks either in parallel, in series, or in cascade are of particular interest. The signal and noise matrix of interconnections of two two-port network are discussed next.

2.4.1 Series Connection of Networks

Figure 2.13 shows a series connection consisting of two two-port networks, and the individual networks are shown in impedance matrix representation, where I_1', I_2', V_2', and I_2' are the currents and voltages of the sub-network N_1, V_1'', I_1'', V_2'', and I_2'' are the currents and voltages of the sub-network N_2, V_1, I_1, V_2, and I_2 are the currents and voltages of the resulting network at input and output ports.

The currents and voltages of the resulting network at input and output ports can be expressed as follows:

$$V_1 = V_1^{'} + V_1^{''} \tag{2.178}$$

$$V_2 = V_2^{'} + V_2^{''} \tag{2.179}$$

$$I_1 = I_1^{'} = I_2^{''} \tag{2.180}$$

$$I_2 = I_2^{'} = I_2^{''} \tag{2.181}$$

Therefore, the Z matrix of the resulting network in series network can be expressed the sum of each Z matrix of the sub-network:

$$Z = Z_1 + Z_2 \tag{2.182}$$

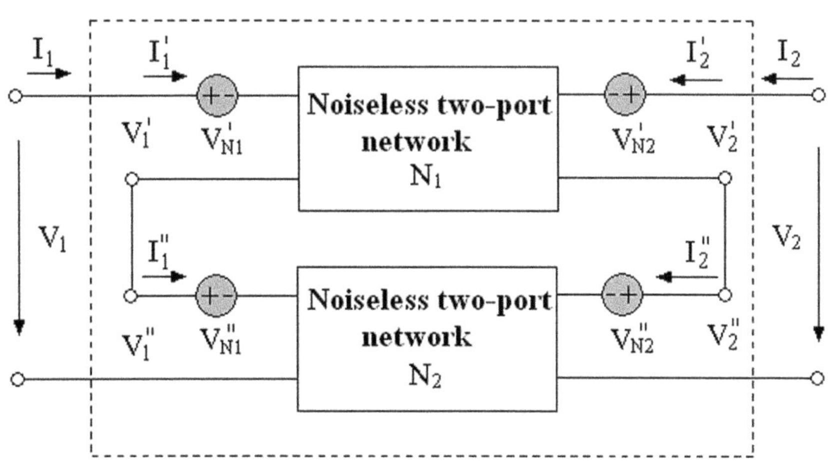

Figure 2.13 Series connection of two two-port networks

The noise voltages V_{N1} and V_{N2} can be expressed as follows:

$$V_{N1} = V_{N1}^{'} + V_{N1}^{''} \tag{2.183}$$

$$V_{N2} = V_{N2}^{'} + V_{N2}^{''} \tag{2.184}$$

where $V_{N1}^{'}$ and $V_{N1}^{''}$, $V_{N2}^{'}$, and $V_{N2}^{''}$ are the independent noise sources, respectively:

$$\left\langle V_{N1}^{'} \cdot (V_{N1}^{''})^{*} \right\rangle = \left\langle V_{N2}^{'} \cdot (V_{N2}^{''})^{*} \right\rangle = 0 \tag{2.185}$$

Based on the definition of the impedance noise matrix, we have

$$\begin{aligned} C_{Z11} &= \left\langle V_{N1} \cdot V_{N1}^{*} \right\rangle = \left\langle (V_{N1}^{'} + V_{N1}^{''}) \cdot (V_{N1}^{'} + V_{N1}^{''})^{*} \right\rangle \\ &= \left\langle V_{N1}^{'} \cdot (V_{N1}^{'})^{*} \right\rangle + \left\langle V_{N1}^{''} \cdot (V_{N1}^{''})^{*} \right\rangle \\ &= C_{Z11}^{'} + C_{Z11}^{''} \end{aligned} \tag{2.186}$$

Similarly, we have

$$C_{Z22} = C_{Z22}^{'} + C_{Z22}^{''} \tag{2.187}$$

$$C_{Z12} = C_{Z12}^{'} + C_{Z12}^{''} \tag{2.188}$$

$$C_{Z21} = C_{Z21}^{'} + C_{Z21}^{''} \tag{2.189}$$

Therefore, the resulting impedance noise matrix for series connection is the sum of respective the noise matrix in impedance of the original two-ports:

$$C_{Z} = C_{Z}^{'} + C_{Z}^{''} \tag{2.190}$$

2.4.2 Parallel Connection of Networks

Figure 2.14 shows a parallel connection consisting of two two-port networks, and the individual networks are shown in admittance matrix representation where I_1', I_2', V_2', and I_2' are the currents and voltages of the sub-network N_1, V_1'', I_1'', V_2'', and I_2'' are the currents and voltages of the sub-network N_2, V_1, I_1, V_2, and I_2 are the currents and voltages of the resulting network at input and output ports. I_{N1}' and I_{N2}' are the short-circuit noise sources of network N_1, and I_{N1}'' and I_{N2}'' are the short-circuit noise sources of network N_2.

The currents and voltages of the resulting network at input and output ports can be expressed as follows:

$$I_1 = I_1' + I_1'' \tag{2.191}$$

$$I_2 = I_2' + I_2'' \tag{2.192}$$

$$V_1 = V_1' = V_2'' \tag{2.193}$$

$$V_2 = V_2' = V_2'' \tag{2.194}$$

Therefore, the Y matrix of the resulting network in series network can be expressed the sum of each Y matrix of the sub-network:

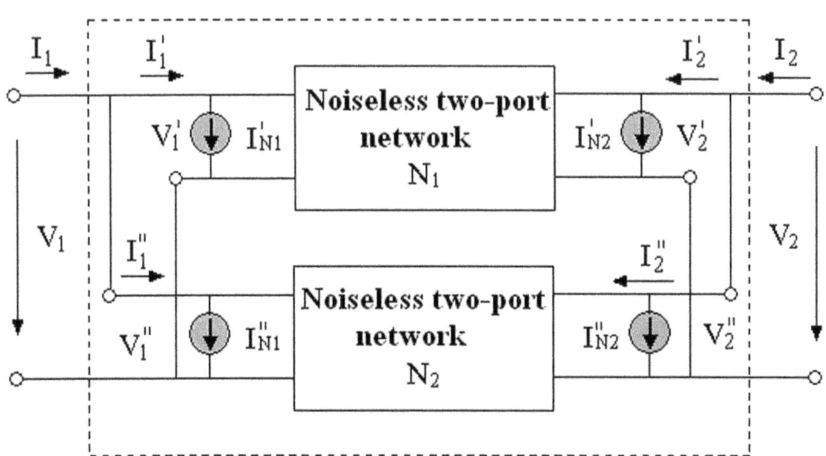

Figure 2.14 Parallel connection of two two-port networks

$$Y = Y_1 + Y_2 \tag{2.195}$$

The noise currents I''_{N1} and I''_{N2} can be expressed:

$$I_{N1} = I'_{N1} + I''_{N1} \tag{2.196}$$

$$I_{N2} = I'_{N2} + I''_{N2} \tag{2.197}$$

where I'_{N1} and I''_{N1}, I'_{N2} and I''_{N2} are independent noise sources, respectively:

$$\left\langle I'_{N1} \cdot (I''_{N1})^* \right\rangle = \left\langle I'_{N2} \cdot (I''_{N2})^* \right\rangle = 0 \tag{2.198}$$

Based on the definition of the admittance noise matrix, we have

$$\begin{aligned} C_{Y11} &= \left\langle I_{N1} \cdot I^*_{N1} \right\rangle = \left\langle (I'_{N1} + I''_{N1}) \cdot (I'_{N1} + I''_{N1})^* \right\rangle \\ &= \left\langle I'_{N1} \cdot (I'_{N1})^* \right\rangle + \left\langle I''_{N1} \cdot (I''_{N1})^* \right\rangle \\ &= C'_{Y11} + C''_{Y11} \end{aligned} \tag{2.199}$$

Similarly,

$$C_{Y22} = C'_{Y22} + C''_{Y22} \tag{2.200}$$

$$C_{Y12} = C'_{Y12} + C''_{Y12} \tag{2.201}$$

$$C_{Y21} = C'_{Y21} + C''_{Y21} \tag{2.202}$$

Therefore, the resulting impedance noise matrix for parallel connection of networks is the sum of respective the noise matrix in impedance of the original two-ports:

$$C_Y = C'_Y + C''_Y \tag{2.203}$$

2.4.3 Cascading Networks

As mentioned already, the chain matrix is most suitable when cascading networks, as depicted in Figure 2.15. In this case the current on the output of the first network is equal in value, but opposite in sign to the input current of the second network. The voltage drop across the output port of the first network is equal to the voltage across the input port of the second network. The resulting chain matrix can be written with the following relations:

$$\begin{bmatrix} V_1 \\ I_1 \end{bmatrix} = \begin{bmatrix} V_1^{'} \\ I_1^{'} \end{bmatrix} = [A_1]\cdot\begin{bmatrix} V_1^{''} \\ I_1^{''} \end{bmatrix} = [A_1]\cdot[A_2]\cdot\begin{bmatrix} V_2^{''} \\ -I_2^{''} \end{bmatrix} = [A_1]\cdot[A_2]\cdot\begin{bmatrix} V_2 \\ -I_2 \end{bmatrix} \tag{2.204}$$

Therefore over all system chain matrix is equal to the product of the chain matrices of the individual networks:

$$[A] = [A_1]\cdot[A_2] \tag{2.205}$$

By using incident and reflection wave instead of the current and voltage:

$$\begin{bmatrix} a_1 \\ b_1 \end{bmatrix} = \begin{bmatrix} a_1^{'} \\ b_1^{'} \end{bmatrix} = [T_1]\cdot\begin{bmatrix} b_2^{'} \\ a_2^{'} \end{bmatrix} = [T_1]\cdot\begin{bmatrix} a_1^{''} \\ b_1^{''} \end{bmatrix} = [T_1]\cdot[T_2]\cdot\begin{bmatrix} a_2 \\ b_2 \end{bmatrix} \tag{2.206}$$

Similarly, over all system chain matrix is equal to the product of the chain scattering matrices of the individual networks:

$$[T] = [T_1]\cdot[T_2] \tag{2.207}$$

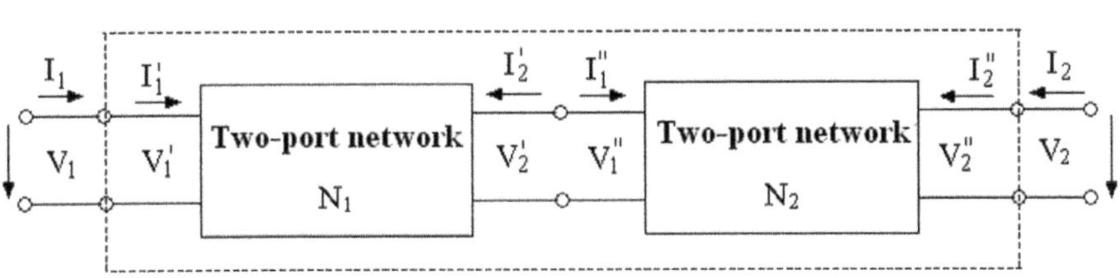

Figure 2.15 Cascading connection of two two-port networks

The S-parameters of the resulting network are as follows:

$$S_{11}^{T} = S_{11}^{A} + \frac{S_{11}^{B}S_{12}^{A}S_{21}^{A}}{1 - S_{11}^{B}S_{22}^{A}} \tag{2.208}$$

$$S_{21}^{T} = \frac{S_{21}^{A}S_{21}^{B}}{1 - S_{11}^{B}S_{22}^{A}} \tag{2.209}$$

$$S_{12}^{T} = \frac{S_{12}^{A}S_{12}^{B}}{1 - S_{11}^{B}S_{22}^{A}} \tag{2.210}$$

$$S_{22}^{T} = S_{22}^{B} + \frac{S_{12}^{B}S_{21}^{B}S_{22}^{A}}{1 - S_{11}^{B}S_{22}^{A}} \tag{2.211}$$

where A and B denote each of two sub-networks in cascade.

The corresponding noise equivalent circuit is shown in Figure 2.16;: V_N' and I_N' are the input noise sources of sub-network N_1, and V_N'' and I_N'' are the input noise sources of sub-network N_2.

To calculate the noise correlation matrix for cascading connection of two two-port networks,it is necessary to transform the noise sources of sub-network N_2 to the input port of the first sub-network N_1 [8]. The detail procedure is shown in Figure 2.17. The noise sources u'' and i'' can be expressed as follows:

$$u'' = A_{11}V_N'' + A_{12}I_N'' \tag{2.212}$$

$$i'' = A_{21}V_N'' + A_{22}I_N'' \tag{2.213}$$

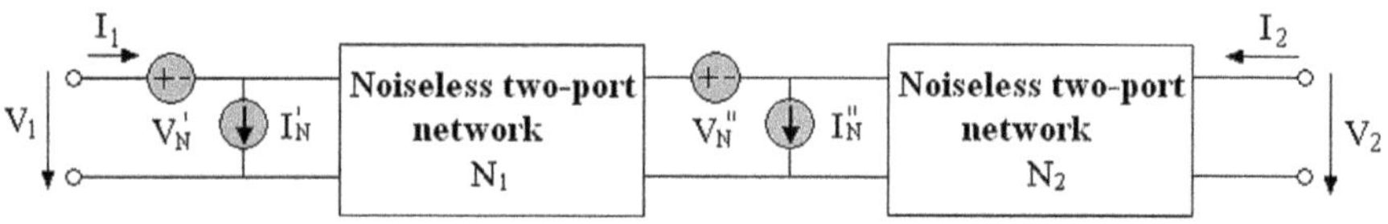

Figure 2.16 Noise circuit of cascading connection of two two-port networks

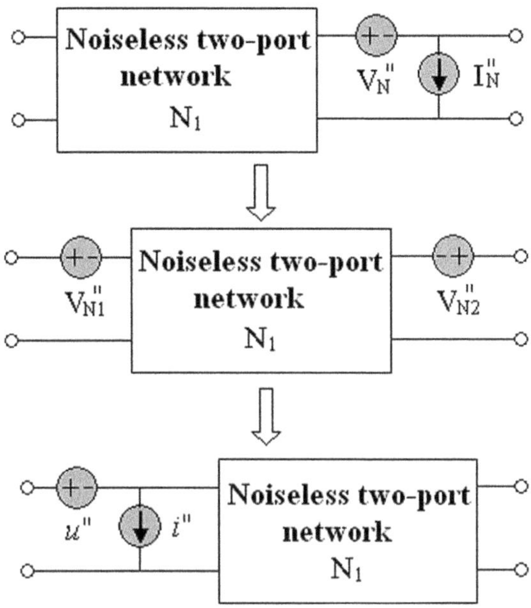

Figure 2.17 Transformation of noise sources of sub-network N_2 to N_1

The resulting impedance noise matrix for cascading connection of networks can be expressed as follows:

$$C_A = C_{A1} + A_1 C_{A2} A_1^+ \tag{2.214}$$

It is noted that a more complicated relation is obtained for cascading connection of networks, which additionally contains the electrical matrix A_1 of the first two-port network.

2.5 Relationship between Three-Port and Two-Port

It is well known that, the microwave active devices such as MESFET and HEMT normally are called two-port network, the reason is that source are perfectly grounded. Actually, MESFET and HEMT are three-port networks. It is necessary to obtained the three-port S-parameters when designing the complex microwave circuit, especially serried and parallel feedback exist. This section focuses on the relationship between three-port and two-port networks.

2.5.1 Conversion S parameters from Three-Port to Two-port

The two-port S-parameters of a three-terminal device, like a transistor, which one terminal grounded can be converted to three-port S-parameters when the grounded terminal is used as a third terminal. Figure 2.18 shows the three-port network and corresponding three two-port networks. For the three-port network, we have [9–10]:

$$\begin{bmatrix} b_1 \\ b_2 \\ b_3 \end{bmatrix} = \begin{bmatrix} S_{11} & S_{12} & S_{13} \\ S_{21} & S_{22} & S_{23} \\ S_{31} & S_{32} & S_{33} \end{bmatrix} \begin{bmatrix} a_1 \\ a_2 \\ a_3 \end{bmatrix} \tag{2.215}$$

For indefinite networks, the three-port S parameters are found by using the following principle:

$$\sum_{i=1}^{3} S_{ij} = 1 \tag{2.216}$$

$$\sum_{j=1}^{3} S_{ij} = 1 \tag{2.217}$$

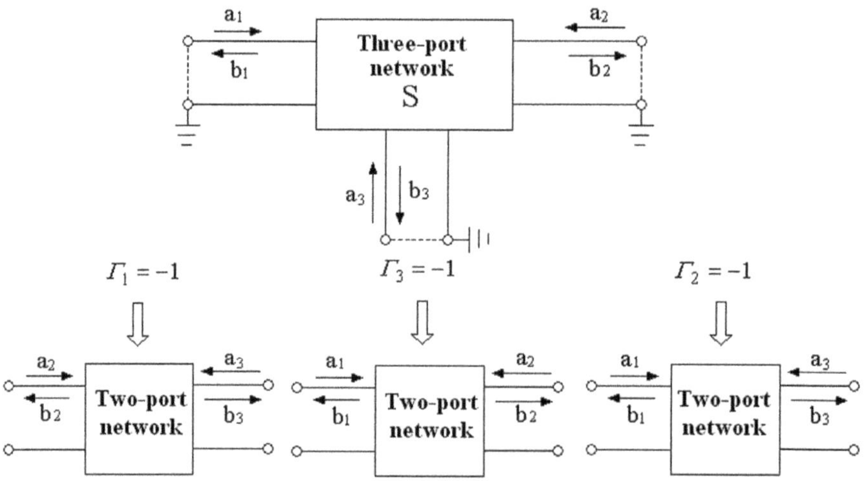

Figure 2.18 Three-port network and corresponding three two-port networks

When terminal three is perfectly ground, we have:

$$\Gamma_3 = \frac{a_3}{b_3} = -1 \tag{2.218}$$

With (2.218) substitute into (2.215), the new two-port circuit is characterized by the following linear equation:

$$\begin{bmatrix} b_1 \\ b_2 \end{bmatrix} = \begin{bmatrix} S_{11}^T & S_{12}^T \\ S_{21}^T & S_{22}^T \end{bmatrix} \begin{bmatrix} a_1 \\ a_2 \end{bmatrix} = \begin{bmatrix} S_{11} - \dfrac{S_{13}S_{31}}{1+S_{33}} & S_{12} - \dfrac{S_{13}S_{32}}{1+S_{33}} \\ S_{21} - \dfrac{S_{31}S_{23}}{1+S_{33}} & S_{22} - \dfrac{S_{23}S_{32}}{1+S_{33}} \end{bmatrix} \begin{bmatrix} a_1 \\ a_2 \end{bmatrix} \tag{2.219}$$

If S parameters of the arbitrary two-port network are known, the corresponding three-port S-parameters can be directly obtained from equation (2.219):

$$S_{33} = \frac{\sum_{i=1}^{2}\sum_{j=1}^{2} S_{ij}^T}{4 - \sum_{i=1}^{2}\sum_{j=1}^{2} S_{ij}^T} \tag{2.220}$$

$$S_{32} = \frac{1+S_{33}}{2}(1 - S_{12}^T - S_{22}^T) \tag{2.221}$$

$$S_{23} = \frac{1+S_{33}}{2}(1 - S_{21}^T - S_{22}^T) \tag{2.222}$$

$$S_{22} = S_{22}^T + \frac{S_{23}S_{32}}{1+S_{33}} \tag{2.223}$$

$$S_{13} = 1 - S_{23} - S_{33} \tag{2.224}$$

$$S_{31} = 1 - S_{33} - S_{32} \tag{2.225}$$

$$S_{13} = 1 - S_{23} - S_{33} \tag{2.226}$$

$$S_{21} = 1 - S_{22} - S_{23} \tag{2.227}$$

$$S_{11} = 1 - S_{21} - S_{31} \tag{2.228}$$

If terminal three is not perfectly grounded, an arbitrary impedance at port three occurs, the matrix (2.219) becomes

$$\begin{bmatrix} b_1 \\ b_2 \end{bmatrix} = \begin{bmatrix} S_{11} + \dfrac{\Gamma_3 S_{13} S_{31}}{1 - \Gamma_3 S_{33}} & S_{12} + \dfrac{\Gamma_3 S_{13} S_{32}}{1 - \Gamma_3 S_{33}} \\ S_{21} \dfrac{\Gamma_3 S_{31} S_{23}}{1 - \Gamma_3 S_{33}} & S_{22} + \dfrac{\Gamma_3 S_{23} S_{32}}{1 - \Gamma_3 S_{33}} \end{bmatrix} \begin{bmatrix} a_1 \\ a_2 \end{bmatrix} \tag{2.229}$$

The corresponding three-port S-parameters are as follows:

$$S_{33} = \frac{\sum_{i=1}^{2}\sum_{j=1}^{2} S_{ij}^T - \Gamma_3 - 1}{1 - 3\Gamma_3 - \sum_{i=1}^{2}\sum_{j=1}^{2} S_{ij}^T} \tag{2.230}$$

$$S_{i3} = \frac{1 - \Gamma_3 S_{33}}{1 - \Gamma_3}(1 - \sum_{j=1}^{2} S_{ij}^T) \quad (i = 1,2) \tag{2.231}$$

$$S_{3j} = \frac{1 - \Gamma_3 S_{33}}{1 - \Gamma_3}(1 - \sum_{i=1}^{2} S_{ij}^T) \quad (j = 1,2) \tag{2.232}$$

$$S_{ij} = S_{ij}^T - \frac{\Gamma_3 S_{i3} S_{3j}}{1 - \Gamma_3 S_{33}} \quad (i,j = 1,2) \tag{2.233}$$

When terminal two is perfectly ground, we have:

$$\Gamma_2 = \frac{b_2}{a_2} = -1 \tag{2.234}$$

Substituting (2.234) into (2.215), the new two-port circuit is characterized by the following linear equation:

$$\begin{bmatrix} b_1 \\ b_3 \end{bmatrix} = \begin{bmatrix} S_{11}^T & S_{13}^T \\ S_{31}^T & S_{33} \end{bmatrix} \begin{bmatrix} a_1 \\ a_3 \end{bmatrix} = \begin{bmatrix} S_{11} - \dfrac{S_{12}S_{21}}{1+S_{22}} & S_{13} - \dfrac{S_{12}S_{23}}{1+S_{22}} \\ S_{31} - \dfrac{S_{32}S_{21}}{1+S_{22}} & S_{33} - \dfrac{S_{32}S_{23}}{1+S_{22}} \end{bmatrix} \begin{bmatrix} a_1 \\ a_3 \end{bmatrix} \tag{2.235}$$

When terminal one is perfectly grounded, we have

$$\Gamma_1 = \frac{b_1}{a_1} = -1 \tag{2.236}$$

Substituting (2.236) into (2.215), the new two-port circuit is characterized by the following linear equation:

$$\begin{bmatrix} b_3 \\ b_2 \end{bmatrix} = \begin{bmatrix} S_{33}^T & S_{32}^T \\ S_{23}^T & S_{22}^T \end{bmatrix} \begin{bmatrix} a_3 \\ a_2 \end{bmatrix} = \begin{bmatrix} S_{33} - \dfrac{S_{31}S_{13}}{1+S_{11}} & S_{32} - \dfrac{S_{31}S_{12}}{1+S_{11}} \\ S_{23} - \dfrac{S_{21}S_{13}}{1+S_{11}} & S_{22} - \dfrac{S_{21}S_{12}}{1+S_{11}} \end{bmatrix} \begin{bmatrix} a_3 \\ a_2 \end{bmatrix} \tag{2.237}$$

From (2.229), (2.235), and (2.237), it can be observed that all *S*-parameters of the two-port network can be expressed as follows:

$$S_{ji}^{T} = S_{ji} - \frac{S_{ki}S_{jk}}{1+S_{kk}} \tag{2.238}$$

where *k* represents the number of the grounded port.

2.5.2 Conversion Noise Waves from Three-Port to Two-Port

Much like *S*-parameters, the noise properties of the three-port network can be characterized by the three noise waves at each port [4–5]. Assuming b_{N1} is the reflected noise power wave at the first port, b_{N2} is the reflected noise power wave at the second port, and b_{N3} is the reflected noise power wave at the third port, the noise wave complex amplitudes at the ports are related by a linear matrix equation for a three-port network:

$$\begin{bmatrix} b_1 \\ b_2 \\ b_3 \end{bmatrix} = \begin{bmatrix} S_{11} & S_{12} & S_{13} \\ S_{21} & S_{22} & S_{23} \\ S_{31} & S_{32} & S_{33} \end{bmatrix} \begin{bmatrix} a_1 \\ a_2 \\ a_3 \end{bmatrix} + \begin{bmatrix} b_{N1} \\ b_{N2} \\ b_{N3} \end{bmatrix} \tag{2.239}$$

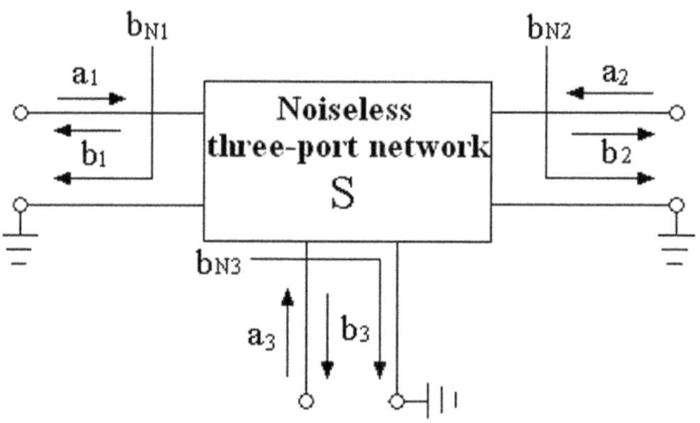

Figure 2.19 A noisy three-port with three outgoing noise waves

Let us consider the general case when terminal three is terminated with an arbitrary impedance as shown in Figure 2.20. The new two-port circuit is then characterized by the following linear equation:

$$\begin{bmatrix} b_1 \\ b_2 \end{bmatrix} = \begin{bmatrix} S_{11} + \dfrac{\Gamma_3 S_{13} S_{31}}{1-\Gamma_3 S_{33}} & S_{12} + \dfrac{\Gamma_3 S_{13} S_{32}}{1-\Gamma_3 S_{33}} \\ S_{21} + \dfrac{\Gamma_3 S_{31} S_{23}}{1-\Gamma_3 S_{33}} & S_{22} + \dfrac{\Gamma_3 S_{23} S_{32}}{1-\Gamma_3 S_{33}} \end{bmatrix} \begin{bmatrix} a_1 \\ a_2 \end{bmatrix} + \begin{bmatrix} c_{N1} \\ c_{N2} \end{bmatrix} \tag{2.240}$$

That is

$$b_1 = a_1 (S_{11} + \frac{\Gamma_3 S_{13} S_{31}}{1-\Gamma_3 S_{33}}) + a_2 (S_{12} + \frac{\Gamma_3 S_{13} S_{32}}{1-\Gamma_3 S_{33}}) + c_{N1} \tag{2.241}$$

$$b_2 = a_1 (S_{21} + \frac{\Gamma_3 S_{31} S_{23}}{1-\Gamma_3 S_{33}}) + a_2 (S_{22} + \frac{\Gamma_3 S_{23} S_{32}}{1-\Gamma_3 S_{33}}) + c_{N2} \tag{2.242}$$

where C_{N1} and C_{N2} are the noise waves and are given by:

$$c_{N1} = b_{N3} \frac{\Gamma_3 S_{13}}{1-\Gamma_3 S_{33}} + a_{N3} \frac{S_{13}}{1-\Gamma_3 S_{33}} + b_{N1} \tag{2.243}$$

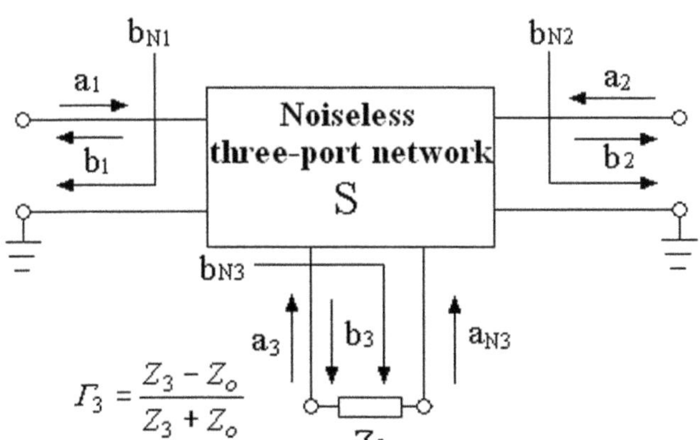

Figure 2.20 A general three-port network with a termination of Γ_3

$$c_{N2} = b_{N3}\frac{\Gamma_3 S_{23}}{1-\Gamma_3 S_{33}} + a_{N3}\frac{S_{23}}{1-\Gamma_3 S_{33}} + b_{N2} \tag{2.244}$$

where a_{N3} is the additional noise wave generated by Z_3:

$$a_{N3} = kT\Delta f\left|1-\left|\Gamma_3\right|^2\right| \tag{2.245}$$

Since the noise waves will eventually be time averaged, we can say that:

$$b_{N1} + b_{N2} + b_{N3} = 0 \tag{2.246}$$

Based on the equations (2.243) through (2.246), the three-port noise waves are found from the two-port noise wave:

$$b_{N1} = C_{N1}(1+\frac{\Gamma_3 S_{13}}{1-\Gamma_3}) + C_{N2}\frac{\Gamma_3 S_{13}}{1-\Gamma_3} + \mathrm{a}_{\mathrm{N3}}\frac{S_{13}}{\Gamma_3 - 1} \tag{2.247}$$

$$b_{N2} = C_{N1}\frac{\Gamma_3 S_{23}}{1-\Gamma_3} + C_{N2}(1+\frac{\Gamma_3 S_{23}}{1-\Gamma_3}) + \mathrm{a}_{\mathrm{N3}}\frac{S_{23}}{\Gamma_3 - 1} \tag{2.248}$$

$$b_{N3} = (C_{N1} + C_{N2})\frac{1-\Gamma_3 S_{33}}{\Gamma_3 - 1} - \mathrm{a}_{\mathrm{N3}}\frac{1-S_{33}}{\Gamma_3 - 1} \tag{2.249}$$

2.6 PI- and T-type Networks

PI- and T-type networks are the most commonly used two-port networks for semiconductor modeling [11–12], and the corresponding diagram of typical PI- and T-type networks are shown as in Figure 2.21, where Z_A, Z_B, and Z_C are the impedances in the PI-type network, and Z_1, Z_2, and Z_3 the impedances in the T-type network.

For the T-type network, the corresponding Z-parameters can be written directly as:

$$Z_{11} = Z_A + Z_B$$

$$Z_{12} = Z_{21} = Z_B \tag{2.250}$$

$$Z_{22} = Z_B + Z_C$$

For the PI-type network (the name of the network comes from the resemblance with the Greek letter π), the corresponding Y-parameters can be written directly as:

$$Y_{11} = \frac{1}{Z_1} + \frac{1}{Z_2}$$

$$Y_{12} = Y_{21} = -\frac{1}{Z_2} \tag{2.251}$$

$$Y_{22} = \frac{1}{Z_3} + \frac{1}{Z_2}$$

Table 2.13 gives the expressions of the microwave network parameters (such as Y, Z, chain, and S-parameters) for T- and PI-type networks. The relationship between T- and PI-type networks can be derived from the Table 2.13:

$$Z_A = \frac{Z_1 Z_2}{Z_1 + Z_2 + Z_3} \tag{2.252}$$

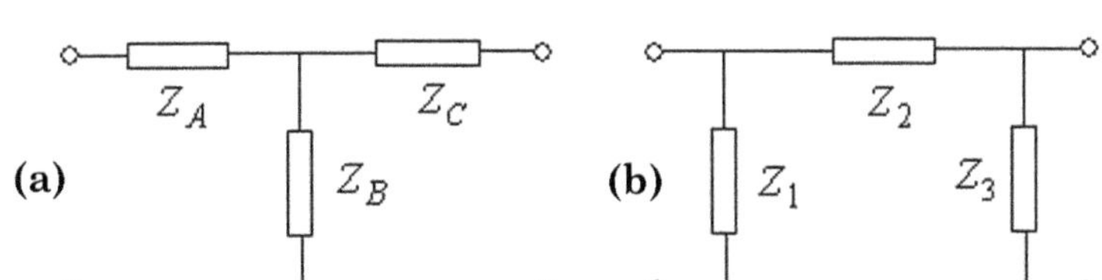

Figure 2.21 T- and PI-type networks

$$Z_B = \frac{Z_1 Z_3}{Z_1 + Z_2 + Z_3} \tag{2.253}$$

$$Z_C = \frac{Z_2 Z_3}{Z_1 + Z_2 + Z_3} \tag{2.254}$$

$$Z_1 = \frac{Z_A Z_B + Z_B Z_C + Z_A Z_C}{Z_C} \tag{2.255}$$

$$Z_2 = \frac{Z_A Z_B + Z_B Z_C + Z_A Z_C}{Z_B} \tag{2.256}$$

$$Z_3 = \frac{Z_A Z_B + Z_B Z_C + Z_A Z_C}{Z_A} \tag{2.257}$$

For passive two-port networks, impedance and admittance noise correlation matrix can be expressed as follows:

$$C_Z = 4kT\,\mathrm{Re}\{Z\} \tag{2.258}$$

$$C_Y = 4kT\,\mathrm{Re}\{Y\} \tag{2.259}$$

They are completely determined by the temperature T and the real part of their electrical matrices in impedance and admittance representation, respectively. For passive T-type networks, the impedance noise correlation matrix is as follows:

$$C_Z = 4kT\,\mathrm{Re}\left\{\begin{matrix} Z_A + Z_B & Z_B \\ Z_B & Z_B + Z_C \end{matrix}\right\} = 4kT\begin{bmatrix} R_A + R_B & R_B \\ R_B & R_B + R_C \end{bmatrix} \tag{2.260}$$

where R_A, R_B, and R_C are the real parts of the Z_A, Z_B, and Z_C, respectively.

The corresponding admittance noise correlation matrix is as follows:

$$C_Y = \frac{4kT}{|N|^2}\begin{bmatrix} D & R_B \\ R_B & D \end{bmatrix} \tag{2.261}$$

where

$$D = |Z_C|^2 (R_A + R_B) + |Z_B|^2 (R_A + R_C) + \mathrm{Re}\{Z_A Z_B Z_C^*\} + \mathrm{Re}\{Z_A Z_C Z_B^*\}$$

$$N = Z_A Z_B + Z_A Z_C + Z_B Z_C$$

For passive PI-type networks, the admittance noise correlation matrix is as follows:

$$C_Y = 4kT\,\mathrm{Re}\begin{Bmatrix} Y_1 + Y_2 & -Y_2 \\ -Y_2 & Y_2 + Y_3 \end{Bmatrix} = 4kT\begin{bmatrix} G_1 + G_2 & -G_2 \\ -G_2 & G_2 + G_3 \end{bmatrix} \tag{2.262}$$

where G_1, G_2, and G_3 are the real parts of admittances Y_1, Y_2, and Y_3.

2.7 Summary

Having defined the signal and noise parameters of two-port networks in this chapter, and the relationship between various signal parameters, relationship between various noise parameters (i.e., *Z*, *Y*, *ABCDS*- and *T*-parameters) are presented.

Table 2.13 Microwave Network Parameters for T- and PI-Type Networks

	T-Type Network	PI-Type Network
Z	$Z_{11} = Z_A + Z_B$ $Z_{12} = Z_{21} = Z_B$ $Z_{22} = Z_B + Z_C$	$Z_{11} = \dfrac{Z_1(Z_2 + Z_3)}{Z_1 + Z_2 + Z_3}$ $Z_{12} = Z_{21} = \dfrac{Z_1 Z_3}{Z_1 + Z_2 + Z_3}$ $Z_{22} = \dfrac{Z_3(Z_1 + Z_2)}{Z_1 + Z_2 + Z_3}$
Y	$Y_{11} = \dfrac{Z_B + Z_C}{Z_A Z_B + Z_B Z_C + Z_A Z_C}$ $Y_{21} = Y_{12}$ $= -\dfrac{Z_B}{Z_A Z_B + Z_B Z_C + Z_A Z_C}$ $Y_{22} = \dfrac{Z_A + Z_B}{Z_A Z_B + Z_B Z_C + Z_A Z_C}$	$Y_{11} = Y_1 + Y_2$ $Y_{12} = Y_{21} = -Y_2$ $Y_{22} = Y_3 + Y_2$ $Y_1 = \dfrac{1}{Z_1} \quad Y_2 = \dfrac{1}{Z_2} \quad Y_3 = \dfrac{1}{Z_3}$
A	$A = 1 + \dfrac{Z_A}{Z_B}$ $B = Z_C + Z_A(\dfrac{Z_C}{Z_B} + 1)$ $C = \dfrac{1}{Z_B} \quad D = 1 + \dfrac{Z_C}{Z_B}$	$A = 1 + \dfrac{Z_2}{Z_3} \quad B = Z_2$ $C = \dfrac{1}{Z_1}(1 + \dfrac{Z_2}{Z_3}) + \dfrac{1}{Z_3}$ $D = 1 + \dfrac{Z_2}{Z_1}$
S	$S_{11} = \dfrac{-Z_o^2 + (Z_A - Z_C)Z_o + T}{Z_o^2 + (Z_A + Z_C + 2Z_B)Z_o + T}$ $S_{21} = S_{12}$ $= \dfrac{2Z_o Z_B}{Z_o^2 + (Z_A + Z_C + 2Z_B)Z_o + T}$	$S_{11} = \dfrac{Y_o^2 - (Y_1 - Y_3)Y_o - T}{Y_o^2 + (Y_1 + Y_3 + 2Y_2)Y_o + T}$ $S_{21} = S_{12}$ $= \dfrac{2Y_o Y_2}{Y_o^2 + (Y_1 + Y_3 + 2Y_2)Y_o + T}$

Table 2.13 Microwave Network Parameters for T- and PI-Type Networks (Continued)

S	$S_{22}=\dfrac{-Z_o^2-(Z_A-Z_C)Z_o+T}{Z_o^2+(Z_A+Z_C+2Z_B)Z_o+T}$ $T=Z_AZ_B+Z_AZ_C+Z_BZ_C$	$S_{22}=\dfrac{Y_o^2+(Y_1-Y_3)Y_o-T}{Y_o^2+(Y_1+Y_3+2Y_2)Y_o+T}$ $T=Y_1Y_2+Y_1Y_3+Y_2Y_3$
T	$T_{11}=\dfrac{Z_o^2+(Z_A+Z_C+2Z_B)Z_o+\Delta Z}{2Z_oZ_B}$ $T_{21}=\dfrac{-Z_o^2+(Z_A-Z_C)Z_o+\Delta Z}{2Z_oZ_B}$ $T_{12}=\dfrac{Z_o^2+(Z_A-Z_C)Z_o-\Delta Z}{2Z_oZ_B}$ $T_{22}=\dfrac{-Z_o^2+(Z_A+Z_C+2Z_B)Z_o-\Delta Z}{2Z_oZ_B}$ $\Delta Z=Z_{11}Z_{22}-Z_{12}Z_{21}$	$T_{11}=\dfrac{Y_o^2+(Y_1+Y_3+2Y_2)Y_o+\Delta Y}{2Y_oY_2}$ $T_{21}=\dfrac{Y_o^2-(Y_1-Y_3)Y_o-\Delta Y}{2Y_oY_2}$ $T_{12}=\dfrac{-Y_o^2-(Y_1-Y_3)Y_o+\Delta Y}{2Y_oY_2}$ $T_{22}=\dfrac{Y_o^2+(Y_1+Y_2+2Y_3)Y_o+\Delta Y}{2Y_oY_2}$ $\Delta Y=Z_{11}Z_{22}-Z_{12}Z_{21}$

Chapter 3

Microwave and RF Measurement Techniques

Microwave and RF measurement techniques are the basis of characterization of the microwave and RF devices and circuits, especially for semiconductor devices such as BJT and FET etc. The microwave and RF integrated circuits (ICs) also need verification by using microwave and RF measurements. It is noted that unlike the coarse measurement, the microwave and RF measurements techniques are the highly accurate measurements; for example, a small error will cause a large discrepancy for the semiconductor device modeling and parameter extraction, and the corresponding RF ICs designed by using the device model. Therefore, before introducing the FET modeling and parameter extraction techniques, first the basis of the microwave and RF measurement techniques are provided.

Microwave and RF measurements can be classified in two distinct but often overlapping categories [1–2]: signal measurements and network measurements. Signal measurements include observation and determination of the characteristics of waves and waveforms. These parameters can be obtained in time, frequency, or modulation domain. Network measurement determines the terminal and signal transfer characteristics of device and systems with any number of ports. Because of this book focuses on the semiconductor modeling and parameter extraction, this chapter strongly emphasizes on the three detail measurements in the frequency domain as follows:

1. S-parameters measurement technique (belong to network measurement)
2. Noise measurement technique (belong to signal measurement)
3. Nonlinear measurement technique (belong to signal measurement)

3.1 *S*-parameters Measurement

As mentioned in Chapter 2, when moving to higher and higher frequencies, it is difficult to directly measure the impedance *Z*-, admittance *Y*-, and *ABCD*- parameters, which based on the voltages and currents at the input and output ports. The reasons are as follows:

1. Equipment is not readily available to measure total voltage and total current at the ports of the network.
2. Short and open circuits are difficult to achieve over a broad band of frequencies.
3. Active devices, such as transistors and tunnel diodes, very often will not be short- or open-circuit stable.

To overcome these problems, the logical variables to use at these frequencies are traveling waves rather than total voltages and currents. The scattering (*S*) parameters are currently the easiest parameters to measure at frequencies above a few tens of MHz as they are measured with 50 or 75 Ω network analyzers. In actuality, the 50 Ω reference impedance was selected from a trade-off between the lowest loss (77 Ω) and maximum power-handling dimension (30 Ω) for an air line coaxial cable.

3.1.1 *S*-parameters Measurement System

Figure 3.1 shows the basic diagram for *S*-parameters measurement system; the major components consists of the vector network analyzer (VNA), device under test (DUT), power supply, bias tee, and RF cables, and bias cables connect between them.

RF cables are used to connect the VNA and DUT. The insertion loss and voltage standing wave ratio (VSWR) are required as low as possible for qualified RF cables. To achieve this, the cable's dielectric core has a low dielectric constant and loss tangent. The outer diameter of the center conductor provides plenty of conductive surface, diminishing its inductance. There are two types of the RF cables: semi-rigid and flexible. The commonly used dimensions are as follows:

1. 3.5 mm Operating frequency up to 26.5 GHz
2. 2.9 mm Operating frequency up to 40 GHz
3. 2.5 mm Operating frequency up to 50 GHz
4. 1.0 mm Operating frequency up to 110 GHz

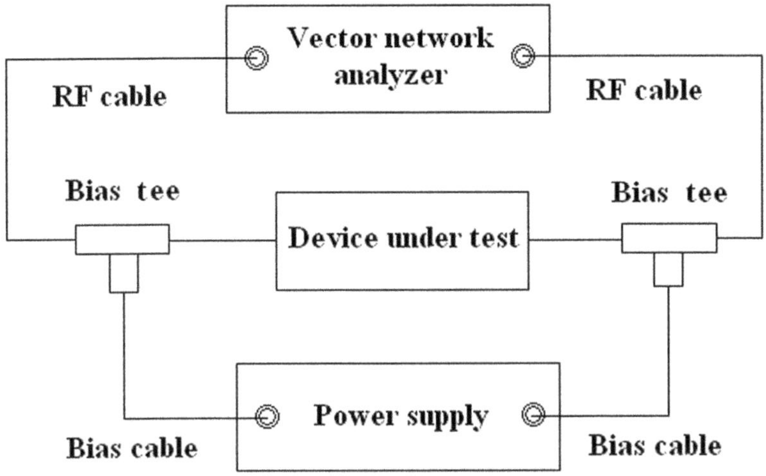

Figure 3.1 *S*-parameters measurement system diagram

Bias cables connect the bias tee and power supply source; the commonly used bias cables are triaxial cable which includes three concentric conductors: force/sense, guard, and common. Voltage and current are applied on the force and measured with sense. With low resistance and negligible current, the triaxial sense line precisely reads the force voltage or current, thereby capturing any voltage drop along the fore line. At the power supply end of the sense line, a high impedance prevents current from flowing into the power supply. The structure of the bias cables is shown as Figure 3.2(a).

The bias network provides a means of supplying DC bias to the center conductor of a coaxial line of a component or device while blocking the dc bias to the RF input port. The bias tee includes two RF ports and another two ports for DC power supply; the RF signal is fed into the input port of the bias tee, and RF signal and DC depart from the output port as shown in Figure 3.2(b). Figure 3.3 shows a schematic diagram of the bias network. The circuit is a "tee" in which the capacitor in the left arm acts as a DC block / high-pass filter. The vertical arm, with its series inductance and shunt capacitance, acts as a low-pass filter. The high-pass filter keeps DC from flowing into the VNA, while the low-pass filter keeps the applied RF from disturbing the power supply.

3.1.2 Vector Network Analyzer

Network analyzers are traditionally used to measure the transmission and reflection characteristics of components, circuits, and devices. The

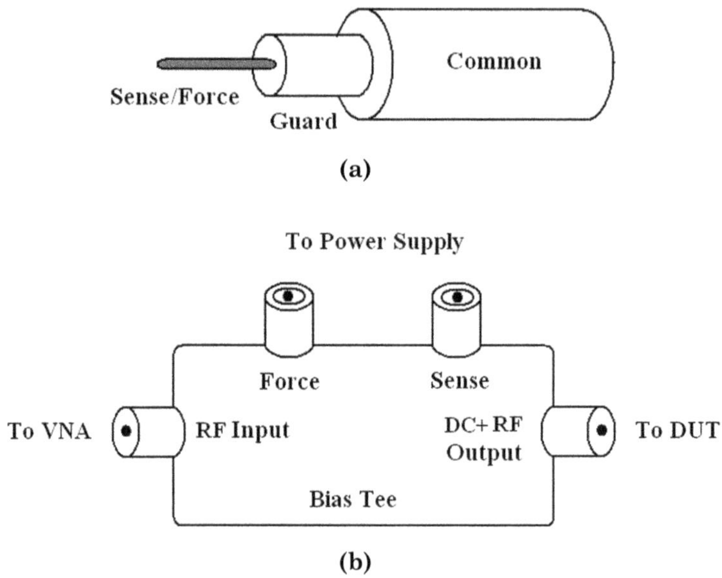

Figure 3.2 Diagram of bias tee and bias cable

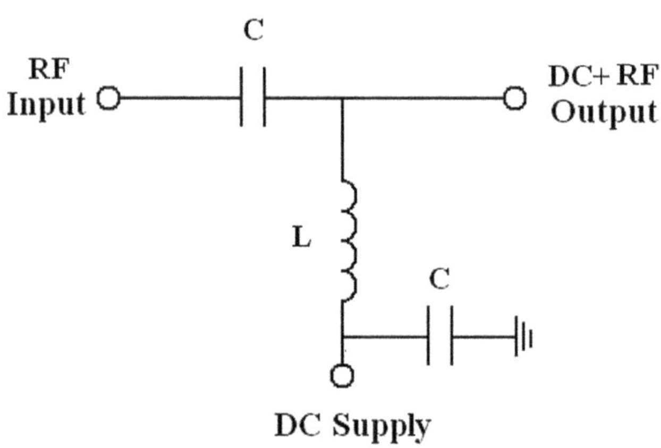

Figure 3.3 Equivalent circuit model of bias tee

network analyzer measures the magnitude, phase, and group delay of two-port networks to characterize their linear behavior. Optionally, the some of network analyzers are also capable of displaying a network's time domain response to an impulse or a step waveform by computing the inverse Fourier. With unmatched accuracy and convenience, the vector network analyzer normally makes broadband measurements from 45 MHz to 50 GHz in 2.4 mm coax, from 45 MHz to 110 GHz in 1.0 mm coax and from 33 GHz to 110 GHz in waveguide bands.

The commonly used network analyzers are the scalar network analyzer and vector network analyzer. Figure 3.4(a) shows the basic network analyzer diagram of the scalar network analyzer system, which normally consists of a source for stimulus, signal-separation devices, receivers for signal detection, and display/processing circuitry for reviewing results [1]. The source is usually a built-in phase-locked (synthesized) voltage-controlled oscillator. The RF signal from the swept oscillator is fed into the signal-separation device. Signal-separation hardware allows measurements of a portion of the incident signal to provide a reference for ratio measurements, and it separates the inci-

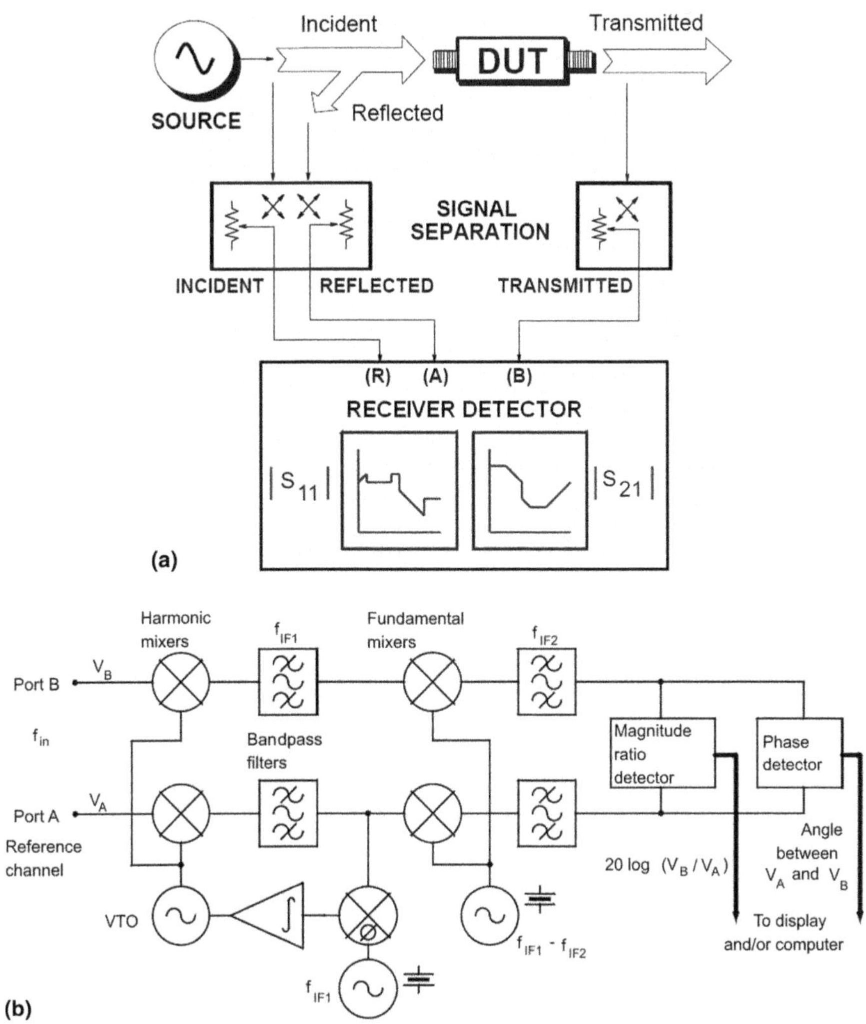

Figure 3.4 Basic network analyzer diagram: (a) scalar (b) vector

dent (forward) and reflected (reverse) signals present at the input of the DUT. Hardware for this purpose includes power dividers (which are resistive and broadband, but have high insertion loss), directional couplers (which have low loss but are usually limited in bandwidth), and directional bridges (which are useful for measuring reflected signals over a broad bandwidth, but may also have significant loss). The dc voltage V_R detected by detector R is proportional to the actual power input to the DUT. A portion of the signal power fed to the DUT is reflected by the DUT to the separation device. The corresponding dc voltage V_A detected by detector A is proportional to the reflected power. The dc voltage V_B detected by detector B is proportional to the output power of the DUT.

Therefore the magnitude of input reflection coefficient S_{11} can be expressed by the ratio of V_A and V_R:

$$|S_{11}| = \sqrt{\frac{V_A}{V_R}} \tag{3.1a}$$

The magnitude of transmission coefficient ration S_{21} can be expressed by the ratio of V_B and V_R:

$$|S_{21}| = \sqrt{\frac{V_B}{V_R}} \tag{3.1b}$$

Similarly, by exchanging the input and output port, the magnitudes of output reflection coefficient S_{22} and feedback coefficient S_{12} can be obtained.

Note that scalar network analyzer can only establish the magnitude of the reflection coefficient so that an absolute impedance cannot be measured. To establish the impedance of a device, the phase angle of the reflected wave relative to the incident wave must be known. To measure the phase difference between the forward and reflected wave, a phase meter or vector network analyzer is used.

The simplified block diagram of a typical multi-channel VNA is shown in Figure 3.4(b). There are two channels fed from the test set. The inputs are converted first to a low intermediate frequency such as 20 MHz and then to 100 kHz before being routed to phase detectors. The first conversion oscillator is followed by a comb generator, and the oscillator is phase locked to the mixer output so the unit will frequency track the test source. The mixers convert the test and reference signals to a low frequency (at intermediate frequency, IF) where comparison of phase and magnitude between signals from the two channel are made.

Therefore the phase of input reflection coefficient S_{11} can be expressed by the phase difference between V_A and V_R:

$$\angle S_{11} = \angle V_A - \angle V_R \tag{3.2a}$$

The phase of transmission coefficient ration S_{21} can be expressed by the phase difference between V_B and V_R:

$$\angle S_{11} = \angle V_A - \angle V_R \tag{3.2b}$$

3.1.3 VNA Calibration

A considerable challenge in S-parameter VNA measurements is to define exactly where the measurement system ends and the DUT begins. This location is called *reference plane*. However, any measurement includes not only that of the DUT, but also from the fixture and cables. Note that with increasing frequency, the electrical contribution of the fixture and cables becomes increasingly significant. In addition, practical limitations of the VNA in the form of limited dynamic range, isolation, imperfect source/load match, and other imperfections contribute systematic error to the measurement. This means that all error contributions, inside the VNA and in the cables up to this reference plane, have to be calibrated out.

There are three sources of network analyzer measurement errors [3]:

1. Systematic errors: These include impedance mismatch and leakage terms in the test setup, isolation characteristics between the reference and test signal paths, and system frequency response. These are the repeatable, predictable errors that the system can measure and significantly reduce; thus, the accuracy of measurement can be enhanced. Vector accuracy enhancement applies to these errors.

2. Instrumentation errors: These errors are due to frequency inaccuracy, frequency instability, dynamic instability, resolution, and sensitivity. For high-quality equipment, these error contributions are a very small part of total measurement uncertainly.

3. Random errors: These errors are non-repeatable measurement variations that occur due to noise, environmental changes, frequency drift, and other physical changes in the test setup between calibra-

tion and measurement. These are any errors that the system itself cannot measure or cannot model with an acceptable degree of certainty. Uncertainty of repeatable measurement in test cables and connectors contribute greatly to these errors.

Network analyzer treats everything between the measurement reference planes as a single device. Therefore, our task is reduced to finding a way to calibrate the network analyzer in such a way that it becomes possible to eliminate the effect of all undesired influences or parasitics. The main goal of a calibration procedure is to characterize the error boxes prior to measuring the DUT.

Calibration must be performed within the operating temperature specified for the calibration kit. For all calibration kits the operating temperature is 23°C ±3°C. For a calibration to remain fully verifiable, the temperature of the network analyzer must remain within ±1°C around the initial measurement calibration temperature.

VNAs allow calibration with the following standards:

- **(T)**hrough (direct connection of both measurement planes),
- **(M)**atch (perfect termination in the system-impedance, typically 50 Ω),
- **(S)**hort (perfect short connected to the reference planes),
- **(O)**pen (perfect, eventually modelled open at the reference planes).

The calibration of a VNA is performed by rather complex procedures. Such are Short-Open-Load-Through (SOLT), Through-Reflection-Load (TRL), or Load-Reflection-Match (LRM) (to be presented shortly) and the associated error correction model [4–9].

3.1.3.1 Error Model

The network analyzer and text fixture calibration is usually based on the error models. The common used basic error model with automatic network analyzer is the 12-term error model [4–9]. Each error model term is defined by a complex scattering parameter separately at every frequency. Figure 3.5 shows the typical two-port VNA error model: (a) is the forward model, and (b) is the reverse model. The subscript F denotes a forward direction, and the subscript R denotes a reverse direction according to the source/load switch.

In Figure 3.5, we have

E_{DF}: the error associate with forward directivity of the bridge

E_{RF}: the error associate with forward reflection measurement

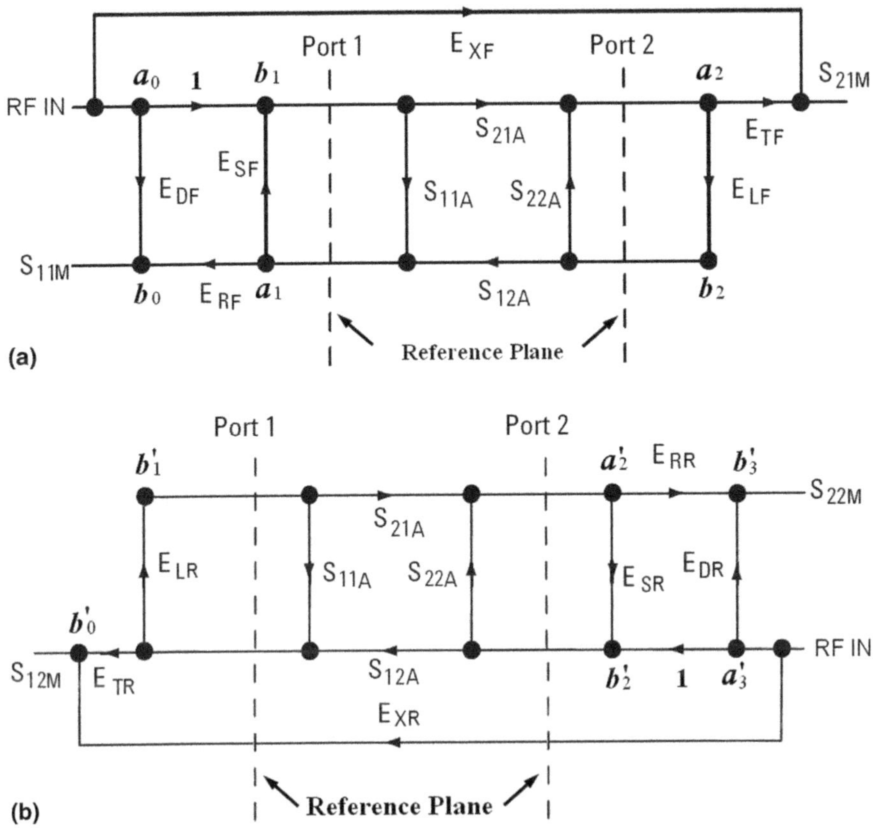

Figure 3.5 Typical two-port VNA error models: (a) forward model; (b) reverse model

E_{XF}: the error associate with forward crosstalk

E_{SF}: the error associate with forward source mismatch

E_{LF}: the error associate with forward load mismatch

E_{TF}: the error associate with forward transmission measurement

Similarly,

E_{DR}: the error associate with reverse directivity of the bridge

E_{RR}: the error associate with reverse reflection measurement

E_{XR}: the error associate with reverse crosstalk

E_{SR}: the error associate with reverse source mismatch

E_{LR}: the error associate with reverse load mismatch

E_{TR}: the error associate with reverse transmission measurement

Based on the signal flow chart, for a forward six-term error model the following can be seen:

$$b_0 = E_{DF}a_0 + E_{RF}a_1$$

$$b_1 = a_0 + E_{SF}a_1 \qquad (3.3a)$$

$$b_2 = E_{LF}a_2$$

$$b_3 = E_{XF}a_0 + E_{TF}a_2$$

where

$$a_1 = S_{11A} \cdot b_1 + S_{12A} \cdot b_2$$

$$a_2 = S_{21A} \cdot b_1 + S_{22A} \cdot b_2$$

Similarly, based on the signal flow chart, for a reverse six-term error model the following can be seen:

$$b_0' = E_{XR}a_3' + E_{TR}a_1'$$

$$b_1' = E_{LR}a_1' \qquad (3.3b)$$

$$b_2' = a_3' + E_{SR}a_2'$$

$$b_3' = E_{DR}a_3' + E_{RR}a_2'$$

where

$$a_1' = S_{11A} \cdot b_1' + S_{12A} \cdot b_2'$$

$$a_2' = S_{21A} \cdot b_1' + S_{22A} \cdot b_2'$$

The relationship between actual and measured S-parameters can be expressed as follows:

$$S_{11M} = E_{DF} + E_{RF} \frac{S_{11A} - E_{LF}\Delta S_A}{1 - E_{SF}S_{11A} - E_{LF}S_{22A} + E_{SF}E_{LF}\Delta S_A} \tag{3.4}$$

$$S_{21M} = E_{XF} + E_{TF} \frac{S_{21A}}{1 - E_{SF}S_{11A} - E_{LF}S_{22A} + E_{SF}E_{LF}\Delta S_A} \tag{3.5}$$

$$S_{12M} = E_{XR} + E_{TR} \frac{S_{12A}}{1 - E_{LR}S_{11A} - E_{SR}S_{22A} + E_{SR}E_{LR}\Delta S_A} \tag{3.6}$$

$$S_{22M} = E_{DR} + E_{RR} \frac{S_{22A} - E_{LR}\Delta S_A}{1 - E_{LR}S_{11A} - E_{SR}S_{22A} + E_{SR}E_{LR}\Delta S_A} \tag{3.7}$$

Two-port error correction yields the most accurate results because it accounts for all of the major sources of systematic error. The error model for a two-port device reveals the four S-parameters measured in the forward and reverse directions.

Once the system error terms have been characterized, the network analyzer utilizes four equations to derive the actual device S-parameters from the measured S-parameters. Because each S-parameter is a function of all four measured S-parameters, a network analyzer must make a forward and reverse test sweep before updating any one S-parameter.

3.1.3.2 SOLT Calibration Method

One of the most commonly used calibration methods for a 12-term error model is that SOLT (Short-Open-Load-Thru) [6]. By using a short, open, and load on each port together with a thru line between ports 1 and 2. The load may be fixed or sliding (used to improve the ultimate residual directivity). This is the most common calibration selection and usually works well for coaxial systems. The algorithm uses models for the standards (particularly the open), requiring characterization of those standards (usually by the manufacturer). The sequence of connection events in a typical SOLT calibration is shown in Figure 3.6.

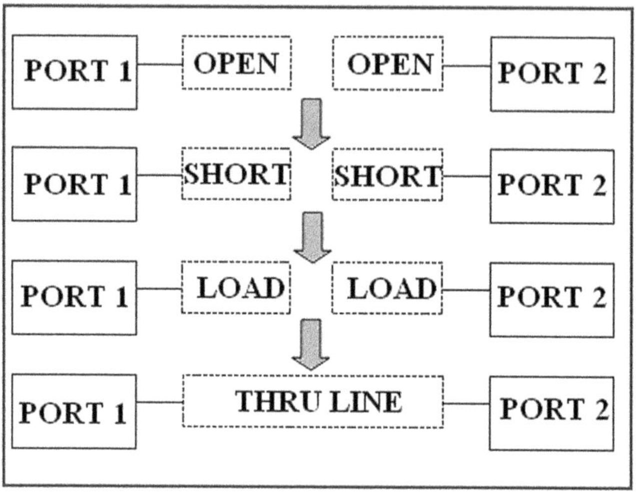

Figure 3.6 Sequence of connection events for SOLT

Determination of the error coefficients requires the use of several standards, although the choice of which standards to use is not necessarily unique. Traditionally, short, open, load, and through standards have been applied, especially in a coaxial medium that facilitates their accurate and repeatable fabrication. The actual device is replaced with known calibration standards, and the S-parameters of the four standards are as follows:

$$S_{short} = \begin{bmatrix} -e^{-2(\alpha+j\beta)l} & 0 \\ 0 & -e^{-2(\alpha+j\beta)l} \end{bmatrix} \tag{3.8}$$

$$S_{open} = \begin{bmatrix} e^{-2(\alpha+j\beta)l} & 0 \\ 0 & e^{-2(\alpha+j\beta)l} \end{bmatrix} \tag{3.9}$$

$$S_{load} = \begin{bmatrix} 0 & 0 \\ 0 & 0 \end{bmatrix} \tag{3.10}$$

$$S_{thru} = \begin{bmatrix} 0 & e^{-(\alpha+j\beta)l} \\ e^{-(\alpha+j\beta)l} & 0 \end{bmatrix} \tag{3.11}$$

The corresponding the transmission line model for short, open, load, and through standards are shown as in Figure 3.7(a), (b), (c), and (d), respectively, where

α = attenuation constant of propagation constant

β = phase constant of propagation constant

l = length of transmission line

$j = \sqrt{-1}$ imaginary unit

C_{open} = open-circuit capacitance

L_{short} = short-circuit inductance

L_{term} = load inductance

After determination of the 12-term error model parameters, the DUT parameters S_{11A} and S_{21A} can be describe as functions of the measured four S-parameters S_{11M}, S_{21M}, S_{12M}, S_{22M}, and the six forward error terms. Likewise, S_{11A} and S_{21A} can be describe as functions of the measured four S-parameters and the six reverse error terms:

$$S_{11A} = \frac{A(1+BE_{SR}) - CDE_{LF}}{(1+AE_{SF})(1+BE_{SR}) - CDE_{LF}E_{LR}} \tag{3.12}$$

$$S_{21A} = \frac{C[1+B(E_{SR} - E_{LF})]}{(1+AE_{SF})(1+BE_{SR}) - CDE_{LF}E_{LR}} \tag{3.13}$$

$$S_{12A} = \frac{D[1+A(E_{SF} - E_{LR})]}{(1+AE_{SF})(1+BE_{SR}) - CDE_{LF}E_{LR}} \tag{3.14}$$

$$S_{22A} = \frac{B(1+AE_{SF}) - CDE_{LR}}{(1+AE_{SF})(1+BE_{SR}) - CDE_{LF}E_{LR}} \tag{3.15}$$

with

$$A = \frac{S_{11M} - E_{DF}}{E_{RF}} E_{SF}$$

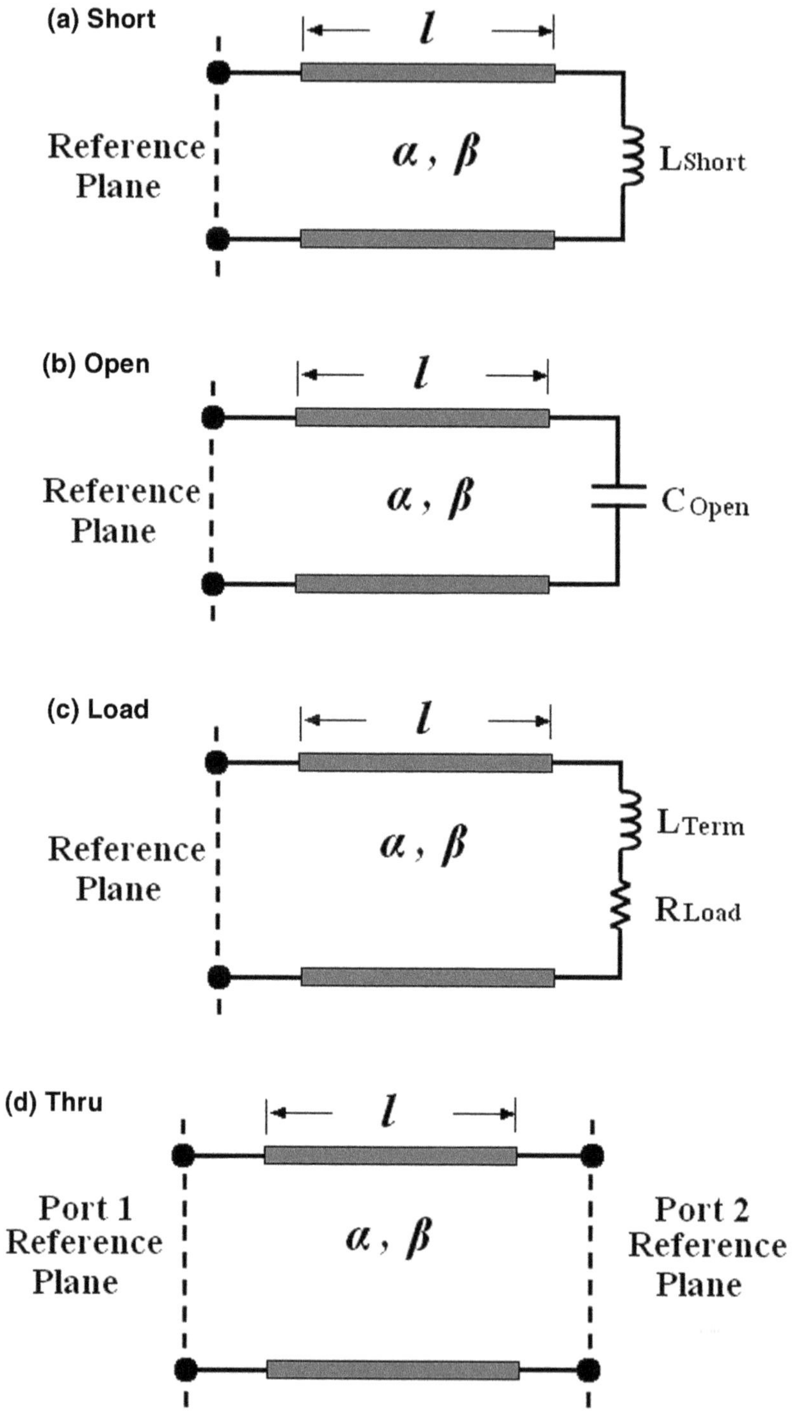

Figure 3.7 Short, open, load, and through standards transmission line model

$$B = \frac{S_{22M} - E_{DR}}{E_{RR}}$$

$$C = \frac{S_{21M} - E_{XF}}{E_{TF}}$$

$$D = \frac{S_{12M} - E_{XR}}{E_{TR}}$$

3.1.3.3 TRL Calibration Method

The simplest calibration method (e.g., SOLT) involves three or more known standards (open, short, and load). The problem with this approach is that such standards are usually imperfect and are likely to introduce errors into the measurement procedures. These errors become especially significant at higher frequencies. To avoid the dependency on the accuracy of calibration standards, several methods (e.g., TRL and LRM) have been developed. The transmission are simple to understand and easy to fabricate. Their physical dimension and the board material decide their characteristic impedance. Because it is based on a transmission standard, TRL is a powerful method.

The sequence of connection events in a typical TRL calibration is shown in Figure 3.8. The through connection is made by directly connecting ports 1 and 2 of the DUT. Next, the reflect (open or short) connection uses as a load with high reflectivity. The reflection coefficient does not have to be known because it will be determined during the calibration process. The line connection is made by connecting ports 1 and 2 via a transmission line matched to the impedance of the error boxes. The TRL technique is potentially very accurate as the quality of the impedance standard is determined primarily by dimensional precision.

3.1.3.4 LRM Calibration Method

In the LRM method, the line and reflect standards are analogous to the through and reflect standards in TRL. The method uses a reflect stan-

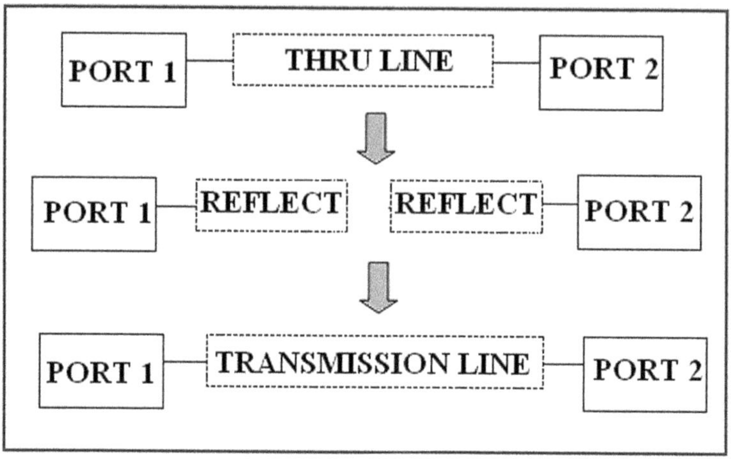

Figure 3.8 Sequence of connection events for TRL

dard (open or short) and a fixed load at each port together with a line between ports 1 and 2. LRM can also be used for on-wafer calibration. The match must be an excellent broadband termination. The LRM technique also permits calibration with fixed probe-probe spacing. The sequence of connection events in a typical TRL calibration is shown in Figure. 3.9.

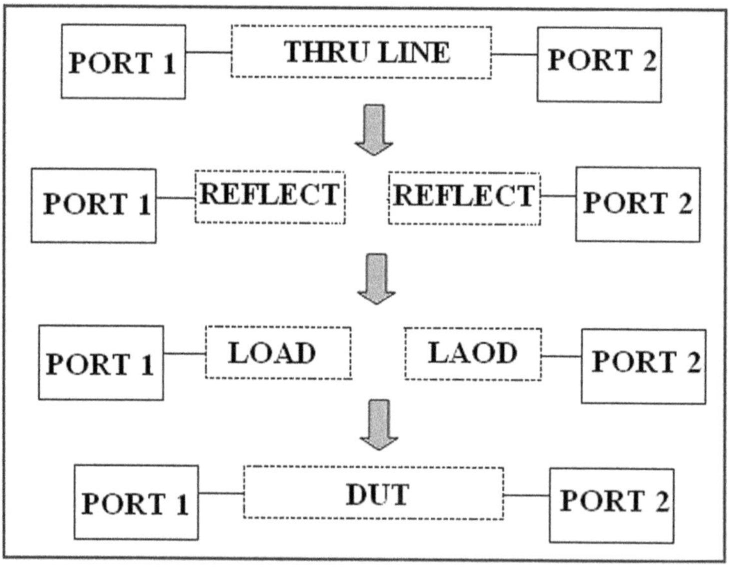

Figure 3.9 Sequence of connection events for LRM

3.1.4 De-embedding Method

As we know, placing the coplanar probes (or PIN) directly on the device to measure it is not possible. Indeed, measurement requires probe pads and interconnection lines leading to the DUT. The disadvantage is that the parasitics of the pads and interconnects can be larger than the device measure it self.

As mentioned already, the contributions from the fixture and cables as well can be calibrated by using SOLT, TRL, and LRM methods. However, the reference plan has been set only at the probe tips, and therefore de-embedding methods are used to move the electrical reference plan from the probe tips to the DUT (as shown in Figure 3.10).

Actually, the purpose of de-embedding is that removing the parasitics caused by pad pattern and feedline (or bond-wire), and de-embedding is the reverse calculation of the interconnections of two-port network [10]. There are three de-embedding techniques for application: parallel, series, and cascade de-embedding. The detailed procedures are discussed next.

3.1.4.1 Parallel De-embedding

Figure 3.11 shows the "OPEN" test structure for parallel de-embedding, and the corresponding equivalent circuit of OPEN test structure with DUT is shown in Figure 3.12.

The Y-parameters of DUT can be expressed by subtracting the effect of pad from measured data:

$$Y_{11}^{DUT} = Y_{11}^{M} - Y_{11}^{PAD} \tag{3.16}$$

$$Y_{12}^{DUT} = Y_{12}^{M} - Y_{12}^{PAD} \tag{3.17}$$

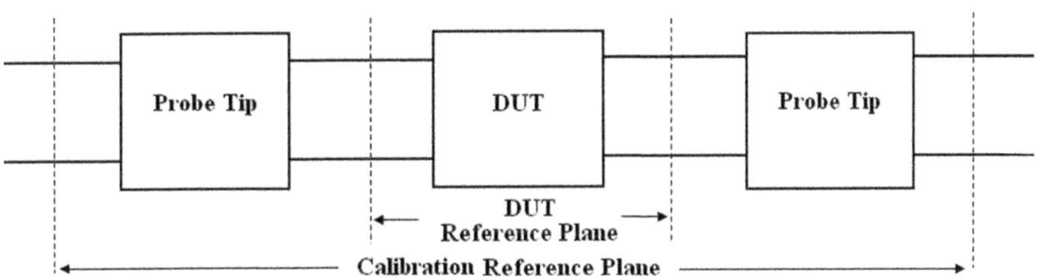

Figure 3.10 Diagram of de-embedding methods

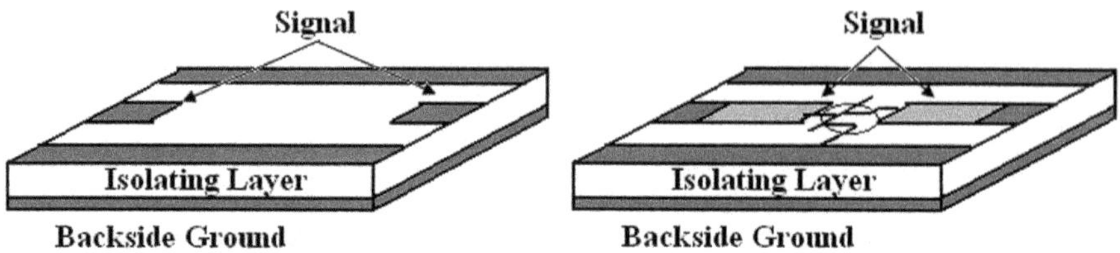

Figure 3.11 OPEN test structure for parallel de-embedding

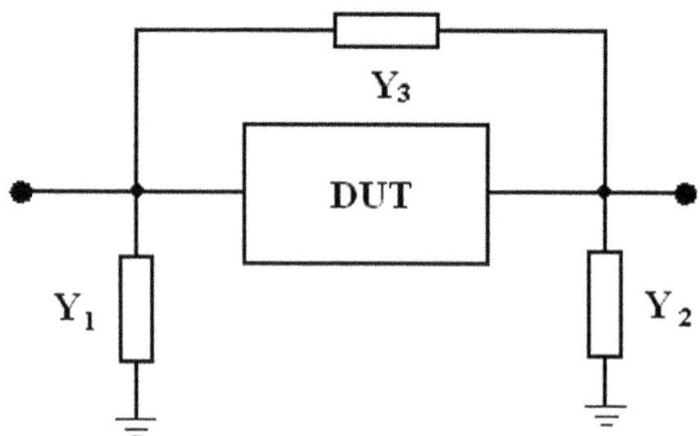

Figure 3.12 Equivalent circuit of OPEN test structure with DUT

$$Y_{21}^{DUT} = Y_{21}^{M} - Y_{21}^{PAD} \tag{3.18}$$

$$Y_{22}^{DUT} = Y_{22}^{M} - Y_{22}^{PAD} \tag{3.19}$$

where Y-parameters of the pad can be expressed as follows:

$$Y_{11}^{PAD} = Y_1 + Y_3 \tag{3.20}$$

$$Y_{12}^{PAD} = -Y_3 \tag{3.21}$$

$$Y_{21}^{PAD} = -Y_3 \tag{3.22}$$

$$Y_{22}^{PAD} = Y_2 + Y_3 \tag{3.23}$$

Superscript *M* denotes the measurement data, *PAD* denotes the OPEN test structure, and *DUT* denotes the device under test.

3.1.4.2 Series De-embedding

The series de-embedding technique means removing the parasitics in series with the DUT, such as the contact resistances between the probe pad and probe tip, as well as series loss of the interconnect. Normally, "SHORT" test structure is used to characterize the series parasitics. Figure 3.13 shows the "SHORT" test structure, and the corresponding equivalent circuit with DUT is shown in Figure 3.14.

The Z-parameters of DUT can be expressed by subtracting the effect of series elements from measured data:

$$Z_{11}^{DUT} = Z_{11}^{M} - Z_{11}^{S} \tag{3.24}$$

$$Z_{12}^{DUT} = Z_{12}^{M} - Z_{12}^{S} \tag{3.25}$$

$$Z_{21}^{DUT} = Z_{21}^{M} - Z_{21}^{S} \tag{3.26}$$

$$Z_{22}^{DUT} = Z_{22}^{M} - Z_{22}^{S} \tag{3.27}$$

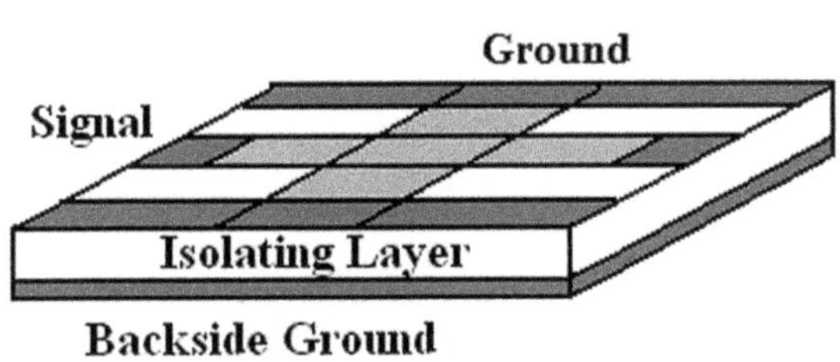

Figure 3.13 SHORT test structure for series de-embedding

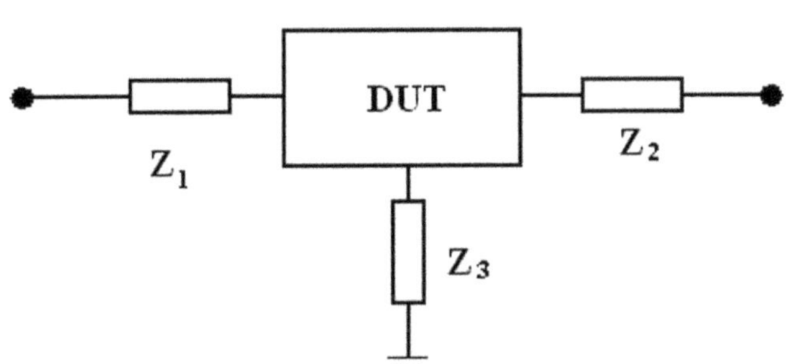

Figure 3.14 Equivalent circuit of SHORT test structure with DUT

where Z-parameters of the series network can be expressed as follows:

$$Z_{11}^{S} = Z_1 + Z_3 \tag{3.28}$$

$$Z_{12}^{S} = Z_3 \tag{3.29}$$

$$Z_{21}^{S} = Z_3 \tag{3.30}$$

$$Z_{22}^{S} = Z_2 + Z_3 \tag{3.31}$$

Superscript *M* denotes the measurement data, *S* denotes the series network, and *DUT* denotes the device under test.

3.1.4.3 Cascading De-embedding

Assuming two parasitic networks with DUT network in cascade, the corresponding circuit is shown in Figure 3.15. The chain parameters can be expressed as follow:

$$A_{DUT} = A_1^{-1} A_M A_2^{-1} \tag{3.32}$$

where

A_{DUT} = chain parameter of DUT

A_1 = chain parameter of parasitic network I

A_2 = chain parameter of parasitic network II

3.1.5 Commercial Measurement System

In terms of the interface between DUT and the test system, there are two kinds of commercial microwave RF measurement systems: coaxial system and on-wafer system.

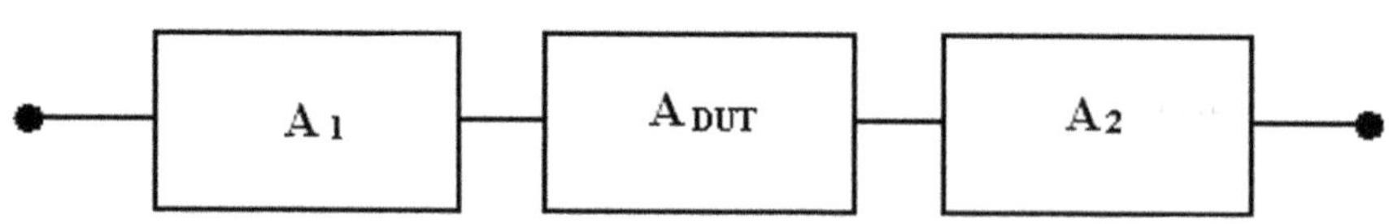

Figure 3.15 Equivalent circuit for cascading de-embedding

3.1.5.1 Coaxial Measurement System

Normally, the final microwave and RF components are not bare wafers but have been wire-bonded and packaged. For packaged wafers the best RF interface to the test system is a test fixture [2–8]. Figure 3.16 shows the coaxial measurement system diagram for measuring the DUT as follows:

1. Packaged semiconductor device (Figure 3.17)
2. Packaged integrated circuit (Figure 3.18)
3. Microwave hybrid integrated circuit (Figure 3.19)

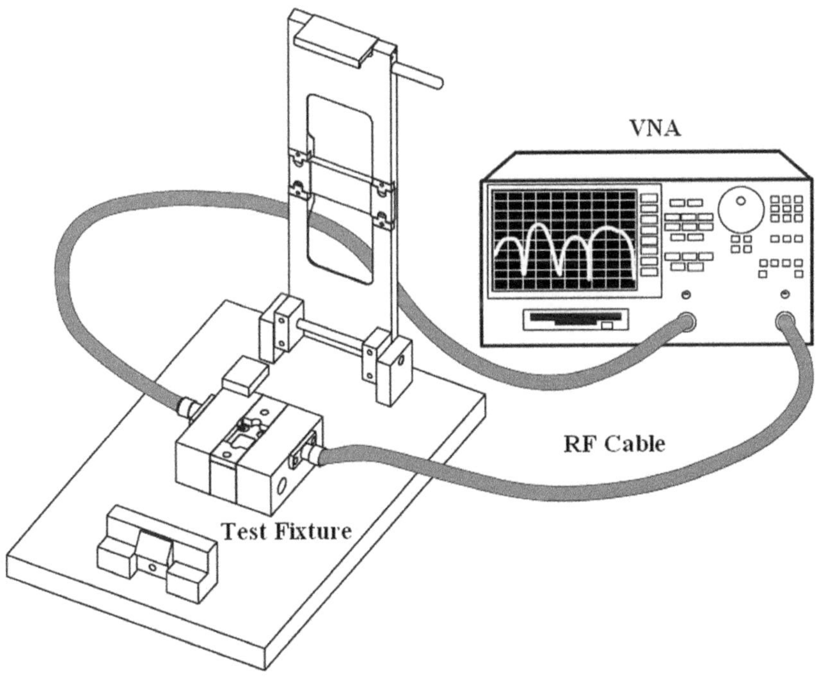

Figure 3.16 Coaxial measurement system diagram

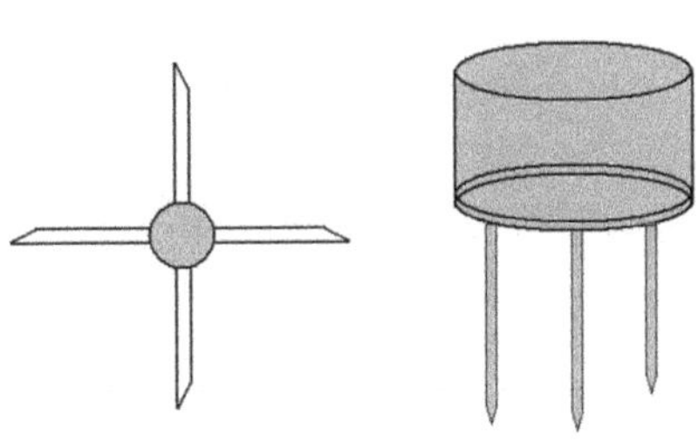

Figure 3.17 Packaged semiconductor devices

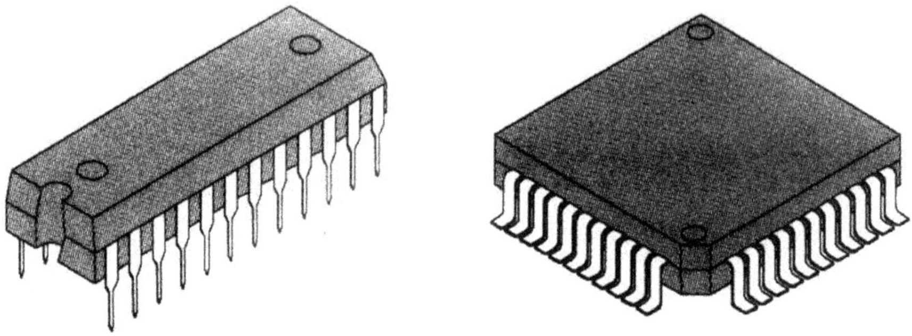

Figure 3.18 Packaged integrated circuits

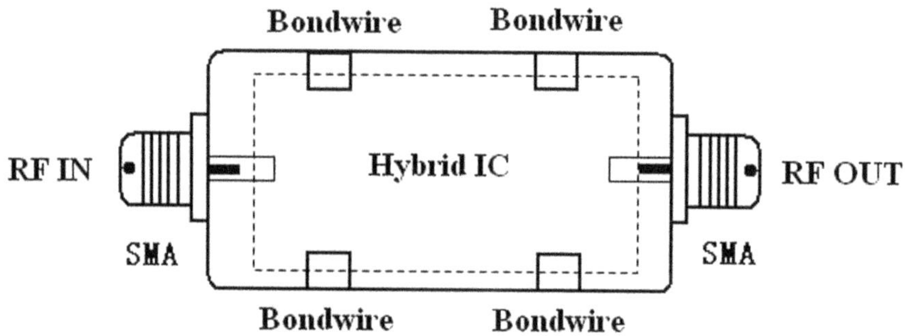

Figure 3.19 Microwave hybrid integrated circuit

3.1.5.2 On Wafer Measurement System

The ever-increasing demand for millimeter-wave circuits and systems has brought about the need to measure components directly on-wafer, that is, at the chip level, to avoid measurement uncertainties associated with packages, connectors, and cables. Access to the devices is normally made through coplanar waveguide probes.

The typical microwave and RF measurement system is composed of a VNA, probe-station, RF cables, bias cables, and the interface to the DUT (see Figure 3.20). The difference between a coaxial system and an on-wafer system is that the interface is the wafer probes, but not test fixture.

The commercial probes may cover a wide range of frequencies (DC-110 GHz), with probes available in single or multi-tip configurations [9]. By maintaining a transmission line impedance all the way to the probe's tips and using a ground-signal-ground coplanar waveguide

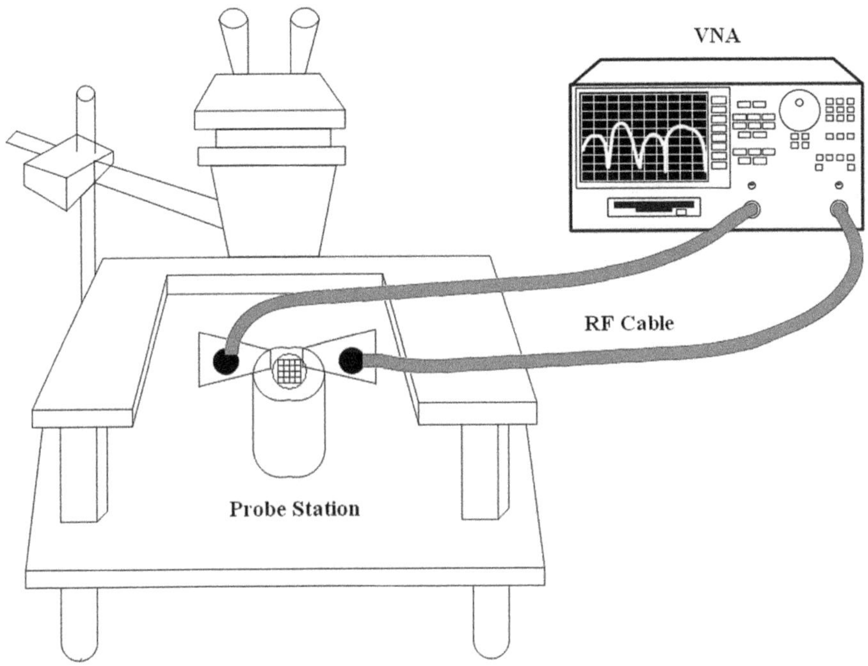

Figure 3.20 On-wafer measurement system diagram

(GSG CPW) launch (Figure 3.22), RF signals are delivered to the wafer with minimal attenuation and excellent impedance control. Wider probe pitch (center-to-center spacing of probe contacts) and single-ground connections can be used at lower frequencies with reduced performance. Variable pitch configurations with flexible signal or ground contacts are not useful for network analyzer measurements since repeatable transition behavior is required for calibration.

In on-wafer measurements, the known calibration standards are provided on Impedance Standard Substrate (ISS). Normal ISS structures include shorting bars, precisely trimmed loads, thru lines, and longer transmission lines. Open standards are most conveniently and reliably created by simply lifting the probe in the air above the ISS.

3.2 Noise Measurement Technique

Noise figure is a figure of merit in many RF systems. A low-noise figure provides improved signal-noise ratio for analog receivers, and reduces bit error rate (BER) in digital receivers. In a development laboratory,

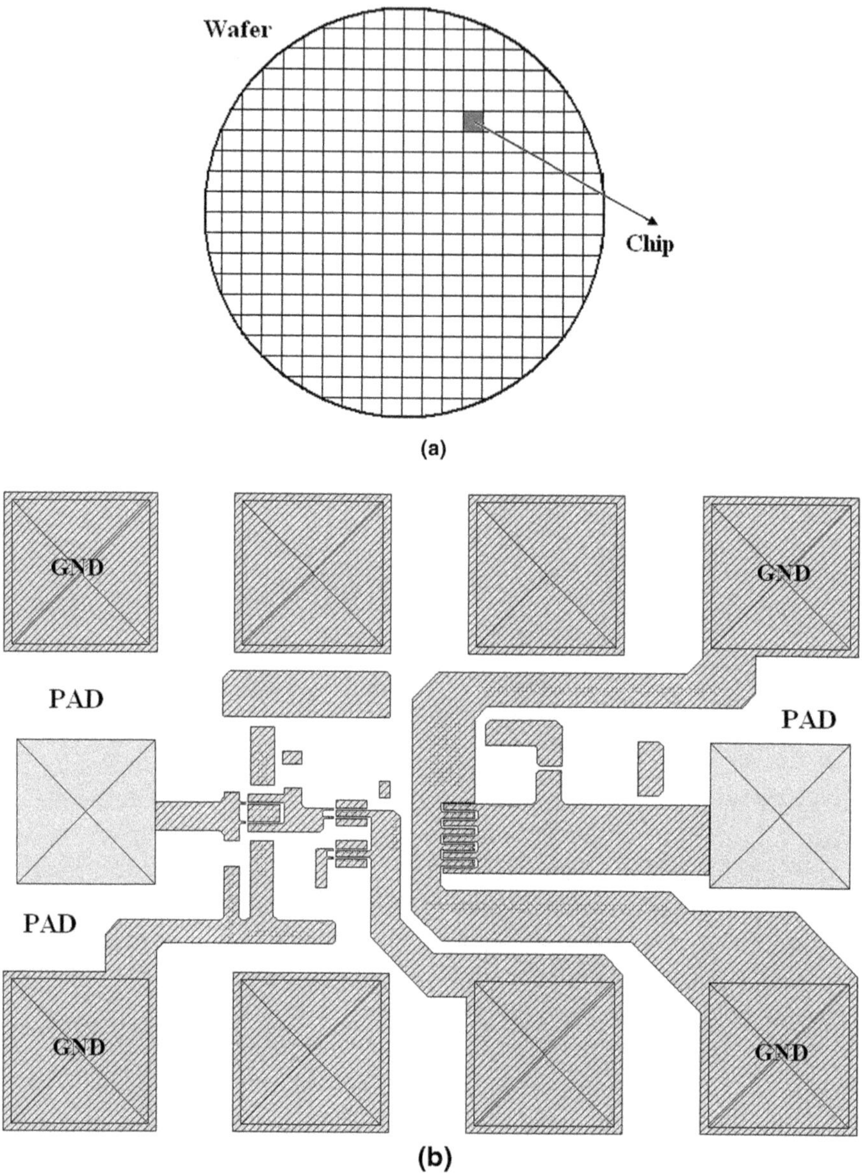

Figure 3.21 Chip diagram: (a) wafer; (b) chip layout

noise figure measurements are essential to verify new designs and support existing equipment. Noise figure is unique in that it is suitable for characterizing the entire system not only, but also the system components such as the pre-amplifier, mixer, and IF amplifier that make up the system. By controlling the noise figure and gain of system components, the designer directly controls the noise figure of the overall sys-

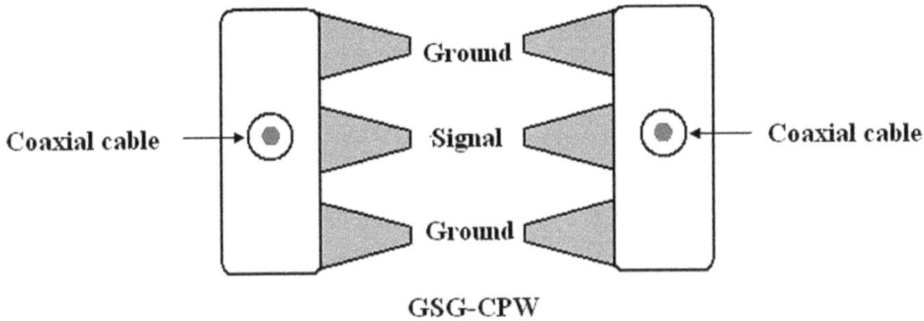

Figure 3.22 GSG Coplanar Probe

tem. Once the noise figure is known, system sensitivity can be easily estimated from system bandwidth [10–15].

There are two kinds of noise sources:

1. Frequency independent noise sources, which include thermal noise and shot noise.
2. Frequency independent noise sources, which include $1/f$ low-frequency noise and high-frequency noise.

3.2.1 Thermal Noise

Thermal noise is called also Johnson or Nyquist noise, and is generated by thermal energy causing random electron motion. Thermal noise can be found in the most of passive and active devices.

For example, the root mean square (rms) thermal noise voltage and current in resistor over a frequency range Δf can be expressed as follows:

$$\overline{v_n^2} = 4kTR\Delta f \tag{3.33}$$

$$\overline{i_n^2} = \frac{4kT\Delta f}{R} \tag{3.34}$$

where T = Kelvin temperature of the resistor (in Kelvins)

K = Boltzmann's constant (=1.38×10^{-23} J/K)

Δf = bandwidth (Hz)

R = resistor (Ω)

A noisy resistor at a temperature can be modeled by an ideal noiseless resistor at 0 K in series with a noise voltage source (or modeled by an ideal noiseless resistor at 0 K in parallel with a noise current source), as shown Figure 3.23.

It is noted that the available noise power (i.e., maximum power available under matched condition) from any arbitrary resistor has been shown by Nyquist to be:

$$P_{\text{out}} = kTB \tag{3.35}$$

It can be found that available noise power increases with increase of the bandwidth, and with increase of the temperature. Therefore, narrower bandwidth amplifiers are less noisy, and cooler devices or amplifiers generate less noise power.

Figure 3.24 shows the available noise power with respect to bandwidth; it can be found that with 1 Hz bandwidth, noise floor is –174 dBm. Thermal noise is also called *white noise*, meaning it has a constant power spectral density with respect to frequency.

Thermal noise sources are very important for semiconductor device modeling; for example, Figure 3.25 (a), (b), and (c) show the thermal noise sources in the BJT, FET, and Hetero-junction bipolar transistor (HBT), respectively.

In Figure 3.25 (a), R_b, R_c, and R_e are the base, collector, and emitter extrinsic resistances of BJT, and v_{nb}^2, v_{nc}^2, and v_{ne}^2 are the corresponding thermal noise voltage sources generated by resistances. In Figure 3.25 (b), R_g, R_d, and R_s are the gate, drain, and source extrinsic resistances of FET, and v_{ng}^2, v_{nd}^2, and v_{ns}^2 are the corresponding thermal noise voltage sources generated by extrinsic resistances. In Figure 3.25 (c), R_{bx}, R_c, and

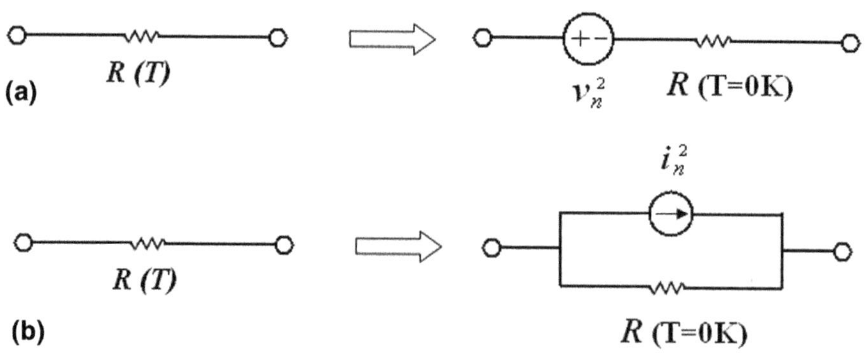

Figure 3.23 Equivalent model of a noisy resistor

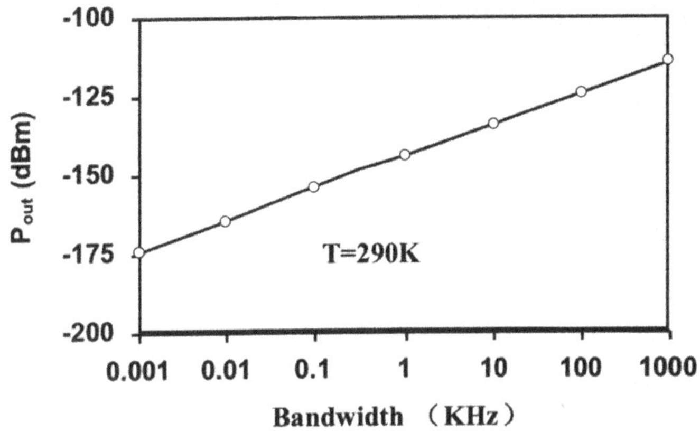

Figure 3.24 Available noise power with respect to bandwidth

R_e are the base, collector, and emitter extrinsic resistances of HBT, and v_{nbx}^2, v_{nc}^2, and v_{ne}^2 are the corresponding thermal noise voltage sources generated by extrinsic resistances. Another thermal noise voltage source v_{nbi}^2 is generated by intrinsic base resistance R_{bi}.

3.2.2 Shot Noise

The fluctuations in the number of electrons emitted from the source constitute the shot noise. Shot noise occur tubes or solid-state devices while crossing a junction or other discontinuities. It is commonly found in a semiconductor device (e.g., in a PN junction or a transistor) and is proportional to square root of dc current. There are two important features:

- It is always associated with a dc current flow in diodes and BJTs.
- It is frequency independent (white noise)

The noise amplitude is represented by the rms value:

$$\overline{i_n} = \sqrt{2qI\Delta f} \tag{3.36}$$

where I is the dc current and q is the electron charge; it can be found that $\overline{i_n^2} / \Delta f$ is proportional to the current I with slope $2q$.

Figure 3.26 shows the equivalent circuit with shot noise source for diode; I_D is the current crossed the pn junction, and the corresponding shot noise $\overline{i_d^2}$ can be expressed as follows:

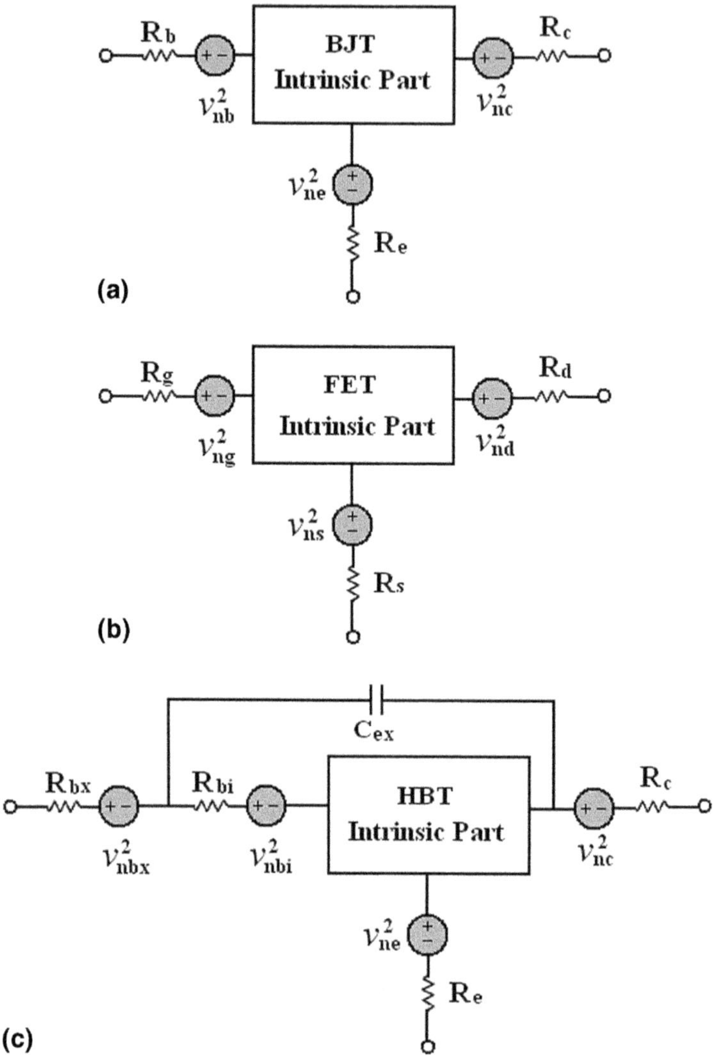

Figure 3.25 Thermal noise sources in semiconductor devices: (a) BJT; (b) FET; (c) HBT

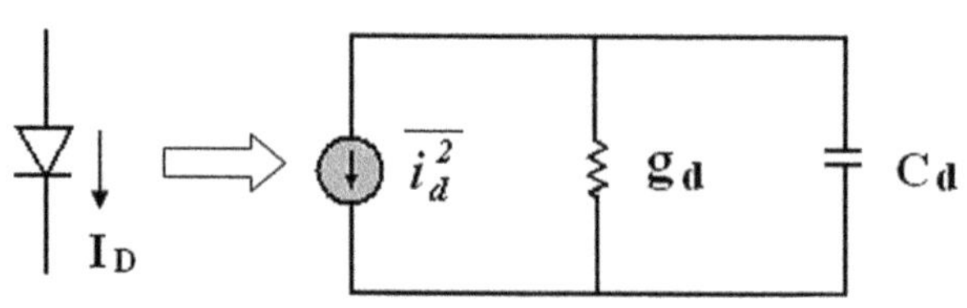

Figure 3.26 Shot noise source in semiconductor diode

$$i_d^2 = 2qI_D\Delta f \quad (3.37)$$

Figure 3.27 shows the equivalent circuit with shot noise sources for BJT and HBT; there are two shot noise sources, $\overline{i_b^2}$ and $\overline{i_c^2}$, which are characterized by their mean quadratic value in a bandwidth Δf centered on the frequency f, and can be given by the following expressions:

$$\overline{i_b^2} = 2qI_B\Delta f \quad (3.38)$$

$$\overline{i_c^2} = 2qI_C\Delta f \quad (3.39)$$

where I_B and I_C are the base and collector dc currents, respectively,

3.2.3 1/*f* Low-Frequency Noise

1/*f* noise is found in many natural phenomena such as nuclear radiation, electron flow through a conductor, or even in the environment. In electrical engineering, it is called also flicker noise. Flicker noise is associated with crystal surface defects in semiconductors and is also found in vacuum tubes. Flicker noise current $i_{1/f}^2$ can be expressed as follows:

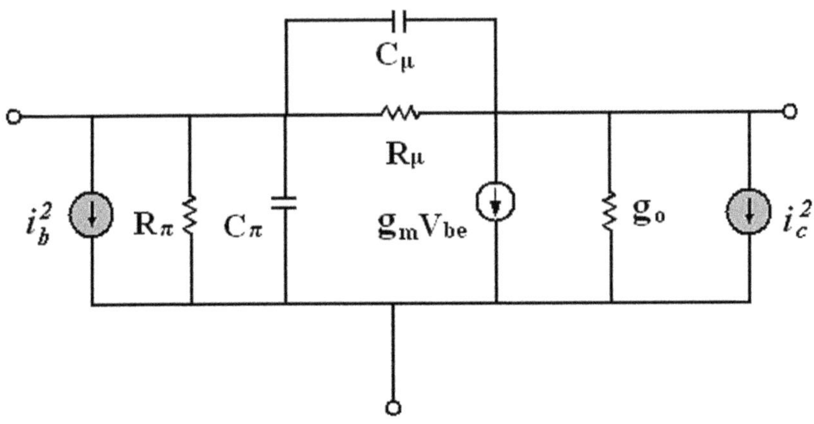

Figure 3.27 Shot noise sources in BJT

$$\overline{i_{1/f}^2} = k_f \frac{I^{\alpha_f}}{f} \Delta f \tag{3.40}$$

or

$$20\log(i_{1/f}) = 10\log(k_f I^{\alpha_f} \Delta f) - 10\log(f) \tag{3.41}$$

where k_f and α_f are the fitting factors.

Figure 3.28 shows the 1/*f* noise current versus frequency. It can be found that the noise power increases with an increase of the bias current; and, unlike thermal and shot noise, flicker noise decreases with frequency, and inverse proportionality with frequency is almost exactly 1/*f* for low frequencies. Flicker noise is essentially random, but because its frequency spectrum is not flat; it is not a white noise. It is often referred to as pink noise because most of the power is concentrated at the lower end of the frequency spectrum.

Figure 3.29 shows 1/*f* noise versus frequency for bipolar and field-effect semiconductor devices, and it can be observed that the Flicker noise is more prominent in FETs but not BJT, and that the corner frequency of FETs is higher than BJT's. This means BJT/HBT devices are better than FET/HEMT for designing low-phase noise components.

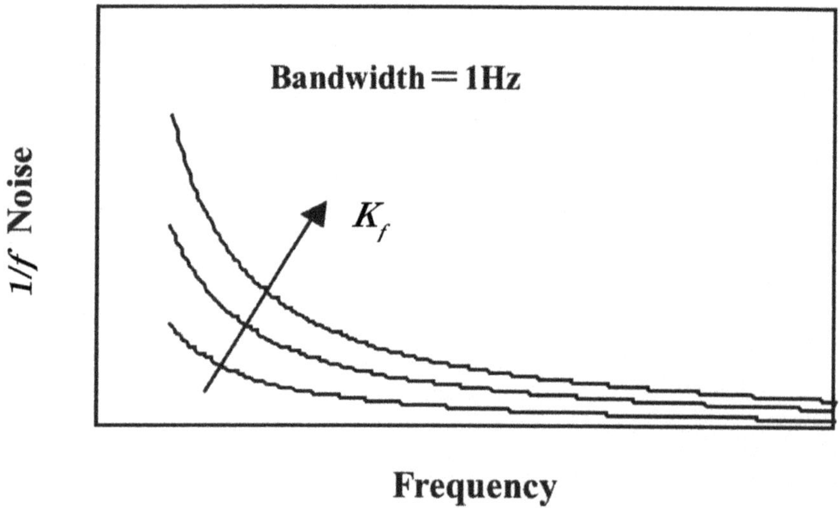

Figure 3.28 1/*f* noise versus frequency

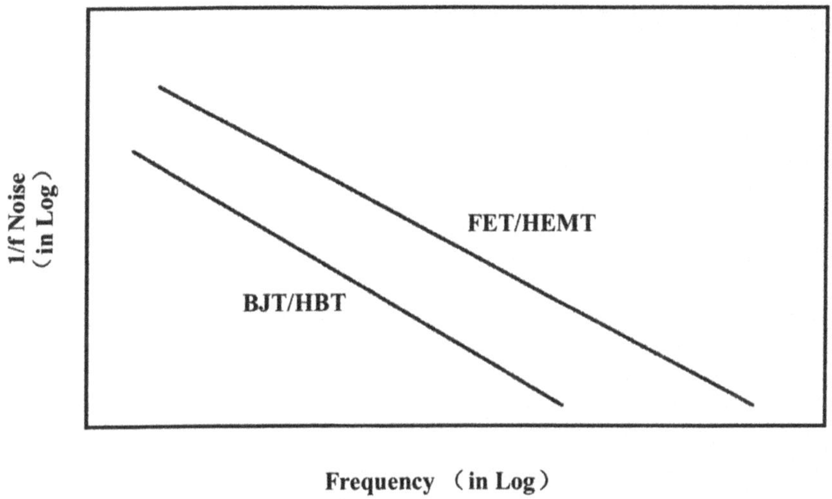

Figure 3.29 1/*f* noise versus frequency for different semiconductor device

3.2.4 High-Frequency Noise

Unlike the thermal and shot noise sources, high-frequency noise just depends on the frequency, and is independent of the DC current. For example, in FET devices, the gate-induced noise is one kind of high-frequency noise and is expressed as follows (see also Chapter 5):

$$\overline{i_g^2} = 4kT\frac{(\omega C_{gs})^2 R}{g_m}\Delta f \tag{3.42}$$

It can be found that high-frequency noise power increases with frequency.

3.2.5 Noise Figure of Two-Port Network

Noise figure (or noise factor) is a figure of merit quantitatively specifying how noisy a component or system is. The most basic definition of noise factor came into popular use in the 1940's when Harold Friis [11] defined the noise figure F of a network to be the ratio of the signal-to-noise power ratio at the input to the signal-to-noise power ratio at the output as shown in Figure 3.30:

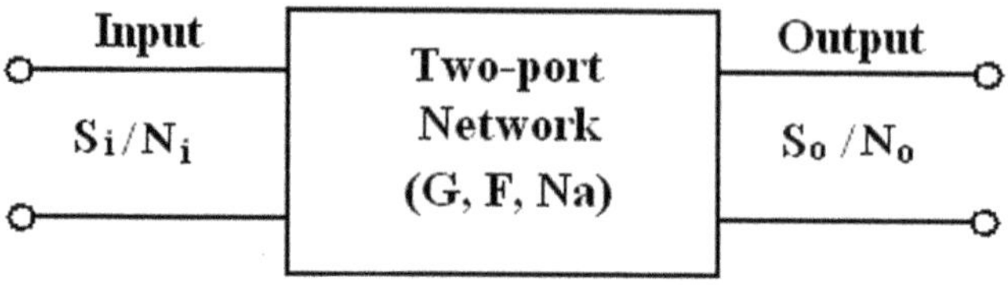

Figure 3.30 Noise figure of two-port network

$$F = \frac{SNR_{in}}{SNR_{out}} = \frac{S_i / N_i}{S_o / N_o} \tag{3.43}$$

The noise figure refers to the noise factor in decibel:

$$NF = 10 \log F \tag{3.44}$$

where SNR_{in} and SNR_{out} are the available signal-to-noise ratio at the input and output ports, respectively, S_i and N_i are the available signal and noise power at the input port, respectively, and S_o and N_o are the available signal and noise power at the output port, respectively.

Thus, the noise figure of a network is the decrease or degradation in the signal-to-noise ratio (SNR) as the signal goes through the network. A perfect amplifier would amplify the noise at its input along with the signal, maintaining the same signal-to-noise ratio at its input and output; therefore the noise factor is 1 or noise figure is 0 dB.

For example, Figure 3.31(a) shows the situation at the input of a noisy two-port network, and Figure 3.31(b) shows the situation at the output of a noisy two-port network. It can be observed that the available signal-to-noise ratio at the input port is 40 dB, and is 30 dB for the available signal-to-noise ratio at the output port; therefore, the two-port has a 10 dB noise figure.

Assuming the available signal and noise power of the two-port network are G and N_a, thus the available noise power at the output port can be expressed as follows:

$$N_o = N_a + GN_i \tag{3.45}$$

where $N_i = KT_0B$. Friis [12] suggested a reference source temperature of 290 K (denoted by T_o), which is equivalent to 16.8°C and 62.3°F. This temperature is close to the average temperature seen by receiving antennas directed across the atmosphere at the transmitting antenna.

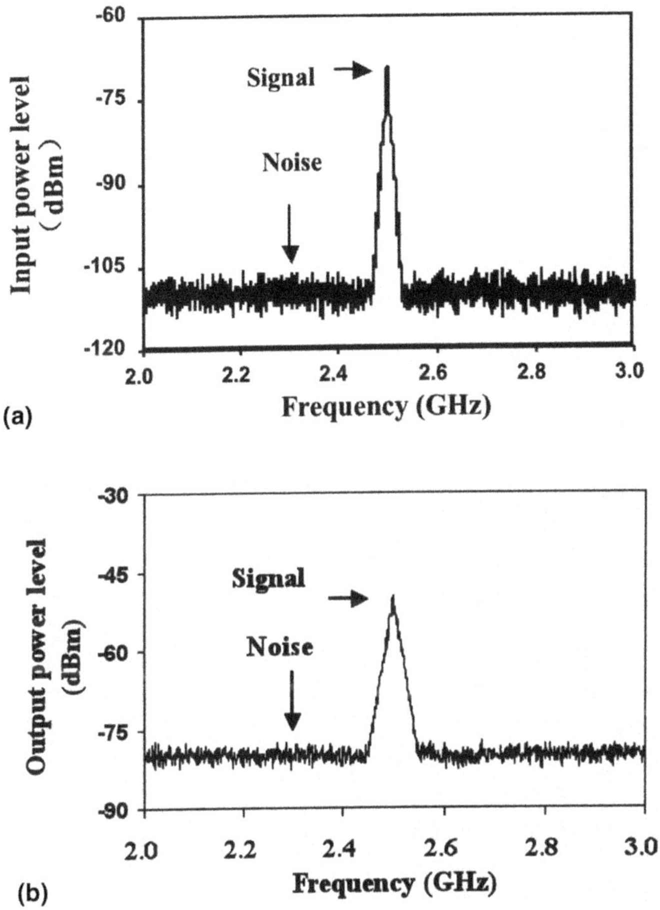

Figure 3.31 Signal and noise levels versus frequency: (a) input; (b) output

Substituting (3.45) into (3.43), we have

$$F = \frac{S_i / N_i}{S_o / N_o} = \frac{N_o}{GN_i} = \frac{S_i / N_i}{GS_i / (N_a + GN_i)} = \frac{N_a + GN_i}{GN_i} \tag{3.46}$$

or

$$F = 1 + \frac{N_a}{kT_0 BG} \tag{3.47}$$

T_e is the equivalent temperature of source impedance into a perfect (noise-free) device that would produce the same added noise, N_a. It is often defined as:

$$T_e = \frac{N_a}{kGB} \tag{3.48}$$

Substituting (3.48) into (3.47)

$$F = \frac{kGB(T_o + T_e)}{kGBT_o} = \frac{T_o + T_e}{T_o} = 1 + \frac{T_e}{T_o} \tag{3.49}$$

or

$$T_e = T_o(F - 1) \tag{3.50}$$

When the two-port network considered earlier is a lossy passive component, such as attenuator or a lossy transmission line, the noise figure is given by

$$F = 1 + \frac{T_e}{T_o} = 1 + (L - 1)\frac{T}{T_o} \tag{3.51}$$

where L is the loss factor or attenuation. If lossy passive two-port networks operate at room temperature, (i.e., $T = T_o$), Equation (3.51) gives

$$F = 1 + L - 1 = L \tag{3.52}$$

Equation (3.52) indicates that the noise figure of a lossy network at room temperature equals the attenuation factor.

Figure 3.32 shows the noise figure of typical low-noise semiconductor device, such as BJT, MESFET and HEMT. It can be found that HEMTs have much lower noise figure than others.

When separate two-port networks are cascaded each having its own gain G_i and noise factor F_i as shown in Figure 3.33, all networks add noise to the signal that travels through them, but the contribution to the over all noise factor from succeeding networks is reduced when pre-

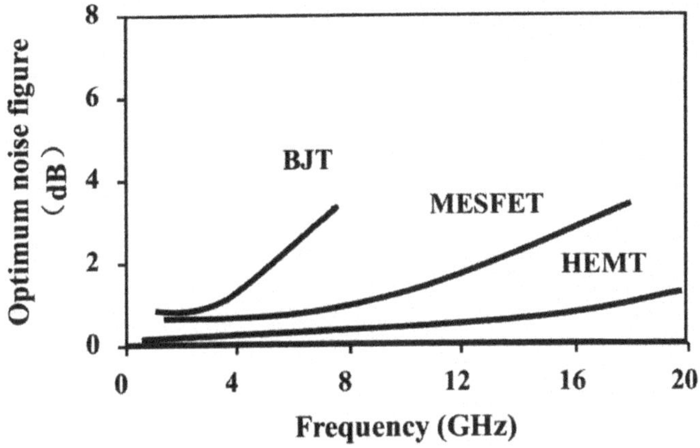

Figure 3.32 Noise figure of semiconductor device

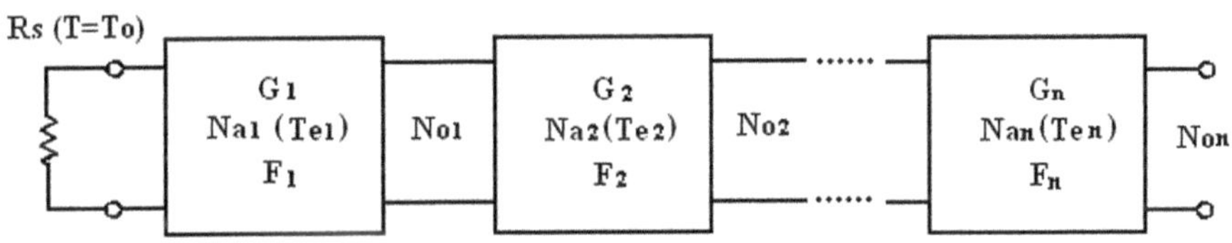

Figure 3.33 Noise factor for a cascade of two-ports

vious stage have amplified the signal. The overall noise factor is calculated using:

$$F = F_1 + \frac{F_2 - 1}{G_1} + \frac{F_3 - 1}{G_1 G_2} + \cdots + \frac{F_n - 1}{G_1 G_2 \cdots G_{n-1}} \tag{3.53}$$

The corresponding effective noise temperature for cascaded network is given by:

$$T_e = T_{e1} + \frac{T_{e2}}{G_1} + \frac{T_{e3}}{G_1 G_2} + \cdots + \frac{T_{en}}{G_1 G_2 \cdots G_{n-1}} \tag{3.54}$$

If all stages are identical, that is

$$G_1 = G_2 = \cdots = G_n = G_a \tag{3.55}$$

$$T_{e1} = T_{e2} = \cdots = T_{en} = T_{ea} \tag{3.56}$$

$$F_1 = F_2 = \cdots = F_n = F_a \tag{3.57}$$

Substituting equations (3.55) through (3.57) into (3.53) and (3.54), we have

$$F = (F_a - 1)\left(\frac{1-(1/G_a)^n}{1-1/G_a}\right) + 1 \tag{3.58}$$

$$T_e = T_{ea}\left(\frac{1-(1/G_a)^n}{1-1/G_a}\right) \tag{3.59}$$

3.2.6 Noise Figure Measurement

There are three ways for measuring noise figure of DUT [13,14]:

1. Signal generator twice-power method
2. Direct noise measurement method
3. Y-factor method

The signal generator twice-power method is a relative simple for measuring the system noise figure (Figure 3.34(a)): (1) measure the output noise power with the input terminated; (2) use a signal generator at the input port, and adjusting its level until a 3 dB increase in output power is attained. Therefore, the signal generator output power is equal to the total output noise power divided by the gain of DUT; thus, the noise figure can be expressed as follows:

$$F = \frac{N_o / G}{N_i} = \frac{P_{SG}}{N_i} = \frac{P_{SG}}{kT_o B} \tag{3.60}$$

The drawback of the signal generator twice-power method is that the gain of DUT is necessary except for bandwidth B.

Also, the noise figure can be determined directly from a measurement of the absolute total noise output power (Figure 3.34(b)). However, it is noted that the noise bandwidth of the power-measuring device must be known, and perhaps a network analyzer is needed.

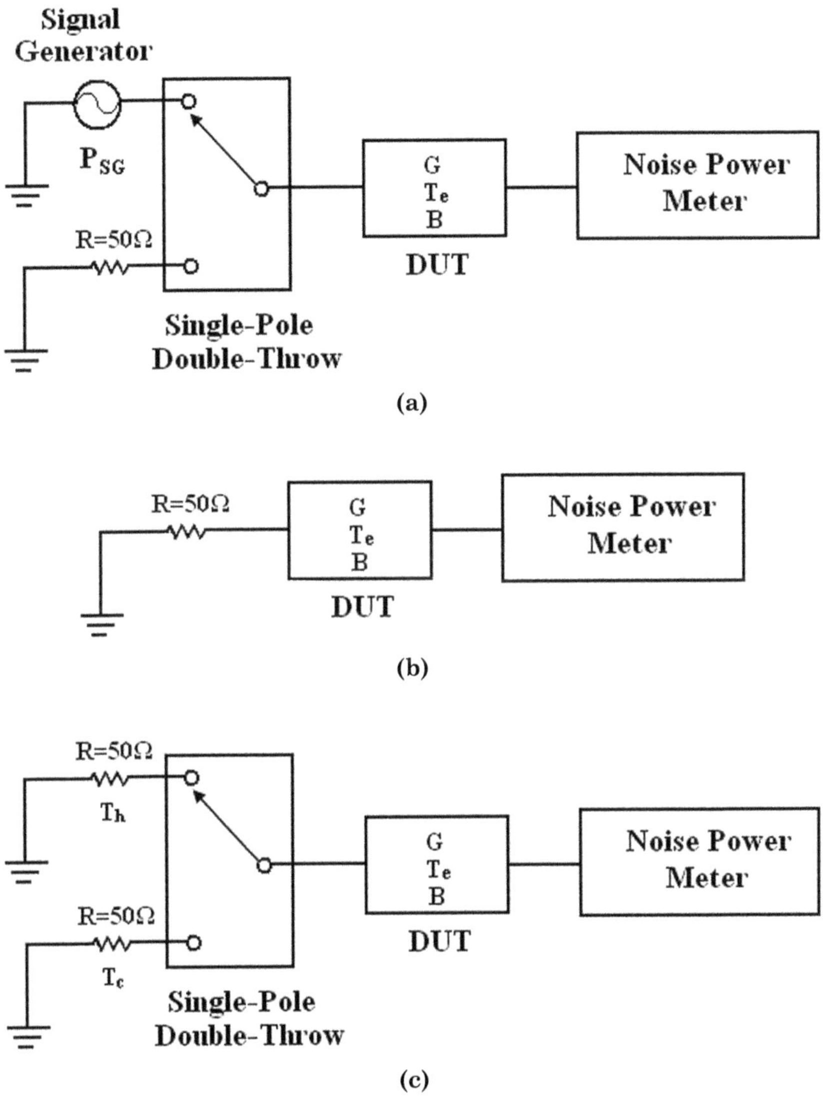

Figure 3.34 Noise figure measurement methods: (a) Signal generator twice-power method; (b) Direct noise measurement method; (c) Y-factor method

The Y-factor method is the basis of most noise figure measurements whether they are manual or automatically performed internally in a noise figure analyzer. With a noise source connected to the DUT, the output power can be measured corresponding to the noise source on and the noise source off (Figure 3.34(c)). The noise figure of the system can be determined from the ratio of these two powers. More details about the Y-factor method are introduced in the next section.

The comparison of noise figure measurement methods previously mentioned is summarized in Table 3.1.

Table 3.1 Comparison of Noise Figure Measurement Methods

Method	Gain of DUT	Bandwidth	Signal Generator	Noise Source
Signal generator twice-power	No	Yes	Yes	No
Direct noise measurement	Yes	Yes	No	No
Y-factor method	No	No	No	Yes

3.2.6.1 Noise Source

One way of determining the noise slope is to apply two different levels of input noise and measure the output power change. A noise source is a device that will provide these two known levels of noise.

The avalanche diode (also known as the solid-state noise source) is the most commonly used noise source and is useful for noise measuring purposes because it is broad and can conveniently be switched between off (or cold) and on (or hot) state. Figure 3.35 shows the equivalent circuit of noise source, the noise source consists of bias tee, matched network, and noise diode. When a large positive bias is applied, the noise diode is reversed biased. Avalanche action occurs where a large DC current plus random current of all frequencies flows through the diode. When a negative bias is applied, no random high-frequency components are generated, and can be regard as a passive resistive network.

To make noise figure measurements, a noise source must have a calibrated output noise level, represented by excess noise ratio (ENR):

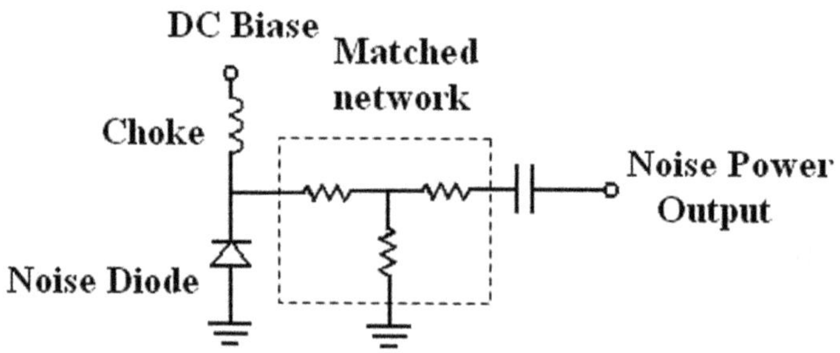

Figure 3.35 Equivalent circuit of noise source

$$ENR(dB) = 10\log 10(\frac{T_h - T_c}{T_o}) \tag{3.61}$$

where T_h is the hot temperature corresponding to the noise source on state, and T_c is the cold temperature corresponding to the noise source off state. Normally T_c is very close to T_o (290 K), that is,

$$ENR(dB) \approx 10\log 10(\frac{T_h - T_o}{T_o}) \tag{3.62}$$

or

$$T_h \approx T_o[10^{ENR(dB)/10} + 1] \tag{3.63}$$

It should be noted that a 0 dB ENR noise source produces a 290 K temperature change between noise source on and off states. When the noise source is used at a different physical temperature, compensation must be applied to the measurement. Usually a noise source comes with a calibration report of T_h or ENR versus frequency within its band of operating frequencies. Figure 3.36 shows the typical ENR and T_h of the noise source versus frequency, it can be found that ENR increases with the increase of frequency; and the corresponding hot temperature T_h increases about 5000 K from 0.1 GHz to 26.5 GHz.

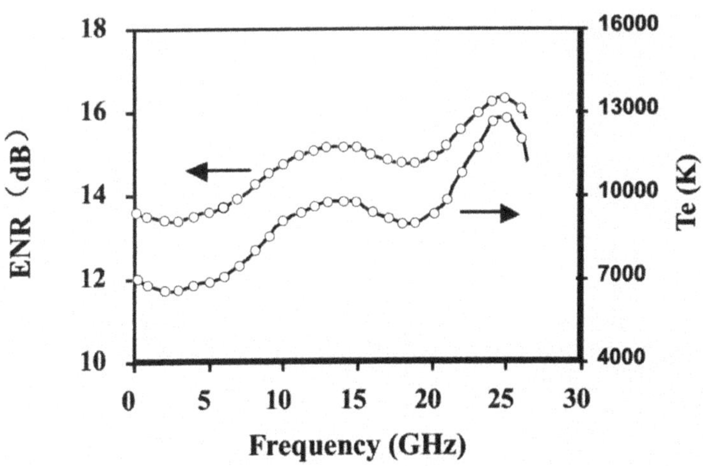

Figure 3.36 Typical ENR and T_h versus frequency

3.2.6.2 Y-Factor Method

The most commonly used noise figure measurement method is the Y-factor method. The corresponding output noise powers with the noise source on and off state are given by:

$$N_h = kBG(T_h + T_e) \tag{3.64}$$

$$N_c = kBG(T_c + T_e) \tag{3.65}$$

where N_h and N_c are the output noise powers with the noise source on and off state, respectively. T_e is the equivalent noise temperature of the DUT. Figure 3.37 shows the output noise power versus source temperature; it can be observed that the slop of the straight line is KGB.

Let Y represent the ration of two noise powers just mentioned

$$Y = \frac{N_h}{N_c} = \frac{T_h + T_e}{T_c + T_e} \tag{3.66}$$

Thus, equivalent noise temperature of the DUT T_e can be obtained from (3.65) directly:

$$T_e = \frac{T_h - YT_c}{Y - 1} \tag{3.67}$$

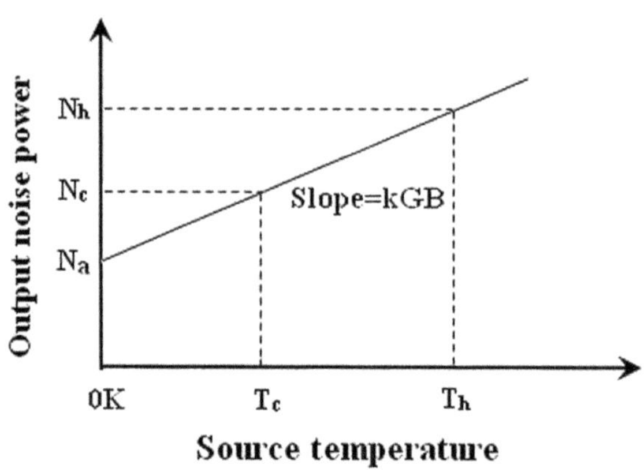

Figure 3.37 Output noise power versus source temperature

Substituting (3.67) into (3.49),

$$F = \frac{(T_h / T_o - 1) - Y(T_c / T_o - 1)}{Y - 1} \tag{3.68}$$

For $T_c \approx T_o$, we have

$$F = \frac{ENR}{Y - 1} \tag{3.69}$$

3.2.6.3 Calibration

The Y-factor method just mentioned is the basis for almost all noise measurement. However, it is unfortunate that all measuring equipment used to measure the noise power will generate the additional noise; therefore, the values of noise figure from equations (3.68) or (3.69) are for the whole noise measurement system, which consists of DUT, noise power meter, noise source, and RF cable, and not the noise figure of DUT only. Therefore, it is necessary to calibrate the system before we detail the measurement methods such as Y-factor method.

Figure 3.38 gives the calibration procedures, which including two steps of noise figure measurement: (1) The noise source is connected directly to the input of the instrument (e.g., noise power meter, Figure 3.38(a)); and (2) the DUT is inserted between the noise source and the instrument (Figure 3.38(b)).

The effective noise temperature of the instrument is given by:

$$T_{e2} = \frac{T_h - Y_2 T_c}{Y_2 - 1} \tag{3.70}$$

where Y_2 is the Y-factor for the first step of the calibration procedure:

$$Y_2 = \frac{N_h}{N_c} = \frac{T_h + T_{e2}}{T_c + T_{e2}} \tag{3.71}$$

The effective noise temperature of the DUT and the instrument in cascade is given by

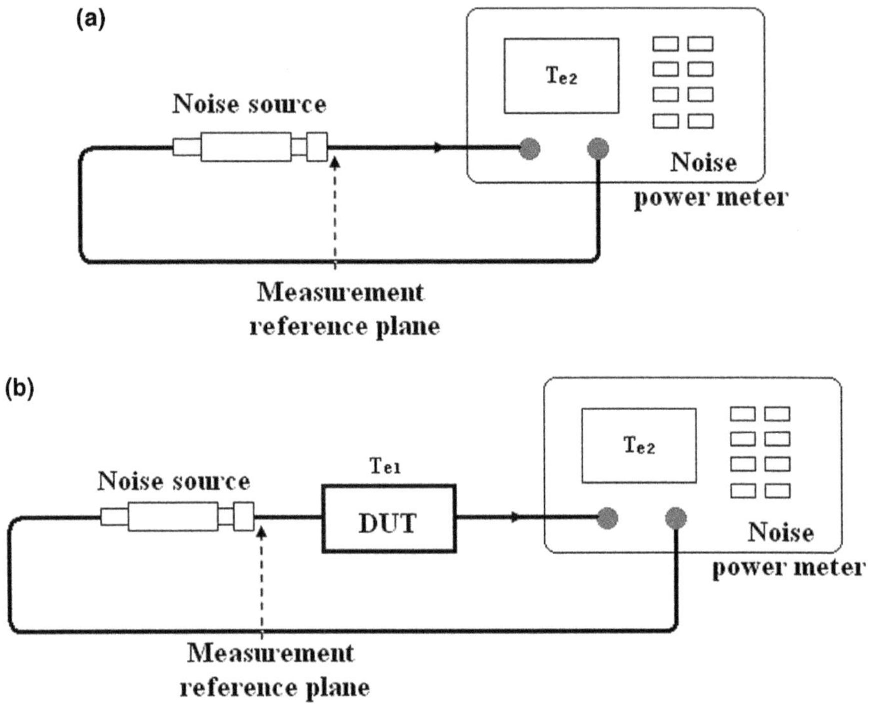

Figure 3.38 Calibration procedure of noise figure measurement: (a) without DUT; (b) with DUT

$$T_{e12} = \frac{T_h - Y_{12}T_c}{Y_{12} - 1} \tag{3.72}$$

where Y_{12} is the Y-factor for the second step of the calibration procedure:

$$Y_{12} = \frac{N'_h}{N'_c} = \frac{T_h + T_{e12}}{T_c + T_{e12}} \tag{3.73}$$

By combining (4.72) and (4.73), the available gain of DUT can be obtained as follows:

$$G = \frac{N'_h - N'_c}{N_h - N_c} \tag{3.74}$$

Thus, the noise figure and corresponding of the DUT are given by

$$T_{e1} = T_{e12} - \frac{T_{e2}}{G} \tag{3.75}$$

$$F = 1 + \frac{T_{e12}}{T_o} - \frac{T_{e2}}{GT_o} \tag{3.76}$$

3.2.6.4 Noise Figure Measurement Error

From the cascade noise figure equation it can be seen that if the DUT gain is large, the measurement system will have little effect on the measurement. The noise figure of high-gain DUTs can be directly measured with the previously discussed methods. When a low-gain DUT is to be measured or the highest accuracy is needed, we discuss here what the uncertainty of the noise figure measurement is [13].

After rearranging the cascade noise figure equation, we can clearly see the dependence of the device noise factor on the other variable:

$$F_{DUT} = F_S - \left(\frac{F_R - 1}{G_{DUT}} \right) \tag{3.77}$$

where subscripts S, R, and DUT denote the system (the DUT and measurement receiver), measurement receiver, and DUT, respectively.

The uncertainty of the noise figure F_{DUT} can be expressed as follows:

$$dF_{DUT} = \frac{\partial F_{DUT}}{\partial F_S} dF_S + \frac{\partial F_{DUT}}{\partial F_R} dF_R + \frac{\partial F_{DUT}}{\partial G_{DUT}} dG_{DUT} \tag{3.78}$$

where

DF_1 = the uncertainty of the DUT's noise factor,

DF_2 = the uncertainty of the measurement receiver's noise factor,

DF_{12} = the uncertainty of the complete system (the DUT and measurement receiver) noise figure.

dG_1 = the uncertainty of the DUT gain.

Equation (3.78) can be rewritten for noise figure in decibels:

$$dNF_1 = \left(\frac{F_{12}}{F_1}\right) dNF_{12} - \left(\frac{F_2}{F_1 G_1}\right) dNF_2 + \left(\frac{F_2 - 1}{F_1 G_1}\right) dG_1 \,(dB) \tag{3.79}$$

The three d terms in the previous equation are due to the NF measurement receiver and the DUT. However, NF instruments rely on a calibrated noise source with a specified ENR. Clearly, there will be an uncertainty associated with this ENR, and this will contribute to the overall uncertainty equation. The final resulting equation for the overall noise figure measurement uncertainty is given by:

$$\begin{aligned} dNF_1 = {} & \left(\frac{F_{12}}{F_1}\right) dNF_{12} - \left(\frac{F_2}{F_1 G_1}\right) dNF_2 + \left(\frac{F_2 - 1}{F_1 G_1}\right) dG_1 \,(dB) \\ & + \left(\frac{F_{12}}{F_1} - \frac{F_2}{F_1 G_1}\right) dENR \end{aligned} \tag{3.80}$$

3.3 Power Measurement System

To verify the accuracy of the semiconductor device nonlinear model, the power performance, which includes the gain compression, harmonic distortion, and inter-modulation distortion (IMD) is necessary. Before discussing the nonlinearity of the RF devices or components, the definition of the various power gain of two-port networks is first introduced [15–19].

3.3.1 Power Gain of Two-Port Network

Consider an arbitrary two-port network connected to source and load impedances Z_S and Z_L, respectively. Figure 3.39 illustrates the corresponding signal flowchart, and different powers used in the gain equations are indicated. In the figure,

P_{in} = power deliver to the input of the two-port network

P_{avs} = power available from the source

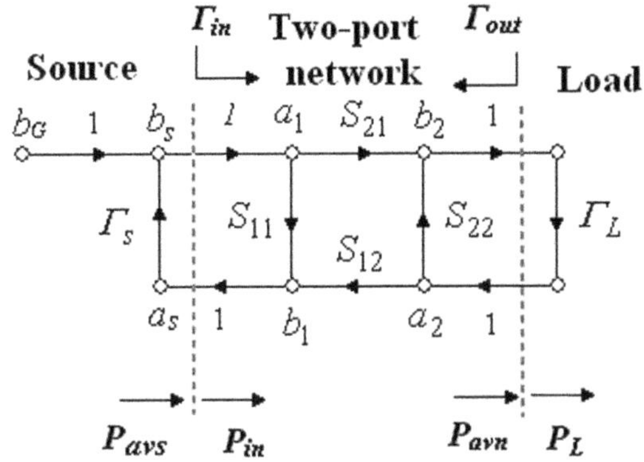

Figure 3.39 Signal flowchart of arbitrary two-port network with input and output network

P_L = power deliver to the load

P_{avn} = power available from the two-port network

The power available from the source, P_{avs}, is the maximum power that can be delivered to the network. This occurs when the input impedance of the terminated network is conjugately matched to the source impedance:

$$P_{avs} = P_{in}\big|_{\Gamma_{in}=\Gamma_S^*} \tag{3.81}$$

where Γ_{in} is the reflection coefficient seen looking into the input port of the network, and Γ_s is the source reflection coefficient.

Based on the signal flow chart (Figure 3.39), P_{avs} can be expressed as follows:

$$P_{avs} = \frac{|b_G|^2}{1-|\Gamma_S|^2} \tag{3.82}$$

Similarly, the power available from the network, P_{avn}, is the maximum power that can be delivered to the load and is given by

$$P_{avn} = P_L\big|_{\Gamma_L=\Gamma_{out}^*} \tag{3.83}$$

Based on the signal flow chart (Figure 3.39), P_{avn} can be expressed as follows:

$$P_{avn} = \frac{|S_{21}|^2 |b_G|^2}{|1 - S_{11}\Gamma_S|^2 (1 - |\Gamma_{out}|^2)} \tag{3.84}$$

where Γ_{out} is the reflection coefficient seen looking into the output port of the network, and Γ_L is the load reflection coefficient.

Several power gain equations appear in the literature and are used in designing microwave amplifiers. The three most commonly used power gains are operating power gain G_p, transducer power gain G_f, and available power gain G_a.

The operating power gain G_p is the ratio of power dissipated in the load to the power delivered to the input of the two-port network, and is defined as follows:

$$G_p = \frac{\text{power deliver to the load}}{\text{power input to the network}} = \frac{P_L}{P_{in}} \tag{3.85}$$

In terms of S-parameters of the whole network, G_p can be expressed

$$G_p = \frac{1}{1 - |\Gamma_{in}|^2} \cdot |S_{21}|^2 \cdot \frac{1 - |\Gamma_L|^2}{|1 - S_{22}\Gamma_L|^2} \tag{3.86}$$

with

$$\Gamma_{in} = S_{11} + \frac{S_{12}S_{21}\Gamma_L}{1 - S_{22}\Gamma_L}$$

When the output port of the two-port network is terminated by a matched load, that is $\Gamma_L = 0$, thus, equation (3.92) can be simplified as follows:

$$G_p = \frac{|S_{21}|^2}{1 - |S_{11}|^2} \tag{3.87}$$

The transducer power gain G_f is defined as the power delivered to a load divided by the power available from a source:

$$G_t = \frac{\text{power deliver to the load}}{\text{power available from the source}} = \frac{P_L}{P_{avs}} \tag{3.88}$$

From the signal flow chart, we have:

$$G_t = \frac{1-|\Gamma_S|^2}{|1-\Gamma_{in}\Gamma_S|^2}|S_{21}|^2 \frac{1-|\Gamma_L|^2}{|1-S_{22}\Gamma_L|^2} \tag{3.89}$$

or

$$G_t = \frac{1-|\Gamma_S|^2}{|1-S_{11}\Gamma_S|^2}|S_{21}|^2 \frac{1-|\Gamma_L|^2}{|1-\Gamma_{out}\Gamma_L|^2} \tag{3.90}$$

The available power gain G_a is the ration of the power available from the two-port network to the power available from the source. This assumes conjugate matching of both the source and load. The definition of the available power gain is as follows:

$$G_a = \frac{\text{power available from the network}}{\text{power available from the source}} = \frac{P_{avn}}{P_{avs}} \tag{3.91}$$

G_a is given by using S parameters:

$$G_a = \frac{1-|\Gamma_S|^2}{|1-S_{11}\Gamma_S|^2}|S_{21}|^2 \frac{1}{1-|\Gamma_{out}|^2} \tag{3.92}$$

When the input port of the two-port network is terminated by a matched source, that is $\Gamma_S = 0$, thus equation (3.94) can be simplified as follows:

$$G_a = \frac{|S_{21}|^2}{1-|S_{22}|^2} \tag{3.93}$$

These definition differ primarily in the way the source and load are matched to the two-port network. In general, transducer power gain G_t is lower than operating power gain G_p and available power gain G_a; that is,

$$G_p > G_t,\ G_a > G_t$$

If the input and output are both conjugately matched to the two-port network, then the gains is maximized, and $G_t = G_a = G_p$.

The definitions of various power gains are summarized in Table 3.2.

3.3.2 Nonlinearity of Two-Port Network

In an ideal system, the output is linearly related to the input; however, in any real device the transfer function is usually a lot more complicated. Semiconductor devices such as diode and transistor are nonlinear components, and this nonlinearity is of great utility for functions such as amplification, detection, and frequency conversion. To analyze the effects of nonlinearities in microwave circuit, one must be able to describe the input-output relationship of signals that pass through them. Nonlinear circuits are generally characterized by input-output relationships called transfer characteristics.

Figure 3.40 shows a general nonlinear network, having an input voltage v_i and an output voltage v_o. In most general cases, the output response of a nonlinear circuit can be modeled as a Taylor series in terms of input signal voltage:

$$v_o = \sum_{k=0}^{n} g_i v_i^k = g_0 + g_1 v_i + g_2 v_i^2 + \cdots\cdots + g_n v_i^n \tag{3.94}$$

To describe the nonlinearity perfectly, an infinite number of terms is required; however, in many practical components, the first four terms

Figure 3.40 A general nonlinear network

Table 3.2 Definitions of Various Power Gains [18]

Transducer power gain in 50 Ω system	$G_T = \|S_{21}\|^2$
Transducer power gain for arbitrary Γ_L and Γ_S	$G_T = \dfrac{\left(1-\|\Gamma_S\|^2\right)\|S_{21}\|^2\left(1-\|\Gamma_L\|^2\right)}{\left\|(1-S_{11}\Gamma_S)(1-S_{22}\Gamma_L)-S_{12}S_{21}\Gamma_S\Gamma_L\right\|^2}$
Unilateral transducer power gain	$G_{TU} = \dfrac{\|S_{21}\|^2\left(1-\|\Gamma_S\|^2\right)\left(1-\|\Gamma_L\|^2\right)}{\|1-S_{11}\Gamma_S\|^2\|1-S_{22}\Gamma_L\|^2}$
Operating power gain with $\Gamma_L = 0$	$G_p = \dfrac{\|S_{21}\|^2\left(1-\|\Gamma_L\|^2\right)}{\|1-S_{22}\Gamma_L\|^2\left(1-\|\Gamma_{in}\|^2\right)} = \dfrac{\|S_{21}\|^2}{1-\|S_{11}\|^2}$
Available power gain with output conjugate match	$G_A = \dfrac{\|S_{21}\|^2\left(1-\|\Gamma_S\|^2\right)}{\|1-S_{11}\Gamma_S\|^2\left(1-\|S_{22}'\|^2\right)} = \dfrac{\|S_{21}\|^2}{1-\|S_{22}\|^2}$ $\Gamma_S = 0$
Maximum available power gain	$G_{a,\max} = \left\|\dfrac{S_{21}}{S_{12}}\right\|\left(k-\sqrt{k^2-1}\right)$
Maximum unilateral transducer power gain	$G_{TU,\max} = \dfrac{\|S_{21}\|^2}{\left(1-\|S_{11}\|^2\right)\left(1-\|S_{22}\|^2\right)}$
Maximum stable power gain	$G_{ms} = \dfrac{\|S_{21}\|}{\|S_{12}\|}$
Unilateral power gain	$U = \dfrac{1/2\|S_{21}/S_{12}-1\|^2}{k\|S_{21}/S_{12}\|-\mathrm{Re}(S_{21}/S_{12})}$

are sufficient to characterize the circuit with a fair degree of accuracy. Therefore, equation (3.96) can be simplified as follows:

$$v_o \approx \sum_{k=0}^{3} g_i v_i^k = g_0 + g_1 v_i + g_2 v_i^2 + g_3 v_i^3 \tag{3.95}$$

When a single frequency sinusoid is applied to the input of a general nonlinear network, that is,

$$v_i = v_a \cos(\omega_c t) \tag{3.96}$$

then (3.97) gives the output voltage as:

$$\begin{aligned} v_o &\approx g_0 + g_1 v_a \cos(\omega_c t) + g_2 v_a^2 \cos^2(\omega_c t) + g_3 v_a^3 \cos^3(\omega_c t) \\ &= g_0 + \frac{g_2 v_a^2}{2} + (g_1 v_a + \frac{3 g_3 v_a^3}{4}) \cos(\omega_c t) + \frac{g_2 v_a^2}{2} \cos(2\omega_c t) \\ &+ \frac{g_3 v_a^3}{4} \cos(3\omega_c t) \end{aligned} \tag{3.97}$$

Figure 3.41 shows the output power versus input power for a general nonlinear network. It can be observed that the fundamental, second and third harmonic signals will be saturation when the amplitude of input signal v_a is large enough. Also along with the fundamental components at a frequency ω_c, there exists a DC component, and harmonic components at integer multiples of ω_c.

Otherwise, for small values of v_a, the g_1 term will dominate the fundamental signal, giving a 1:1 slope when output power is plotted against the input power on a log (in decibel) scale. The second and third harmonic will have a 2:1 and 3:1 slops, respectively. The fundamental, second and third harmonic signals can be approximately expressed as follows:

$$P_{f_c}(dB) \approx 20\log_{10}(g_1 v_o + \frac{3 g_3 v_o^3}{4}) \approx P_{in}(dB) + 20\log_{10}(g_1)(\text{constant}) \tag{3.98}$$

$$P_{2f_c}(dB) \approx 20\log_{10}(\frac{g_2 v_o^2}{2}) = 2P_{in}(dB) + 20\log_{10}(\frac{g_2}{2})(\text{constant}) \tag{3.99}$$

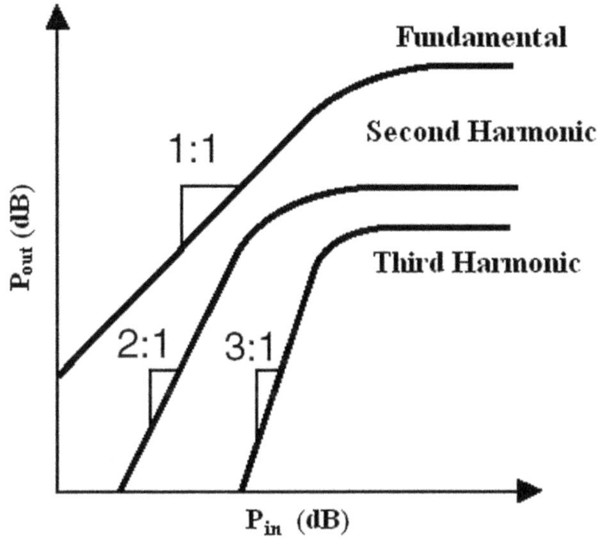

Figure 3.41 Output power versus input power for a general nonlinear network

$$P_{3f_C}(dB) \approx 20\log_{10}(\frac{g_3 v_o^3}{4}) = 3P_{in}(dB) + 20\log_{10}(\frac{g_3}{4})(\text{constant}) \tag{3.100}$$

where P_{in} is the input power level:

$$P_{in}(dB) = 20\log_{10}(v_o) \tag{3.101}$$

When a signal is applied to the input of the two-port network, which consists of two closely spaced frequencies (i.e., two-tone input voltage):

$$v_i = v_1\cos(\omega_1 t) + v_2\cos(\omega_2 t) \tag{3.102}$$

From (3.95) the output is

$$\begin{aligned} v_o = g_0 &+ g_1[v_1\cos(\omega_1 t) + v_2\cos(\omega_2 t)] + g_2[v_1\cos(\omega_1 t) + v_2\cos(\omega_2 t)]^2 \\ &+ g_3[v_1\cos(\omega_1 t) + v_2\cos(\omega_2 t)]^3 \end{aligned} \tag{3.103}$$

Unlike the one-tone input, the response of the two-tone input will generate the combination of two input frequencies, called inter-modulation distortion products. The IMD such as $2\omega_1 - \omega_2$ and $2\omega_2 - \omega_1$ is the figure of merit of the nonlinear network, because of these products located near the original input signal ω_1 and ω_2, and is difficult to filtered by bandpass filter.

The 1-dB gain compression point and third-order intercept point are the two most commonly used figures of merit for testing the linearity of the network. The 1-dB gain compression point is defined as the power level the output power has decreased by 1-dB from the ideal characteristic. Third-order intercept point is the extension intersection of idealized response of the fundamental and third-order IMD. We can view them graphically by plotting the output power for the fundamental and third-order inter-modulation products versus input power on log-log scales, as shown in Figure 3.42.

The relationship between the 1-dB gain compression point and third-order intercept point is as follows:

$$P_i^{IP3}(dB) = P_i^{1dB}(dB) + 9.66(dB) \tag{3.104}$$

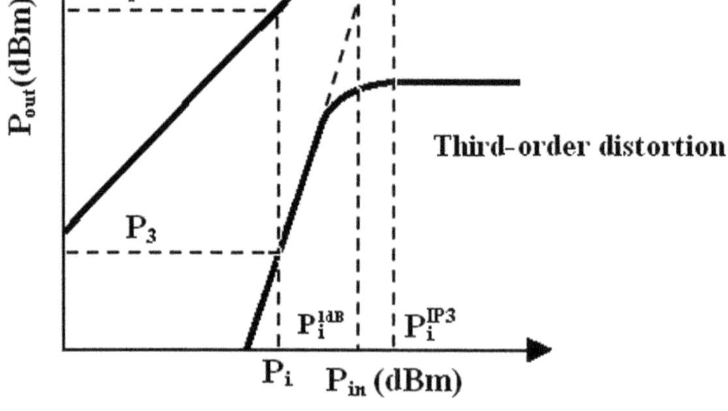

Figure 3.42 1-dB gain compression point and third-order intercept point diagram

or

$$P_o^{IP3}(dB) = P_o^{1dB}(dB) + 9.66(dB) \tag{3.105}$$

The 1-dB gain compression point can be easily obtained from the fundamental output power versus input power measurement. However, the third-order intercept point cannot be measured directly. For an arbitrarily given input power P_i, the fundamental and third-order IMD power are P_1 and P_3, respectively (as shown in Figure 3.42). Thus, the third-order intercept point power is given by:

$$P_i^{IP3} = P_i + \frac{1}{2}(P_1 - P_3) \tag{3.106}$$

or

$$P_o^{IP3} = \frac{1}{2}(3P_1 - P_3) \tag{3.107}$$

When two nonlinear networks are cascaded (Figure 3.43), the gain and third-order intercept point output power of the first network are G_A and P_A^{IP3}. For a given input power P_{in}, the fundamental output power P_A^1 and third-order output power IMD P_A^3 are given by:

$$P_A^1 = P_{in} + G_A \tag{3.108}$$

$$P_A^3 = 3P_A^1 - 2P_A^{IP3} \tag{3.109}$$

Assuming the power gain and third-order intercept point output power of the second network are G_B and P_B^{IP3}, for the given input power P_A^1, the fundamental output power P_B^1 and third-order output power IMD P_B^3 are given by:

$$P_B^1 = P_A^1 + G_B \tag{3.110}$$

$$P_B^3 = 3P_B^1 - 2P_B^{IP3} \tag{3.111}$$

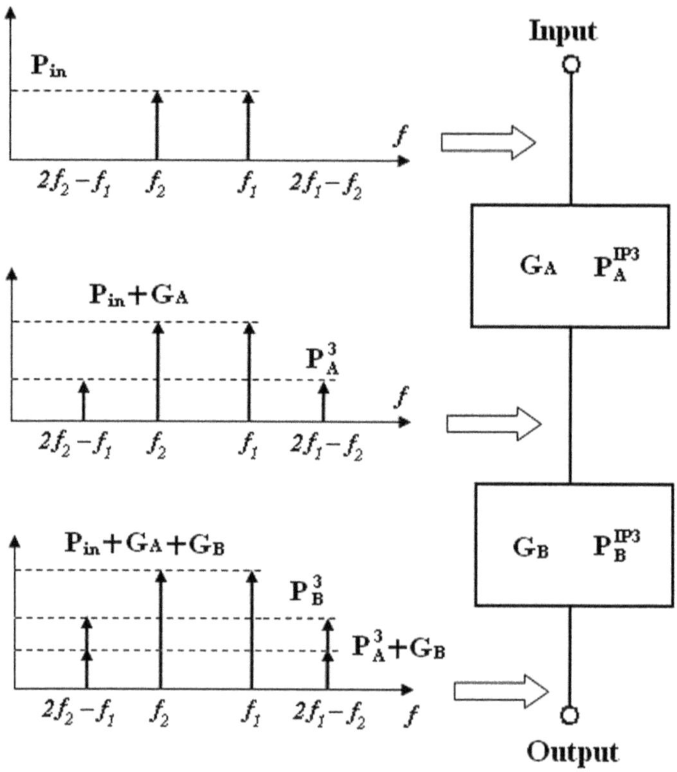

Figure 3.43 Calculation of IMD of two networks in cascade

The third-order IMD output power of the resulting network can be expressed as follows [1]:

$$P_{tot}^{3} = 10\log_{10}[10^{P_B^3/10} + 10^{(P_A^3+G_B)/10}] \tag{3.112}$$

3.3.3 Power Measurement Technique

Power measurement technique includes power gain compression, harmonic distortion, and IMD measurements. Figure 3.44 and Figure 3.45 show the on-wafer and coaxial gain compression and harmonic distortion measurement systems; it can be found that the systems consists of signal generator, spectrum analyzer, RF cables, RF connectors, and probes. An attenuator may be needed to protect the spectrum analyzer from overload. Note that the power level present at the spectrum analyzer input includes all harmonics, not just the ones displayed on the

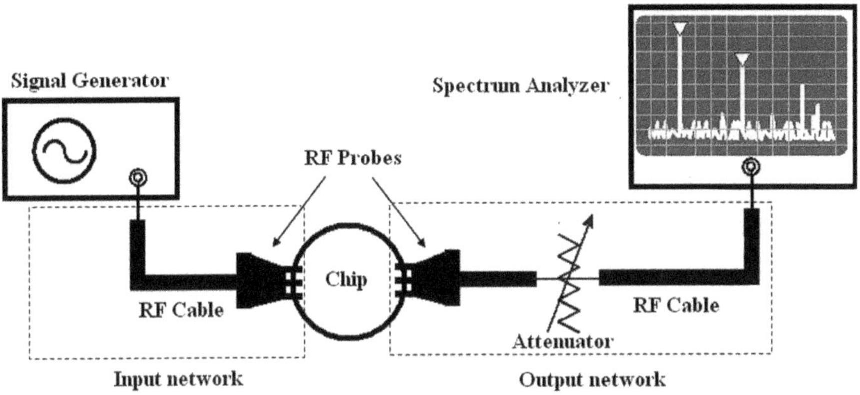

Figure 3.44 On-wafer gain compression and harmonic distortion measurement system

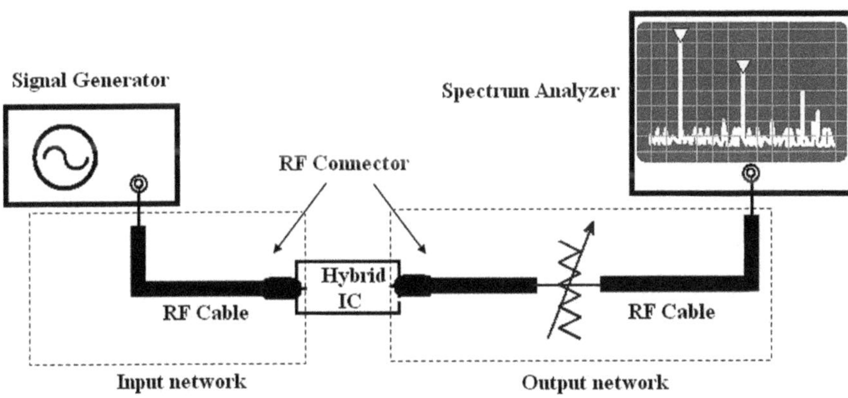

Figure 3.45 Coaxial gain compression and harmonic distortion measurement system

screen. It is important to note that spectrum analyzers have their own nonlinear characteristics that depend on the level input to the instrument. It is sometimes difficult to ascertain whether measured harmonic distortion is being generated within the device or with the test instrument. One method to do this is to use a step attenuator at the output of the device and step up and down. If distortion is being generated with the spectrum analyzer, the harmonic levels will change with different attenuator settings.

To obtain the power performance of DUT, it is necessary to de-embed the effect of the input and output networks (Figure 3.46). The detailed procedure is as follows:

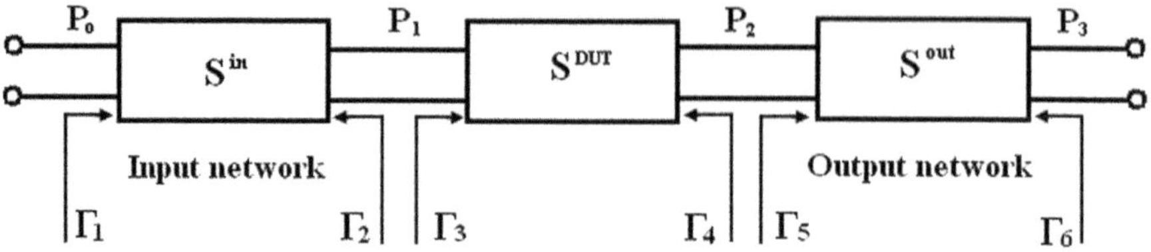

Figure 3.46 De-embedding technique of power gain measurement

1. Measurement of each S-parameters of the input network, DUT, and output network, S^{in}, S^{DUT}, and S^{out}
2. Measurement of the power gain of the system G_p^m
3. Calculation of the power gain of the input network G_p^{in},

$$G_p^{in} = \frac{P_1}{P_o} = \frac{1}{1-|\Gamma_1|^2} \cdot |S_{21}^{in}|^2 \cdot \frac{1-|\Gamma_3|^2}{|1-S_{22}^{in}\Gamma_3|^2} \tag{3.113}$$

with

$$\Gamma_1 = S_{11}^{in} + \frac{S_{12}^{in}S_{21}^{in}\Gamma_3}{1-S_{22}^{in}\Gamma_3}$$

4. Calculation of the power gain of the output network G_p^{out},

$$G_p^{out} = \frac{P_3}{P_2} = \frac{1}{1-|S_{11}^{out}|^2} \cdot |S_{21}^{out}|^2 \tag{3.114}$$

5. Calculation of the power gain of the output network G_p^{DUT},

$$G_p^{DUT} = \frac{P_2}{P_1} = \frac{G_p^m}{G_p^{in}G_p^{out}} \tag{3.115}$$

Figure 3.47 shows the IMD measurement system. It is observed that two signal generators are needed, and that the corresponding de-embedding technique is more complicated than harmonic.

Automated source and load pull is widely used in power amplifier development to determine device capability and matching network requirements. A typical passive tuner system is shown in Figure 3.48.

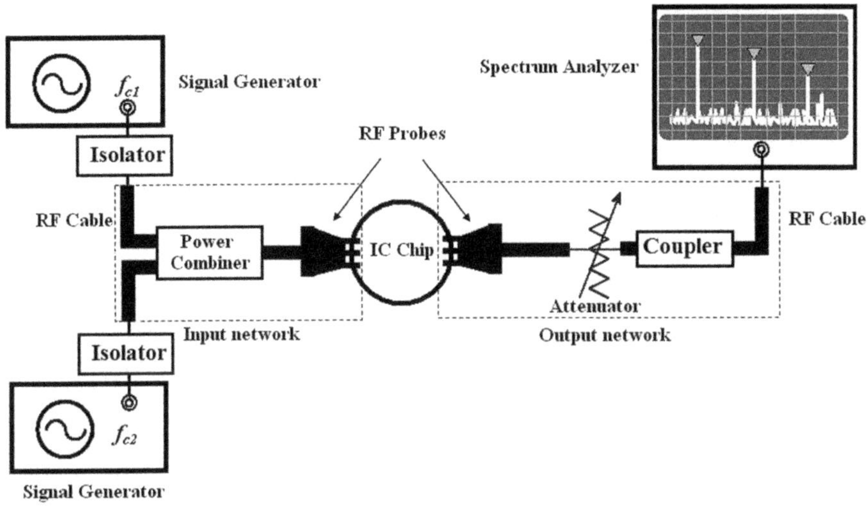

Figure 3.47 IMD measurement system

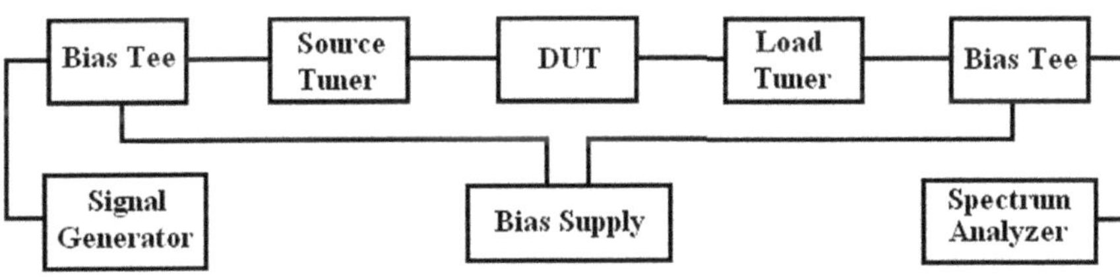

Figure 3.48 Passive tuner system for basic power measurements

Two tuners are used to simultaneously tune the source and load at the fundamental frequency. This is useful for a first-order matching network design, and for validating device models [20]. Fundamental load pull consists of tuning the source and load impedances at the fundamental frequency of operation. The fundamental impedances are the most important by far, which is why fundamental load pull, its accuracy, and repeatability, are most critical.

The tuners are microwave devices generating complex impedance at RF, microwave, and millimeter wave frequencies. Inserted in series before and after DUT, the tuners allow the control of the complex impedance presented to your DUT (e.g., transistors) on the source and load side of your device.

There are two kinds of tuners in commercial system: mechanical and programmable (Figure 4.49). The input and output impedances can be adjusted manually adjustment for the mechanical tuner or by general-purpose interface bus (GPIB) control for the programmable tuner.

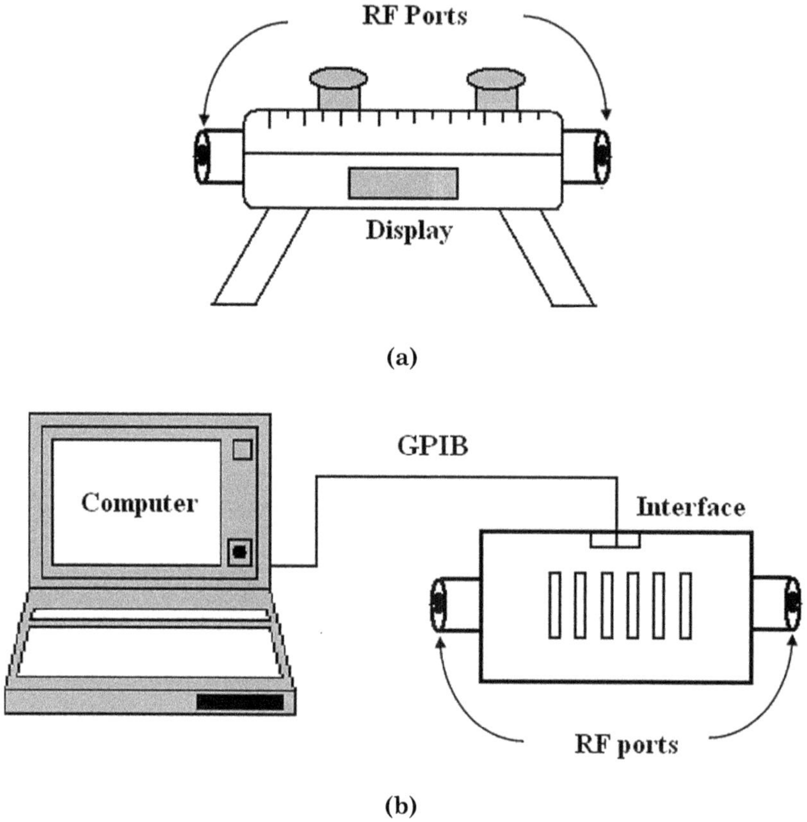

Figure 3.49 Mechanical tuner and programmable tuner

3.4 Summary

Three kinds of microwave and RF measurement techniques are commonly used and have been introduced in this chapter. One is the S-parameters measurement technique for small signal device and circuit characterization. The vector network analyzer is the key instrument, and its calibration can be carried out by using error model with SOLT, TRL and LRM methods. Noise figure is a figure of merit in RF systems and components, and can be determined from noise measurement technique by using the Y-factor method. Then the power measurement technique, which includes the power gain compression, harmonic distortion and inter-modulation distortion measurements is introduced.

Chapter 4

FET Small Signal Modeling and Parameter Extraction

As we known, the accurate on-wafer parameter measurements are of great importance in the modeling and characterization of RF devices (e.g., microwave transistors and passive components). Therefore, the analysis of experimental S-parameters plays an important role in the design of devices and circuits used for RF/microwave applications [1–12]. With fast growth in the RF wireless communications market, the demand for high-performance but low-cost RF solutions is rising. This advanced performance of FETs (e.g., MESFET and HEMT) is attractive for high-frequency circuit design in view of a system-on-a-chip realization, where digital, mixed-signal baseband, and RF transceiver blocks would be integrated on a single chip.

In the high-frequency characterization of microwave transistors, small-signal models are often used to parameterize complicated behaviors with relatively simple equations. A small-signal model is preferably designed so that the model parameters represent something physical in the transistor. This can provide important information to optimize the test structures layout, to perform the simulation of the complete structure using an equivalent circuit, and to study the sensitivity of the device under test (DUT) parameters to a given de-embedding procedure. For RF products, time to market and design cycle reduction depend greatly on the capability of the design environment to predict circuit performance accurately using simulation. To have an efficient design environment, design tools with accurate models for devices and interconnect parasitics are essential. For analog and RF applications, the accuracy of circuit simulation is strongly determined by device models. Accurate device models become crucial to predict the circuit performance correctly. The manufacture of microwave products will use the services of a foundry or specify semiconductor from their suppliers.

In this chapter, we introduce small-signal modeling and parameter extraction technique for FETs, an III-V compound semiconductor device pseudomorphic high electron-mobility transistor (PHEMT) is used as a example. The parameter extraction includes pad capacitances, extrinsic inductances, extrinsic resistances, and intrinsic elements extractions.

4.1 HEMT Device

The derivative device technologies of GaAs MESFETs are HFET and PHEMT. These devices utilize advanced epitaxial material growth technology (e.g., molecular beam epitaxy [MBE] or molecular organic chemical vapor deposition [MOCVD]) and bandgap engineering techniques to achieved high-speed and low-noise performance. HFET is also called modulated doped FET (MODFET) [1,2].

There are different variations of HFETs. One of the most commonly used versions is called high electron mobility transistor. A typical device structure is shown in Figure 4.1. A large bandgap doped material (e.g., Aluminium gallium arsenide [AlGaAs]) is grown heteroepitaxially on an undoped lower bandgap material (e.g., GaAs). The undoped GaAs provides a high-mobility two-dimensional (2-D) channel for carriers supplied from AlGaAs. An undoped AlGaAs layer is used to avoid electronic interaction and to increase mobility. An HFET is also called a high electron mobility transistor, modulation-doped FET), or 2-D electron gas FET (TEGFET) due to its initial development by various groups.

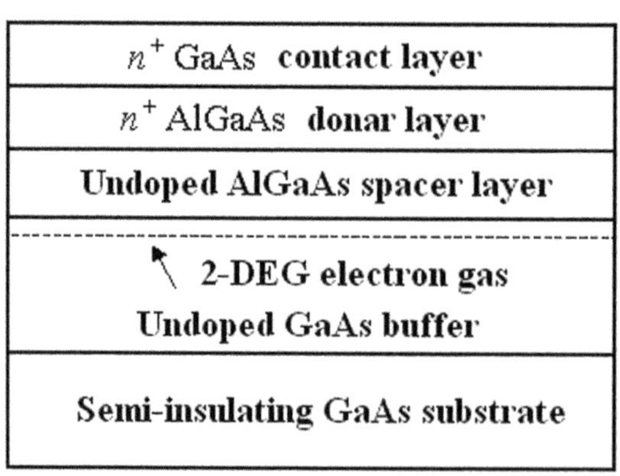

Figure 4.1 HEMT device structure

To improve GaAs-HFET/HEMT performance, an Indium gallium arsenide (InGaAs) strained layer is used as the electron gas channel material instead of GaAs. The more indium percentage incorporated in InGaAs material, the higher the electron drift velocity is. The PHEMT is then developed to enhance the electron mobility in 2-D electron gas (2DEG) layer (as shown Figure 4.2). The lattice mismatched InGaAs layer can increase carrier mobility and improve carrier confinement in the conducting channel. The larger conduction bandgap difference at the AlGaAs/In- GaAs heterointerface allows higher sheet charge density and hence higher current density and transconductance.

4.2 Small Signal Modeling

Figure 4.3 shows the chosen small-signal equivalent circuit model for PHEMTs [1–12], which equivalent circuit model can be divided into two parts:

1. The intrinsic elements: g_m, g_{ds}, C_{gs}, C_{gd}, C_{ds}, R_i, and τ
2. The extrinsic elements: L_g, L_d, L_s, R_g, R_s, R_d, C_{pg}, C_{pd}, and C_{pdg}

The various components in the model are defined in the following list:

Extrinsic elements:

C_{pg}: gate pad capacitance

C_{pd}: drain pad capacitance

C_{pdg}: isolation capacitance between gate and drain pad

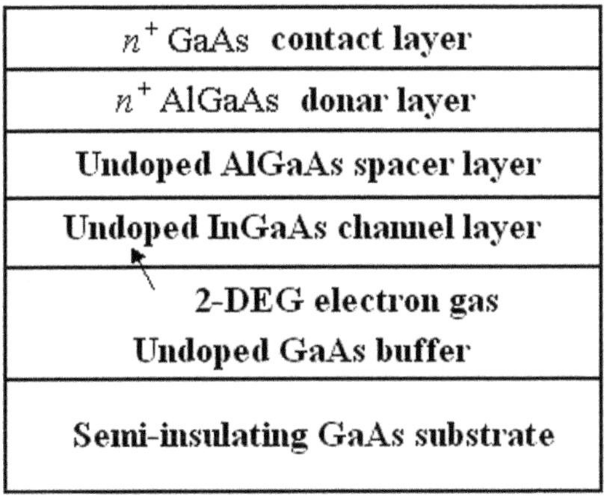

Figure 4.2 PHEMT device structure

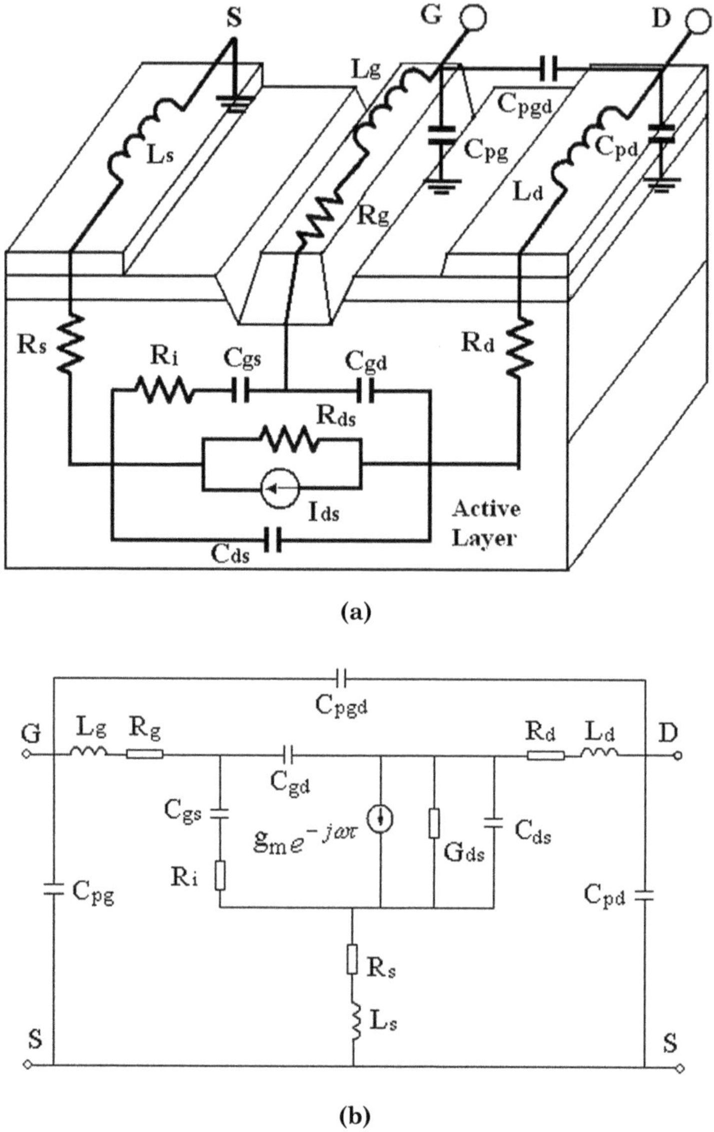

Figure 4.3 FET small signal equivalent circuit model: (a) cubic diagram; (b) planar diagram

L_g: inductance of the gate feedline

L_d: inductance of the drain feedline

L_s: inductance of the source feedline

R_g: distributed gate resistance

R_d: Drain to channel resistance, including contact resistance

R_s: Source to channel resistance, including contact resistance

Intrinsic elements:

g_m: transconductance

g_{ds}: drain conductance

C_{gs}: gate-to-source capacitance

C_{gd}: gate-to-drain capacitance

C_{ds}: drain-to-source capacitance

R_i: channel resistance

τ: time delay associated with transconductance

The Y matrix of complete device model can be written as:

$$Y = Y_{PAD} + [Z_{RL} + Y_{INT}^{-1}]^{-1} \tag{4.1}$$

where Y_{PAD} represents pad capacitance part

$$Y_{PAD} = \begin{bmatrix} j\omega(C_{pg} + C_{pgd}) & -j\omega C_{pgd} \\ -j\omega C_{pgd} & j\omega(C_{pd} + C_{pgd}) \end{bmatrix} \tag{4.2}$$

Z_{RL} represents parasitic resistance and inductance part:

$$Z_{RL} = \begin{bmatrix} R_g + R_s + j\omega(L_g + L_s) & R_s + j\omega L_s \\ R_s + j\omega L_s & R_d + R_s + j\omega(L_g + L_s) \end{bmatrix} \tag{4.3}$$

The intrinsic part is characterized by the Y- parameters:

$$Y_{INT} = \begin{bmatrix} \dfrac{j\omega C_{gs}}{1 + j\omega R_i C_{gs}} + j\omega C_{gd} & -j\omega C_{gd} \\ \dfrac{g_m e^{-j\omega\tau}}{1 + j\omega C_{gs} R_i} - j\omega C_{gd} & g_{ds} + j\omega(C_{ds} + C_{gd}) \end{bmatrix} \tag{4.4}$$

There are two most important figures of merit for FETs: (1) the cutoff frequency f_T, and (2) the maximum oscillation frequency f_{max}. The cutoff frequency f_T is defined as the frequency at which the device has unity current gain. The frequency at which the maximum available gain (MAG) is unity signifies the maximum frequency of operation f_{max}.

Forward current gain h_{21} with output port short-circuited is given by:

$$h_{21} = \frac{Y_{21}}{Y_{11}} \tag{4.5}$$

By neglecting all the parasitics, the current gain h_{21} can be rewritten as:

$$|h_{21}| = \left|\frac{Y_{21}}{Y_{11}}\right| \approx \frac{g_m}{2\pi f C_{gs}} \tag{4.6}$$

Since f_T is defined as the point at which $|h_{21}| = 1$, f_T is given by

$$f_T = \frac{g_m}{2\pi C_{gs}} \tag{4.7}$$

The maximum available gain is given by [1]:

$$MAG = \left(\frac{f_T}{f}\right)^2 \frac{1}{4R/R_{ds} + 4\pi f_T C_{gd}(R + R_g + \pi f_T L_s)} \tag{4.8}$$

where

$$R = R_g + R_i + R_s + \pi f_T L_s$$

Since f_{max} is defined as the point at which $MAG = 1$, therefore f_{max} is given by:

$$f_{\max} = \frac{f_T}{\sqrt{4R/R_{ds} + 4\pi f_T C_{gd}(R + R_g + \pi f_T L_s)}} \tag{4.9}$$

Accurate small-signal equivalent circuit of FETs (e.g., MESFETs and PHEMTs) is a supposition for the device performance analysis (e.g., noise, gain) in the design of microwave circuits and the characterization of the device technological process. The most commonly used small-signal parameter extraction technique is the numerical optimization of the model-based S parameters to fit the measured data. However, the accuracy of the numerical optimization methods that minimize the difference between measured and modeled S-parameters versus frequency can vary depending on the optimization method and starting values, and may result in nonphysical and non-unique values of the equivalent circuit elements. The analytical approach for the extraction of the equivalent circuit parameters of FETs has been addressed. The intrinsic elements can be determined using closed-form expressions after a de-embedding process where the influence of the parasitic elements is removed. Therefore, for the accurate determination of the bias-dependent intrinsic elements, firstly all extrinsic parasitic elements have to be determined precisely.

4.3 PHEMT Device Structure

The AlGaAs/InGaAs/GaAs PHEMTs with 0.25 μm mushroom gates were grown and fabricated using Nanyang Technical University's (NTU's) in-house developed process technology. The layer structure of the wafer, from bottom to top, is described in the following list:

1. Semi-isolating GaAs substrate
2. Undoped buffer GaAs layer
3. Si δ-dopin plane, 2×10^{12} cm^{-2}
4. $Al_{0.25}Ga_{0.75}As$ layer, 40 Å
5. Undoped $In_{0.22}Ga_{0.78}As$ strained layer, 140 Å
6. $Al_{0.25}Ga_{0.75}As$ spacer layer, 40 Å
7. Si δ-dopin plane, 5×10^{12} cm^{-2}
8. $Al_{0.25}Ga_{0.75}As$ layer220 Å
9. $n^{+} - GaAs$ cap layer, 450 Å

The double heterojunction δ-doped PHEMT structure considered for our studies is shown in Figure 4.4. The PHEMT has been used here, which has 2 × 40 μm gate width (number of gate fingers × unit gate

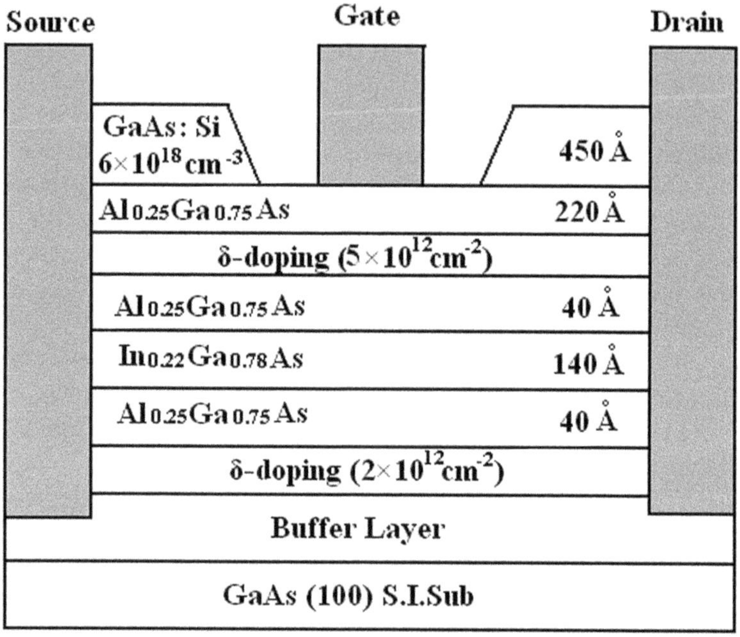

Figure 4.4 PHEMT device layer structure

width) and pinch-off voltage of –0.8 V. The corresponding device layout is shown in Figure 4.5.

4.4 Extraction Method of Pad Capacitances

The three capacitance elements C_{pg}, C_{pd}, and C_{pdg} model the capacitive effects of the measurement probe contacts. The two kinds of most commonly used extraction methods for pad capacitances are the open test structure method and the pinch-off cold-FET method. These procedures are discussed in more detail in the following sections.

4.4.1 Open Test Structure Method

The pad capacitances are determined by measuring an open structure, which consists of only the pads [3,4]. Measurements of the open test structure are modeled as a PI-type network of capacitances. Figure 4.6 shows the open test structure layout with the corresponding equivalent circuit model. The simplicity of this model is a direct consequence of a semi-insulating GaAs substrate and proper device isolation down to the

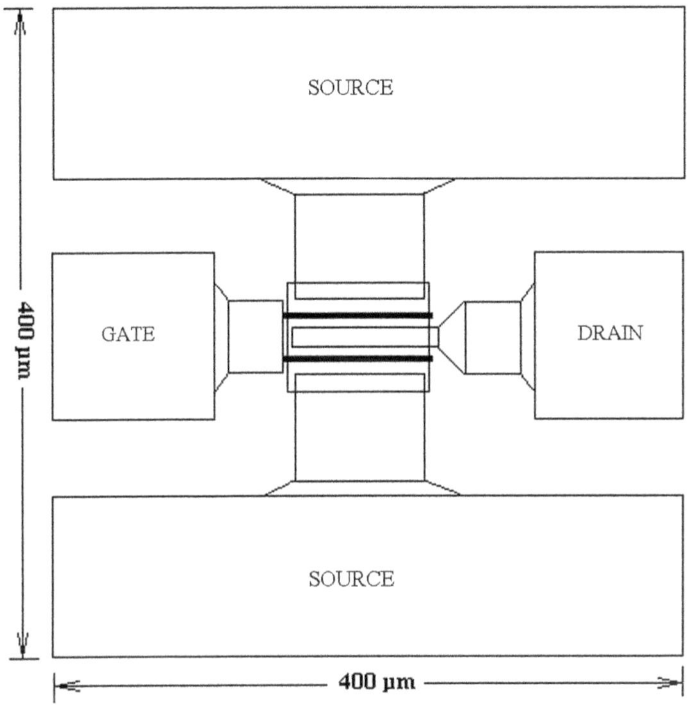

Figure 4.5 PHEMT device layout

semi-insulating substrate. In contrast, suitable pad-parasitic correction for a silicon device is more complex since the modeling of a conducting substrate requires several additional elements.

The Y parameters of open test structure can be expressed as follows:

$$\mathrm{Im}(Y_{11}) = j\omega(C_{pg} + C_{pgd}) \tag{4.10}$$

$$\mathrm{Im}(Y_{12}) = \mathrm{Im}(Y_{21}) = -j\omega C_{pgd} \tag{4.11}$$

$$\mathrm{Im}(Y_{22}) = j\omega(C_{pd} + C_{pgd}) \tag{4.12}$$

From (4.10)–(4.12), the pad capacitances C_{pg}, C_{pd}, and C_{pdg} can be directly obtained:

$$C_{pg} = \frac{1}{\omega}\mathrm{Im}(Y_{11} + Y_{12}) \tag{4.13}$$

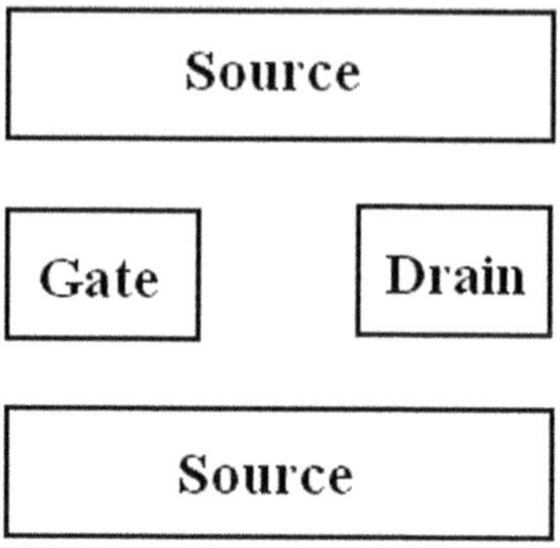

(a)

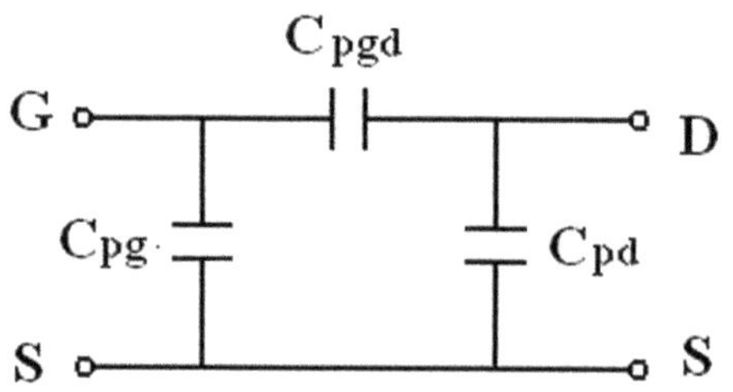

(b)

Figure 4.6 Open test structure (a) and equivalent circuit model (b)

$$C_{pd} = \frac{1}{\omega} \mathrm{Im}(Y_{22} + Y_{12}) \tag{4.14}$$

$$C_{pgd} = -\frac{1}{\omega} \mathrm{Im}(Y_{12}) = -\frac{1}{\omega} \mathrm{Im}(Y_{21}) \tag{4.15}$$

Figure 4.7 shows the frequency dependence of the pad capacitances. Constant values are observed from 50 MHz to 40 GHz with the deviations from the mean values being less than 5%. That means the pad capacitances are frequency independent and, of course, also bias independent. The pad capacitance values are C_{pd} = 17.5 fF, and C_{pgd} = 2.1 fF. The isolation between pads is below 30 dB. The extracted pad capacitances and corresponding dispersions are summarized in Table 4.1.

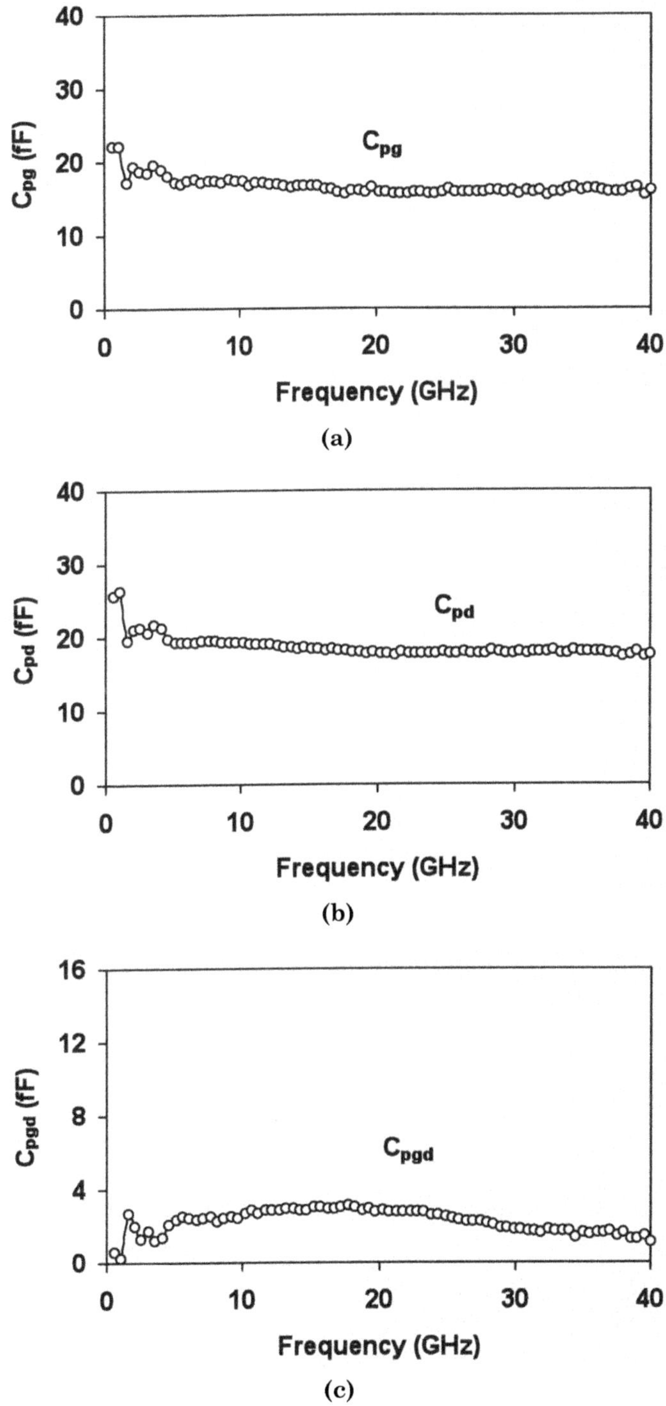

Figure 4.7 Frequency dependence of pad capacitance

Table 4.1 Extracted Pad Capacitances and Dispersions

Capacitance	Values (fF)	Dispersions (fF)
C_{pg}	17.5	±1
C_{pd}	19.5	±1
C_{pdg}	2.1	±0.6

4.4.2 Pinch-Off Method

For a complete small-signal model, the pad capacitances of the coplanar feeding structure matched to on-wafer measurement should also be taken into account. The pad capacitance can be extracted by using the open test structure method [3,4]. However, this method requires special test structures for each device size on the wafer, and the non-uniformity across the wafer has to be ignored. Alternatively, the pinch-off cold-FET method has been extensively used for the extraction of the parasitic gate and drain capacitances C_{pg} and C_{pd} [3–9]. In [5], two identical capacitances C_b are used to describe the depletion-layer extension under the gate. In [6], three identical capacitances C_b are used to describe the depletion-layer extension under pinch-off bias condition. An improved method is proposed by [7] to extract the parasitic capacitances, whereby two identical capacitances C_b are employed to describe the symmetrical nature of the gate-to-source and gate-to-drain geometry, and a capacitance C_c is introduced to account for drain-to-source depletion-layer extension. In this approach, the derived circuit equations are simplified by neglecting some terms depending on the frequency range (low and high frequencies) where the model parameters are extracted. All the conventional cold-FET methods [3–8] are based on a common assumption: The gate-to-source and gate-to-drain depletion-layer capacitances are equal for a symmetrical device structure under pinch-off cold-FET bias condition. Unfortunately, this assumption is not generally valid for the parasitic gate and drain capacitances C_{pg} and C_{pd} of MESFETs and PHEMT, although their difference can be very small.

To overcome these difficulties, a new pinch-off cold-FET method is proposed for the determination of the parasitic capacitance for PHEMTs based on a general scalable equivalent circuit model [10]. In contrast with previous publications [3–9], this method has the following advantages:

1. No restrictions or assumptions are imposed on the depletion-layer capacitances under pinch-off bias condition. Three different capacitances are used to describe the depletion-layer extension.
2. No complex derivation and extraction procedures are needed.
3. Four scalable PHEMT devices with same pad structures are employed to determine the parasitic capacitances.
4. All parasitic capacitances C_{pg}, C_{pd}, and C_{pdg} can be extracted simultaneously.

The PHEMT small-signal equivalent circuit model under pinch-off bias condition is shown in Figure 4.8. Pinch-off bias condition for PHEMTs is defined as the condition when both junctions are zero or reverse biased. Under such a condition, DC current is zero; hence g_m would be extremely small and the device behaves like a passive circuit ($Z_{12} = Z_{21}$). Compared with the conventional cold-FET method, the

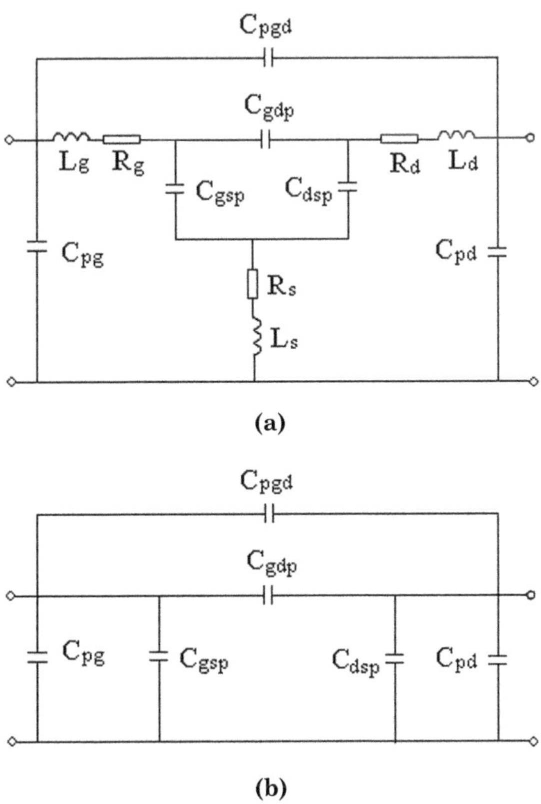

Figure 4.8 Small-signal equivalent circuit model of PHEMT under pinch-off bias condition: (a) at high frequency; (b) at low frequency

pinch-off equivalent circuit model used in this book is more general and consistent with the equivalent circuit model in the active region.

At low frequency, the influence of the parasitic resistances and inductances can be neglected. The PHEMT equivalent circuit of Figure 4.8(a) exhibits a pure capacitive behavior and is simplified as shown in Figure 4.8(b). The imaginary parts of Y-parameters can be written as:

$$\frac{\mathrm{Im}(Y_{11})}{\omega} = C_{pg} + C_{pgd} + C_{gsp} + C_{gdp} \tag{4.16}$$

$$\frac{\mathrm{Im}(Y_{22})}{\omega} = C_{pd} + C_{pgd} + C_{dsp} + C_{gdp} \tag{4.17}$$

$$-\frac{\mathrm{Im}(Y_{12})}{\omega} = C_{pgd} + C_{gdp} \tag{4.18}$$

The scaling formulas for intrinsic capacitances can be written as [10]:

$$C_{gsp}(W) = C_{gspo}W \tag{4.19}$$

$$C_{gdp}(W) = C_{gdpo}W \tag{4.20}$$

$$C_{dsp}(W) = C_{dspo}W \tag{4.21}$$

where C_{gsp}, C_{gdp}, and C_{dsp} are intrinsic capacitances under pinch-off bias condition and C_{gspo}, C_{gdpo}, and C_{dspo} are scaling factors, and W is the gatewidth of the PHEMT.

Substituting (4.19)–(4.21) into (4.16)–(4.18), we have

$$\frac{\mathrm{Im}(Y_{11})}{\omega} = C_{pg} + C_{pgd} + W(C_{gspo} + C_{gdpo}) \tag{4.22}$$

$$\frac{\mathrm{Im}(Y_{22})}{\omega} = C_{pd} + C_{pgd} + W(C_{dspo} + C_{gdpo}) \tag{4.23}$$

$$-\frac{\mathrm{Im}(Y_{12})}{\omega} = C_{pgd} + W(C_{gdpo}) \tag{4.24}$$

By using the three analytical equations (4.22)–(4.24) associated with three linear regression lines, the parasitic and intrinsic capacitances can be determined simultaneously. C_{pg}, C_{pd}, and C_{pdg} can be obtained from the interceptions of the three linear regression lines versus the gatewidth. Therefore, the expressions for parasitic capacitance can be expressed as

$$C_{pg} = \left.\frac{\mathrm{Im}(Y_{11})}{\omega}\right|_{W\to 0} - C_{pgd} \tag{4.25}$$

$$C_{pd} = \left.\frac{\mathrm{Im}(Y_{22})}{\omega}\right|_{W\to 0} - C_{pgd} \tag{4.26}$$

$$C_{pgd} = \left.\frac{\mathrm{Im}(Y_{12})}{\omega}\right|_{W\to 0} \tag{4.27}$$

The intrinsic capacitances also can be solved simultaneously from the slopes of the linear regression lines:

$$C_{gsp} = \frac{\mathrm{d}\left[\mathrm{Im}(\mathrm{Y}_{11})/\omega\right]}{\mathrm{dW}} - C_{gdp} \tag{4.28}$$

$$C_{gdp} = \frac{\mathrm{d}\left[\mathrm{Im}(\mathrm{Y}_{22})/\omega\right]}{\mathrm{dW}} - C_{gdp} \tag{4.29}$$

$$C_{gdp} = -\frac{\mathrm{d}\left[\mathrm{Im}(\mathrm{Y}_{12})/\omega\right]}{\mathrm{dW}} \tag{4.30}$$

The measured pinch-off cold-FET S-parameters for four differently sized PHEMTs are firstly transformed to Y-parameters; then the imaginary parts of Y parameters $\mathrm{Im}(Y_{ij})/\omega$, (i, j = 1,2) can be extracted at low frequencies. The DC bias condition is $V_{gs} = -3$ V, $V_{ds} = 0$ V.

Figure 4.8 shows the imaginary parts of the *Y*-parameters $\mathrm{Im}(Y_{ij})/\omega$, (i,j = 1,2) versus the gatewidth of the PHEMTs. C_{pg}, C_{pd}, and C_{pdg} can be determined from the interceptions of the three straight lines in Figure 4.9 (a), (b), and (c). The values of the parasitic capacitances are as follows: C_{pg} = 25.5 fF, C_{pd} = 28 fF, and C_{pdg} = 4.5 fF. The corresponding intrinsic capacitances under pinch-off bias condition are summarized in Table 4.2. It is noted that the gate-to-source and gate-to-drain depletion-layer capacitances (C_{gsp} and C_{gdp}) are not equal under pinch-off cold-FET bias condition for all the devices of different sizes. It also can be found that the values of the intrinsic capacitances match the scaling rules very well.

Table 4.2 Extracted Results for Intrinsic Capacitances under Pinch-off Bias Condition

Device	C_{gsp} (fF)	C_{gdp} (fF)	C_{dsp} (fF)
2×20	9.5	11	4.5
2×40	18.5	22.5	8
2×60	28	34	13
2×100	44.5	55.5	24

Table 4.3 shows a comparison of parasitic capacitances extracted by using the proposed method and the open test structure method, respectively. It can be seen that the values of the parasitic capacitances based on the new method are larger than that one of the open test structure method. The main reason is the influence of the feedlines and transitions between the pads and feedlines, which have been taken into account by the proposed method and neglected by the test structure method.

Table 4.3 Comparison of Parasitic Capacitances by Using Proposed Method and Open Test Structure Method

Parasitic Capacitances	Proposed Method	Test Structure Method
C_{pg}	25.5	17.5
C_{pd}	28.0	19.5
C_{pdg}	4.5	2.1

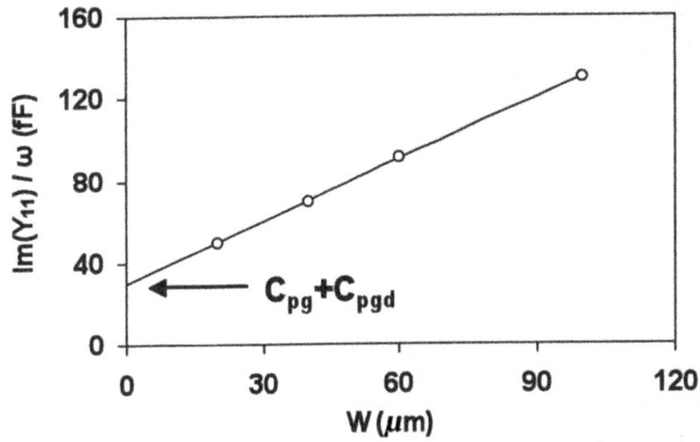

(a)

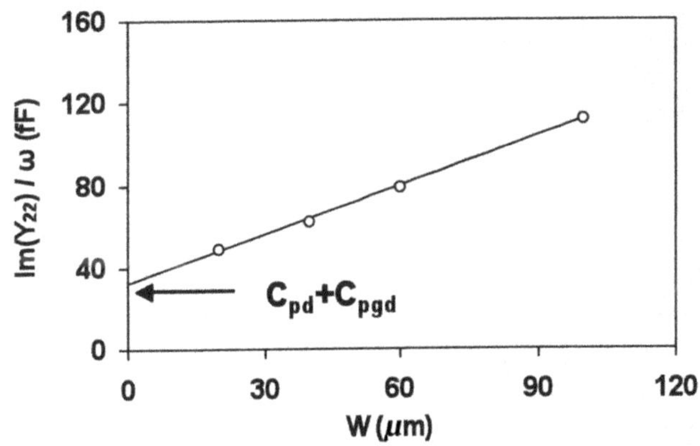

(b)

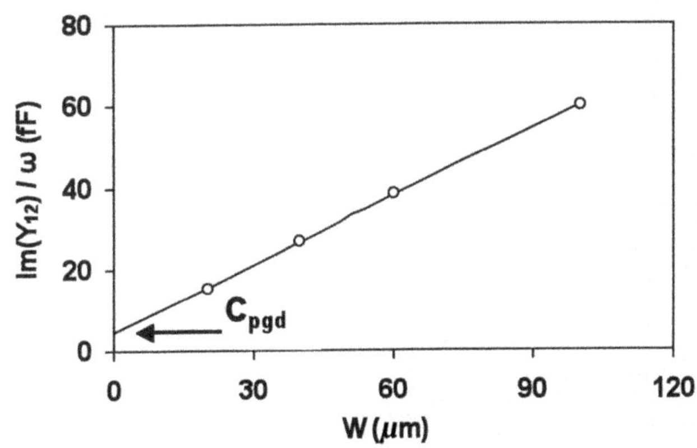

(c)

Figure 4.9 Frequency response of measured $\mathrm{Im}(Y_{ij})/\omega$, (i,j = 1,2) versus gatewidth of the PHEMTs

All the conventional pinch-off cold-FET method [3–8] are based on an assumption that the gate-to-source and gate-to-drain depletion-layer capacitances are equal under pinch-off cold-FET bias condition. The parasitic gate capacitance can be expressed as:

$$C_{pg} = \frac{\mathrm{Im}(Y_{11} + 2Y_{12})}{\omega} \tag{4.31}$$

Figure 4.10 shows the extracted C_{pg} versus gatewidth by using the conventional pinch-off cold-FET method [3–8]. It can be found that C_{pg} decreases with increasing device gatewidth, which means it is not independent of the device size. Figure 4.10 also compares the extracted results of C_{pg} by using the proposed method and conventional pinch-off cold-FET method, and it is very clear that the extracted results based on the conventional pinch-off cold-FET method are underestimated. Because the conventional pinch-off cold-FET methods are based on the assumption that the gate-to-source and gate-to-drain depletion-layer capacitances are equal, the correct capacitance C_{pg} cannot be obtained. As a consequence, the extracted C_{pd} values are also underestimated and overestimated, respectively. (Figure 4.11). The extracted parasitic gate and drain capacitances C_{pg} and C_{pd} are difficult consistent by using the conventional pinch-off cold-FET method. From Figure 4.10 and Figure 4.11, it can also be found that the extracted C_{pg} and C_{pd} are dependent on the device size.

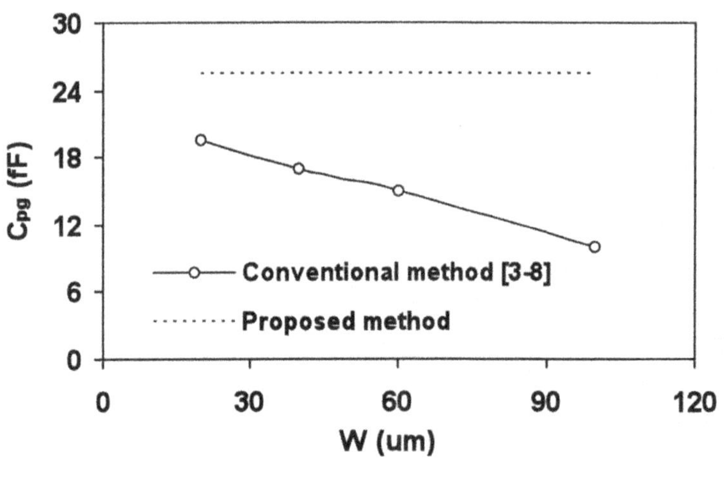

Figure 4.10 Comparison of C_{pg} by using proposed method and conventional pinch-off cold-FET method.

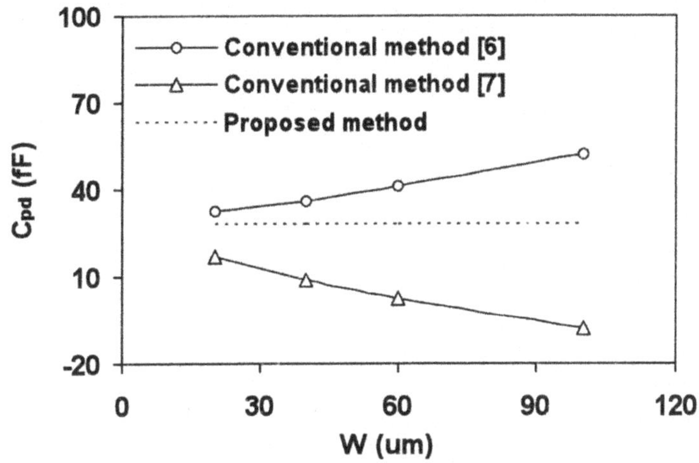

Figure 4.11 Comparison of C_{pd} by using proposed method and conventional pinch-off cold-FET method.

4.5 Extraction Method of Extrinsic Inductances

The extrinsic inductances are caused by the feedlines between the device under test and the test pads. There are three commonly used methods for determinating extrinsic inductances L_g, L_d, and L_s: short test structure method, cold-FET method, and cutoff method.

4.5.1 Short Test Structure Method

The parasitic device-connection impedances can be determined by measuring a test pattern, which consists of the pads, the device feeds, and a short replacing the transistor [3]. The short test structure is modeled as a T-network of series resistors and inductors. Figure 4.12 shows the shorted test structure and corresponding equivalent circuit model.

The extrinsic inductances and feedline losses can be directly determination from Z parameters of the short test structure:

$$L_{ps} = \frac{1}{\omega} I_m(Z_{12}) = \frac{1}{\omega} I_m(Z_{21}) \tag{4.32}$$

$$L_{pg} = \frac{1}{\omega}(Z_{11} - Z_{12}) \tag{4.33}$$

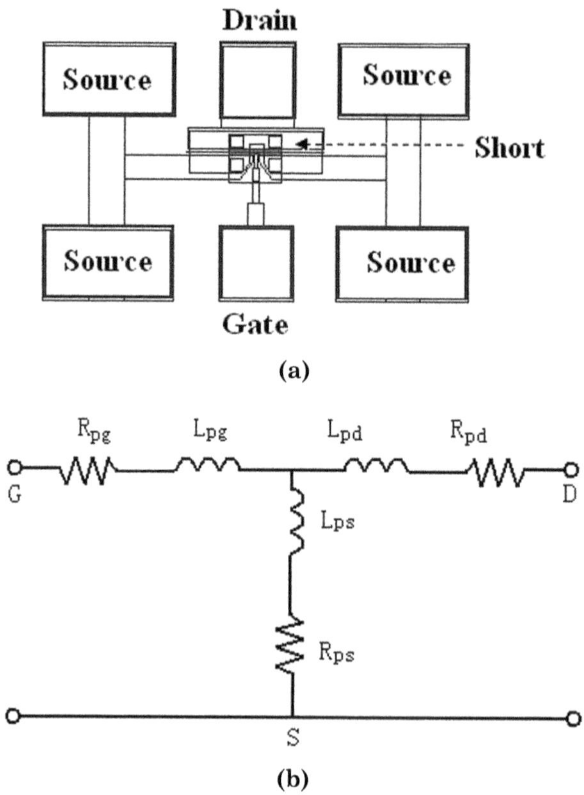

Figure 4.12 Short test structure (a) and equivalent circuit model (b)

$$L_{pd} = \frac{1}{\omega}(Z_{22} - Z_{21}) \tag{4.34}$$

$$R_{pg} = R_e(Z_{11} - Z_{12}) \tag{4.35}$$

$$R_{pd} = R_e(Z_{11} - Z_{21}) \tag{4.36}$$

$$R_{ps} = R_e(Z_{12}) = R_e(Z_{12}) \tag{4.37}$$

Figure 4.13 shows the frequency dependence of the extrinsic inductances, and the feedline losses are shown in Figure 4.14. Constant values are observed from 50 MHz to 40 GHz with the deviations from the mean values being less than 5%. That means the feedline inductances are frequency independent and of course, also bias independent. The

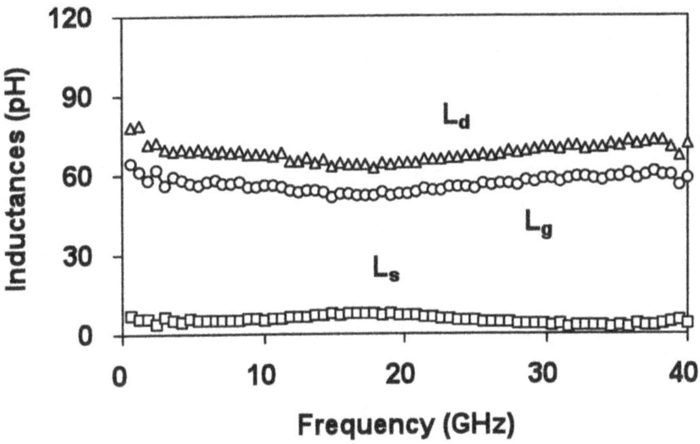

Figure 4.13 Frequency dependence of the extrinsic inductances

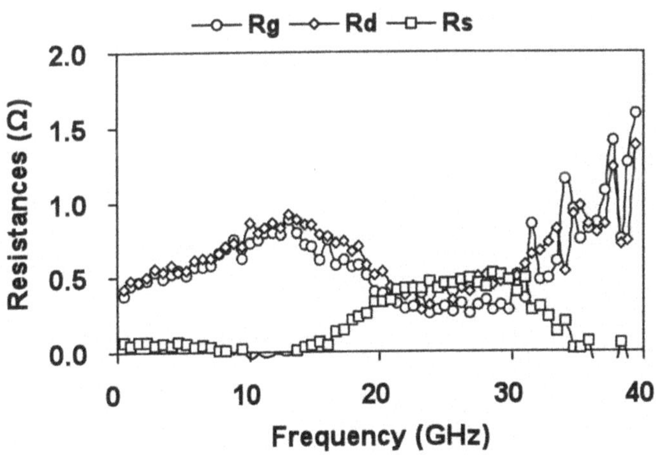

Figure 4.14 Frequency dependence of the feedline losses

extrinsic inductance values are L_g = 60 pH, L_d = 70 pH, and L_s = 4 pH. It can be found that the feedline losses are very small (less than 1 Ω normally), therefore can be neglected compared with the contact resistances.

4.5.2 Forward Biased Cold-FET Method

The conventional cold-FET method is defined as the condition when applying a strong forward bias to the gate of FET, that is, when the gate-to-source voltage is larger than threshold voltage with zero drain-to-source voltage ($V_{gs} > V_{th}$ and $V_{ds} = 0$) [5]. Under such conditions, the device behaves like a passive component and has,

$$g_m = 0\,,\ C_{gs} = C_{gd} = C_{ds} = 0$$

Therefore the equivalent circuit model becomes much simpler as shown in Figure 4.15. The extrinsic inductances can be directly determination from Z parameters:

$$L_g = \frac{\mathrm{Im}(Z_{11} - Z_{12})}{\omega} \tag{4.38}$$

$$L_d = \frac{\mathrm{Im}(Z_{22} - Z_{12})}{\omega} \tag{4.39}$$

$$L_s = \frac{\mathrm{Im}(Z_{12})}{\omega} \tag{4.40}$$

Figure 4.16 shows the frequency dependence of the extrinsic inductances for differently sized devices. The results show that the extracted parameters remain almost constant with frequency.

4.5.3 Reverse Biased Cutoff Method

To extract extrinsic parameters, the conventional cold-FET methods apply a strong forward bias to the gate of FET. However, in the case of

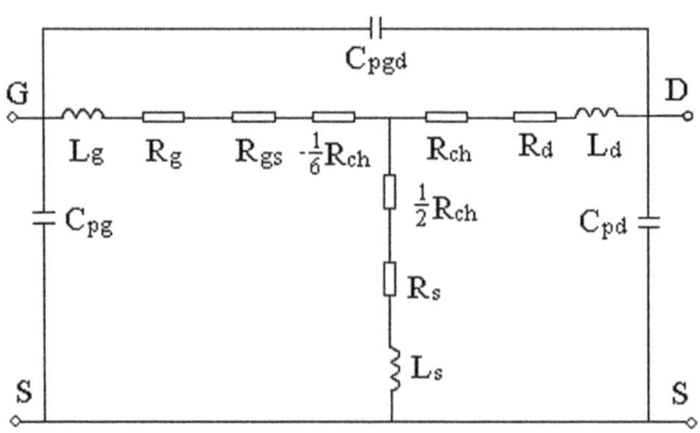

Figure 4.15 Equivalent circuit model under forward cold-FET condition

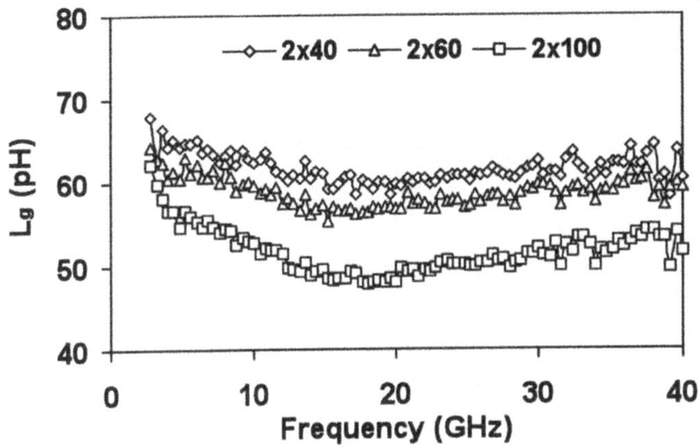

(a)

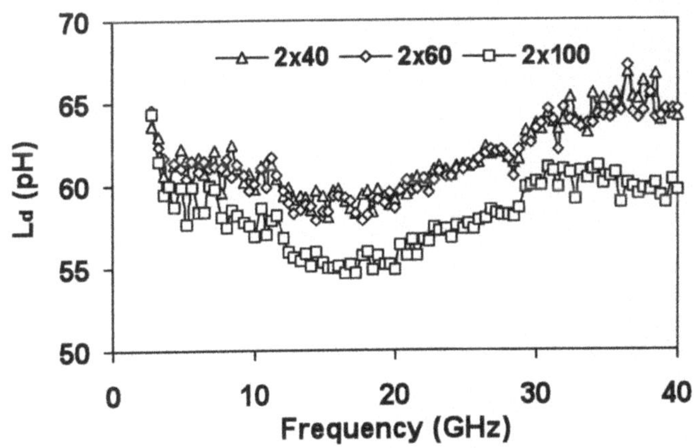

(b)

Figure 4.16 Frequency dependence of the extrinsic inductances for differently sized devices

PHEMT, the method may cause gate degradation due to the large gate current running through the Schottky junction. Here, all parasitic parameters are extracted by using pinch-off bias condition with zero drain voltage only. The DC bias condition used is $V_{gs} = -3\text{V}, V_{ds} = 0\text{V}$ [11].

For frequencies up to a few gigahertz (typically $F < 5$ GHz), the parasitic resistances and inductances can be neglected. The capacitances C_{gsp}, C_{gdp}, and C_{dsp} (see Figure 4.8(a)) are determined by linear regression of $\mathrm{Im}(Y_{ij}^L)$ after de-embedding the pad capacitances C_{pg}, C_{pd}, and C_{pdg}, which can be written as:

$$C_{gdp} = -\frac{\mathrm{Im}(Y_{21}^L)}{\omega} = -\frac{\mathrm{Im}(Y_{12}^L)}{\omega} \tag{4.41}$$

$$C_{gsp} = \frac{\mathrm{Im}(Y_{11}^L + 2Y_{12}^L)}{\omega} \tag{4.42}$$

$$C_{dsp} = \frac{\mathrm{Im}(Y_{22}^L + Y_{12}^L)}{\omega} \tag{4.43}$$

Because parasitic resistances and inductances are very sensitive to S parameters at lower frequencies, so for frequencies beyond a certain high gigahertz (for our PHEMT device typically $F > 25$ GHz), the parasitic resistances and inductances can be directly calculated by:

$$L_s = \frac{\mathrm{Im}(Z_{12})}{\omega} + \frac{C_{gdp}}{\omega^2 (C_{gsp}C_{gdp} + C_{gsp}C_{dsp} + C_{dsp}C_{gdp})} \tag{4.44}$$

$$L_d = \frac{\mathrm{Im}(Z_{22})}{\omega} + \frac{C_{gdp} + C_{gsp}}{\omega^2 (C_{gsp}C_{gdp} + C_{gsp}C_{dsp} + C_{dsp}C_{gdp})} - L_s \tag{4.45}$$

$$L_g = \frac{\mathrm{Im}(Z_{11})}{\omega} + \frac{C_{gdp} + C_{dsp}}{\omega^2 (C_{gsp}C_{gdp} + C_{gsp}C_{dsp} + C_{dsp}C_{gdp})} - L_s \tag{4.46}$$

Figure 4.17 shows the extracted results of the parasitic resistances and inductances; the results show that the extracted parameters remain almost constant with a frequency above 25 GHz. The flat values again show that the assumptions used are valid.

4.6 Extraction Method of Extrinsic Resistance

Compared with the extrinsic capacitances and inductances, the extrinsic resistances are complicated and also a key issue. The extrinsic resis-

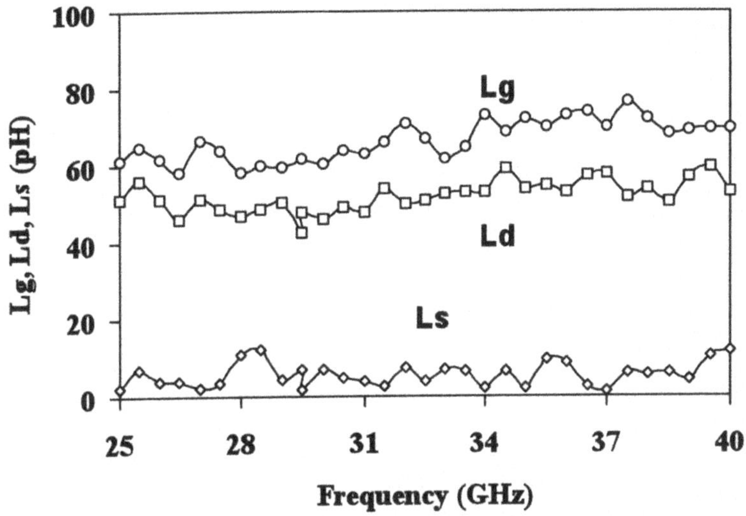

Figure 4.17 Extracted parasitic resistances and inductances results versus frequency for the 2 × 40 μm PHEMT, (bias: V_{gs} = −3 V and V_{ds} = 0 V)

tances significantly affect the DC I-V performance and RF performance. The DC measurement method, cold-FET method, and hot-FET method are the most commonly used ways to determine the three extrinsic resistances R_g, R_d, R_s.

4.6.1 DC Measurement Method

There are several DC measurement methods for determining the three extrinsic resistances R_g, R_d, R_s, [12–16]. The basic principle is the extrinsic resistances can be considered as the series resistances of the Schottky diodes.

Figure 4.18 shows the schematic for the source resistance measurement representing the distributed channel and R_d and R_s [12], it is noted that he gate of FET device is considered as a distributed diode model. The source and drain contact resistances R_d and R_s can be extracted from two different drain current measurements under the assumptions of $I_d >> I_g$, and the specific channel resistance is independent of position. By neglecting the channel resistance, three extrinsic resistances R_g, R_d and R_s can be determined from measurements of $I_G(V_D)$ and $I_G(V_G)$ characteristics at forward bias with floating drain and source [13–15]; the measurements principle is illustrated in Figure 4.19. Another method is proposed in [16], by using two different DC

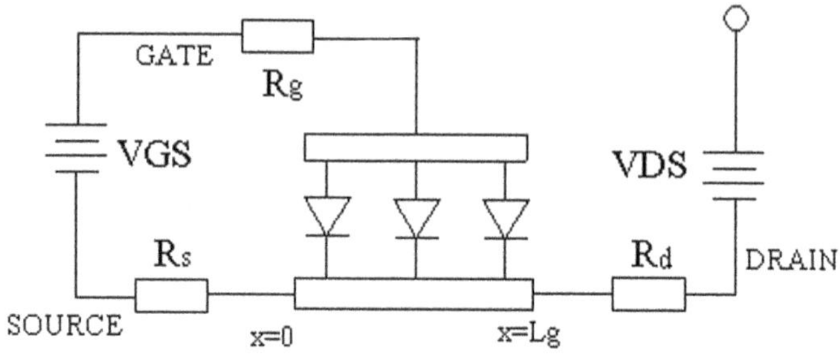

Figure 4.18 The schematic for the source resistance measurement representing the distributed channel and R_d and R_s

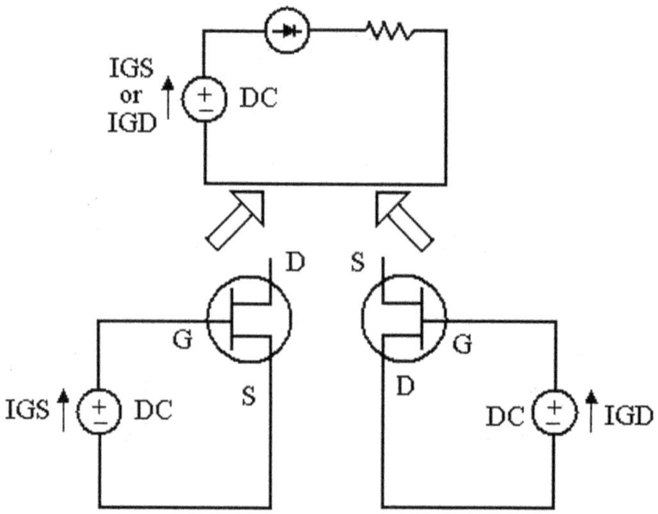

Figure 4.19 Schottky diode model for FET with floating drain or source

measurements, three relations between three unknown extrinsic resistances can be obtained. In a first measurement setup (Figure 4.20(a)), a current I_g is forced through the gate, while at the same time the drain is left open and the source is grounded. A second current source sinks half of this current out of the drain (Figure 4.20(a)).

4.6.2 COLD-FET Method

For cold-FET modeling, all S-parameter measurements are carried out at $V_{ds} = 0$ V. In this case the transistor is a passive symmetric device [17].

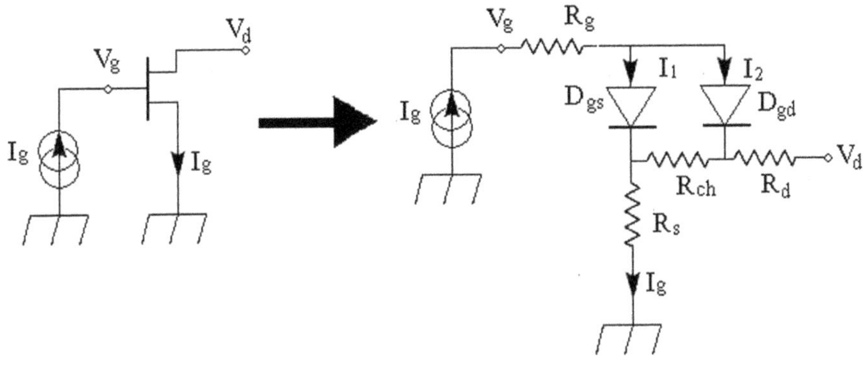

(a)

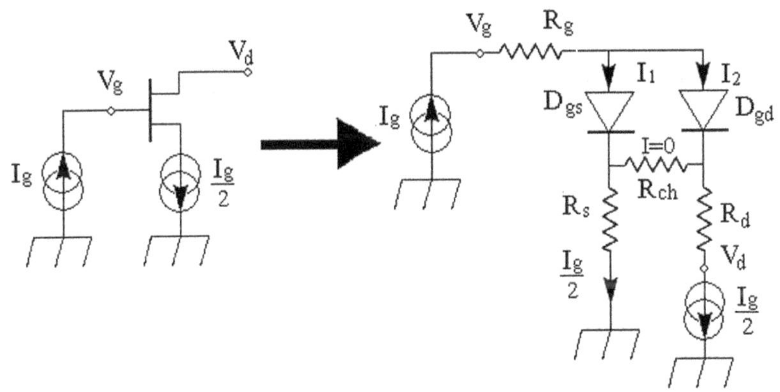

(b)

Figure 4.20 DC measurement setup and equivalent

The parasitic gate, source and drain resistances, and inductances are determined first at a far positively biased gate. If the pad capacitances have been subtracted, the Z parameters for a forward biased cold-FET are:

$$Z_{11} = R_s + R_g + \alpha_g R_c + R_{gs} + j\omega(L_g + L_s) \tag{4.47}$$

$$Z_{12} = Z_{21} = R_s + \alpha R_c + j\omega L_s \tag{4.48}$$

$$Z_{22} = R_s + R_d + 2\alpha R_c + j\omega(L_d + L_s) \tag{4.49}$$

where

R_c = the channel resistance at the applied gate voltage

$R_{gs} = nkT/qI_g$ is the differential resistance of the gate Schottky diode

I_g = the gate current, and n is the ideality factor

α_g = 1/3 and α = 1/2 are dimensionless factors that fall off with increasing gate current

Therefore, the Z- parameters' real parts provide three relations among the four unknowns (R_g, R_s, R_d, and R_c):

$$R_s + R_g + \alpha_g R_c = \mathrm{Re}(Z_{11}) \tag{4.50}$$

$$R_d + R_s + \alpha R_c = \mathrm{Re}(Z_{22}) \tag{4.51}$$

$$R_s + \alpha R_c = \mathrm{Re}(Z_{12}) \tag{4.52}$$

Because the resistances are positive, the variation of R_c can be determined easily from (4.50)–(4.52):

$$R_c = 0 \sim 2\,\mathrm{Re}(Z_{12}) \tag{4.53}$$

However there are five unknown resistors (R_g, R_s, R_d, R_c, and R_{gs}) and only three equations with corresponding real parts; other relations are required. Figure 4.21 shows the real parts of the Z parameters versus frequency; it should be noted that the real part of Z_{11} increases as $1/I_g$ (for low frequency), and R_{gs} can be carried out by determining the slope of $\mathrm{Re}(Z_{11})$ versus $1/I_g$. Table 4.4 gives the real part of Z_{11} (Real Z_{11}) versus V_g and I_g. From Figure 4.22, it can be observed that the Real Z_{11} showed linear dependence on $1/I_g$ in the low frequency ranges (Frequency <10 GHz). Therefore, the Z- parameters' real parts provide three relations among the four unknowns (R_g, R_s, R_d, R_c).

Also, the extrinsic resistances can be determined from reverse biased cold-FET method [18]. Because parasitic series resistances and inductances are very sensitive to S parameters at lower frequencies for frequencies beyond a certain high gigahertz (typically Frequency > 18 GHz), the extrinsic resistances can be directly calculated by:

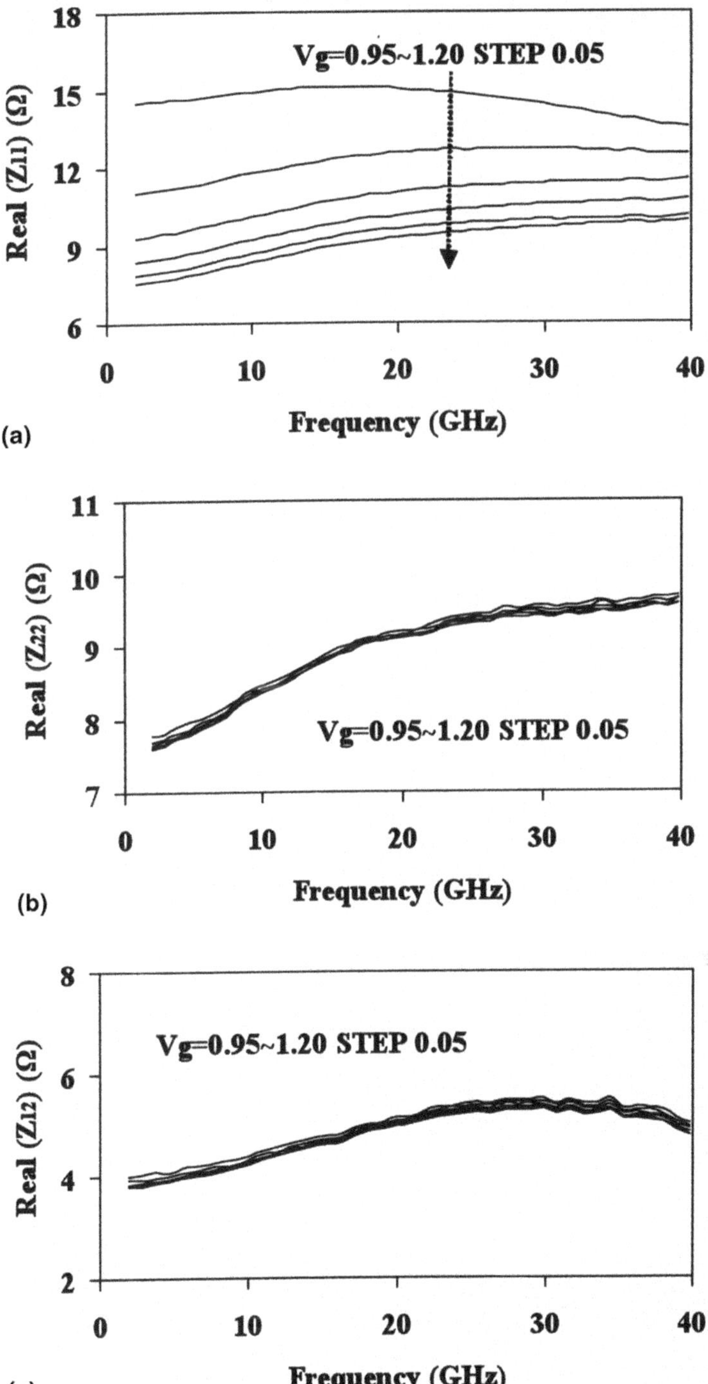

(a) (b) (c)

Figure 4.21 The real parts of the Z parameters versus frequency

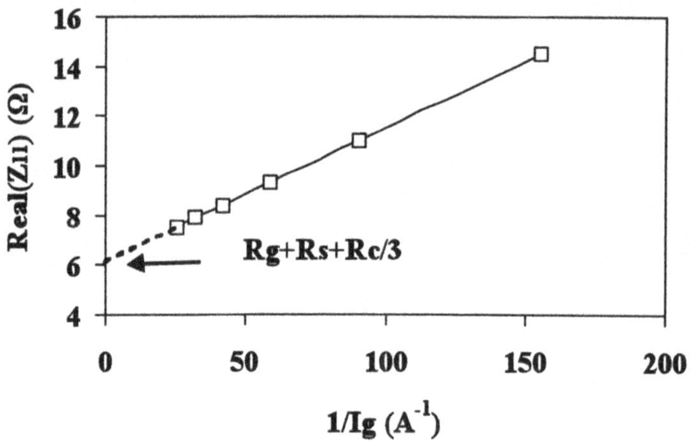

Figure 4.22 The real part of Z_{11} $\mathrm{Re}(Z_{11})$ as a function of $1/I_g$

Table 4.4 The Real Part of V_{11} versus V_g and I_g (Frequency < 10 GHz)

V_g (V)	I_g (mA)	$\mathrm{Re}(Z_{11})$
0.95	6.45	14.5
1.00	11.05	11.05
1.05	16.93	9.32
1.10	23.75	8.4
1.15	31.20	7.89
1.20	39.06	7.5

$$R_g = \mathrm{Re}(Z_{11} - Z_{12}) \tag{4.54}$$

$$R_d = \mathrm{Re}(Z_{22} - Z_{12}) \tag{4.55}$$

$$R_s = \mathrm{Re}(Z_{12}) = \mathrm{Re}(Z_{21}) \tag{4.56}$$

Figure 4.23 shows the extracted parasitic resistances versus frequency for the 2×40 μm PHEMT, and bias condition is $V_{gs} = -3$ V, $V_{ds} = 0$ V.

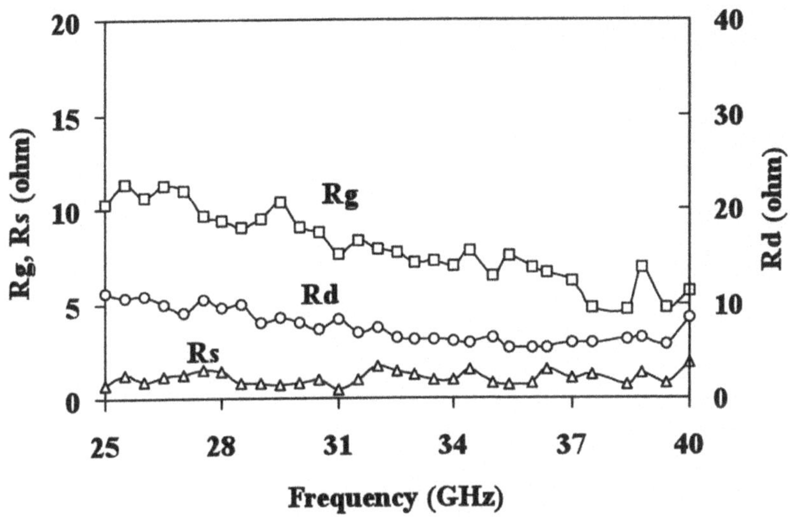

Figure 4.23 Extracted parasitic resistances versus frequency for the 2 × 40 μm PHEMT, bias: $V_{gs} = -3$ V, $V_{ds} = 0$ V

4.6.3 Hot-FET Method

Since usually cold-FET measurements yield only three independent conditions for four resistive elements, an additional condition is essential. If R_c is fixed, the parasitic resistances can be uniquely determined. Unfortunately, cold-FET measurements do not yield a reliable additional independent condition, which means that there is no preferred intrinsic reference plane for in the cold-FET model. A combination of the cold-FET and hot-FET methods is a easy way to determine R_c [19, 20].

Cold parasitic resistances can be expressed as linear functions of R_c:

$$R_a = R_{ch} / 3 + R_g + R_s \tag{4.57}$$

$$R_b = R_{ch} / 2 + R_s \tag{4.58}$$

$$R_c = R_{ch} + R_d + R_s \tag{4.59}$$

Therefore, the active intrinsic admittances can be expressed with extrinsic parameters and cold-FET resistances as follows:

$$\begin{bmatrix} Y_{11} & Y_{12} \\ Y_{21} & Y_{22} \end{bmatrix} = \begin{bmatrix} Z_{11} - R_a + R_{ch}/3 & Z_{12} - R_b + R_{ch}/2 \\ Z_{21} - R_b + R_{ch}/2 & Z_{11} - R_b + R_{ch} \end{bmatrix}^{-1} \tag{4.60}$$

Since the real part of Y_{12} is zero ($\text{Re}(Y_{12}) = 0$), the following equation should be satisfied:

$$\text{Re}[(Z_{12} - R_b + R_{ch}/2)\Delta^*] = 0 \tag{4.61}$$

Equation (4.61) can be arranged as a third-order equation of R_c:

$$a_3 R_{ch}^3 + a_2 R_{ch}^2 + a_1 R_{ch} + a_0 = 0 \tag{4.62}$$

Though equation (4.62) has three algebraic solutions, only one has reasonable magnitude.

4.7 Intrinsic Parameters

Once the extrinsic elements are obtained, the intrinsic elements are determined as follows:

$$d(\omega_i) = \frac{\text{Re}(Y_{11}(\omega_i) + Y_{12}(\omega_i))}{\text{Im}(Y_{11}(\omega_i) + Y_{12}(\omega_i))} \tag{4.63}$$

$$c(\omega_i) = (Y_{21}(\omega_i) - Y_{12}(\omega_i))(1 + j\omega_i d(\omega_i)) \tag{4.64}$$

$$C_{gs}(\omega_i) = \frac{1 + d^2(\omega_i)}{(\omega_i)} \text{Im}(Y_{11}(\omega_i) + Y_{12}(\omega_i)) \tag{4.65}$$

$$R_i(\omega_i) = \frac{d^2(\omega_i)}{(1 + d^2(\omega_i))\text{Re}(Y_{11}(\omega_i) + Y_{12}(\omega_i))} \tag{4.66}$$

$$C_{gd}(\omega_i) = \frac{-\operatorname{Im}(Y_{12}(\omega_i))}{\omega_i} \tag{4.67}$$

$$g_m(\omega_i) = \sqrt{c^2(\omega_i)} \tag{4.68}$$

$$\tau(\omega_i) = -\frac{1}{\omega_i}\tan^{-1}(\operatorname{Im}(c(\omega_i)), \operatorname{Re}(c(\omega_i)) \tag{4.69}$$

$$g_{ds}(\omega_i) = \operatorname{Re}(Y_{22}(\omega_i) + Y_{12}(\omega_i)) \tag{4.70}$$

$$C_{ds}(\omega_i) = \frac{\operatorname{Im}(Y_{22}(\omega_i) + Y_{12}(\omega_i))}{\omega_i} \tag{4.71}$$

where ω_i is the angular frequency and $i(= 0,\ldots,N\!-\!1)$ is the number of sampling points.

Figure 4.24 shows the extracted results of the intrinsic element values; the values and dispersions are summarized in Table 4.5.

Table 4.5 Extracted Intrinsic Elements Values and Dispersions for 2 × 40 μm PHEMT (bias condition: V_{gs} = 0 V, V_{ds} = 2 V).

Intrinsic Elements	Values	Dispersion
C_{gs} (fF)	66	±6
C_{gd} (fF)	23	±1.5
C_{ds} (fF)	11	±3
g_m (mS)	31	±1
g_{ds} (mS)	1.6	±0.3
τ (pS)	0.8	±0.4
R_i (Ω)	0.1	±0.08

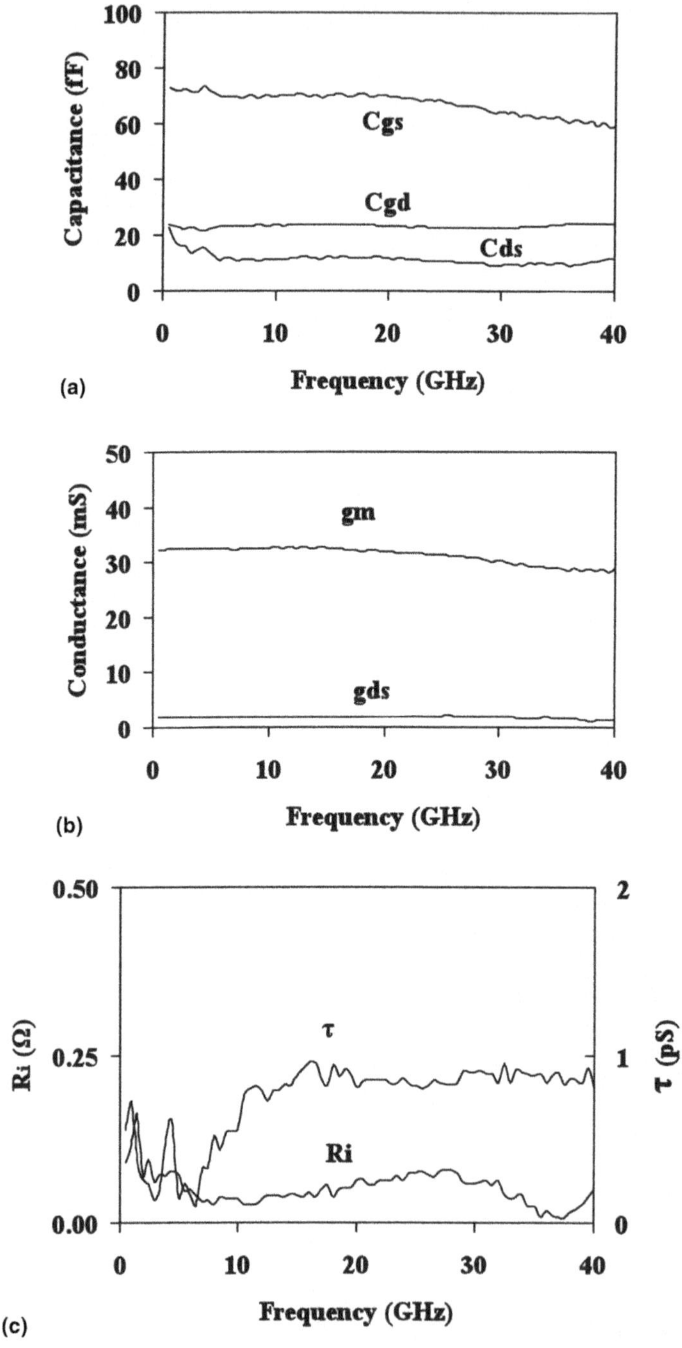

Figure 4.24 Extracted results of intrinsic elements versus frequency for the 2×40 μm PHEMT (bias: $V_{gs} = 0$ V, $V_{ds} = 2$ V).): (a) capacitance C_{gs}, C_{gd}, C_{ds} values; (b) conductance g_m, g_{ds} values; (c) R_i and τ values

4.8 Scalable Small Signal Model

Small signal FET equivalent circuit models are widely used in the design of active linear monolithic microwave integrated circuits (MMICs) and millimeter-wave integrated circuits (MWMICs). However, most of MMICs (e.g., optical receiver and driver) consist of FETs of various size devices and the scalable linear and nonlinear FET models are essential for designing such circuits. The scalable small-signal equivalent circuit model has a number of advantages in the circuit design [10,21]:

1. The small signal model parameters of differently sized devices under the same process conditions can be readily obtained using scalable normalization model parameters, so we can achieve low cost and save a lot of time.
2. It is very useful for studying the scalable nonlinear model and noise model. Combined with DC characteristic and noise parameters microwave measurement, the scalable large-signal model and noise model for FET design can be obtained easily.
3. The scalable small-signal equivalent circuit model is generally more accurate than a linearized large-signal model for predicting the PHEMT S parameters.

In this section, we have developed a scalable small-signal equivalent circuit model for 0.25 μm gatelength double heterojunction δ-doped PHEMTs. It is obvious that the pad capacitances are gatewidth independent. The three extrinsic inductances L_g, L_d, and L_s are dependent on the device size and feedline length but are not scalable. Table 4.6 shows the extracted results of the parasitic inductances.

Table 4.6 Extracted Parasitic Inductance Values for Different Size Devices

Elements	2 × 20 μm	2 × 40 μm	2 × 60 μm	2 × 100 μm
L_g (pH)	65	66	50	56
L_d (pH)	75	60	46	57
L_s (pH)	1.0	0.9	0.7	0.5

Figure 4.25 shows the extracted results of extrinsic resistances versus device gate per finger. It is the seen that R_d and R_s are inversely proportional to the gatewidth of device. The scaling formulas are determined to be as follows:

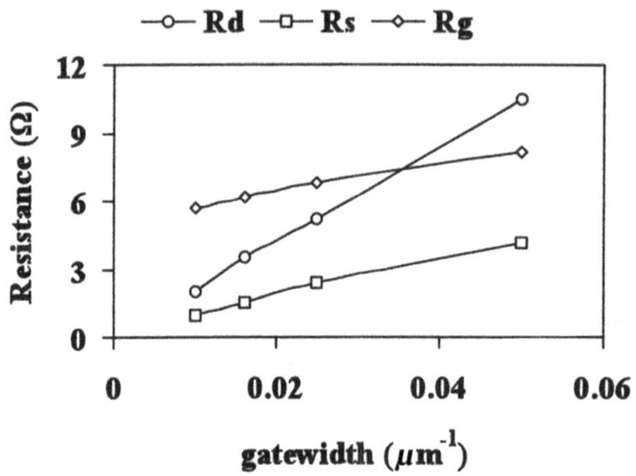

Figure 4.25 Extracted results of parasitic resistances for differently sized devices

$$R_d(W) = R_{do} / W \tag{4.72}$$

$$R_s(W) = R_{so} / W \tag{4.73}$$

where R_{do} and R_{so} are the scaling factors, and W is the gatewidth of the device.

It is also noted that gate resistance R_g consists of two parts: the one is proportional to the size of device and the other is constant. The scaling formulas have the following forms:

$$R_g(W) = R_{go} + R_{g1} / W \tag{4.74}$$

where R_{go} is the constant, and R_{g1} is the scaling factor.

After subtracting the extrinsic elements, the intrinsic elements can be determined directly by the intrinsic Y-parameter expression. On the basis of the equivalent-circuit method developed here, we have investigated the characteristics of the intrinsic small-signal circuit elements at different active bias points. Figure 4.26 shows the extracted results of the intrinsic elements C_{gs}, C_{gd}, C_{ds}, g_m, and g_{ds} for 2×20 μm, 2×40 μm, 2×60 μm and 2×100 μm PHEMTs. It can be found that the intrinsic capacitances, transconductance, and output conductance are proportional to the gatewidth of the device.

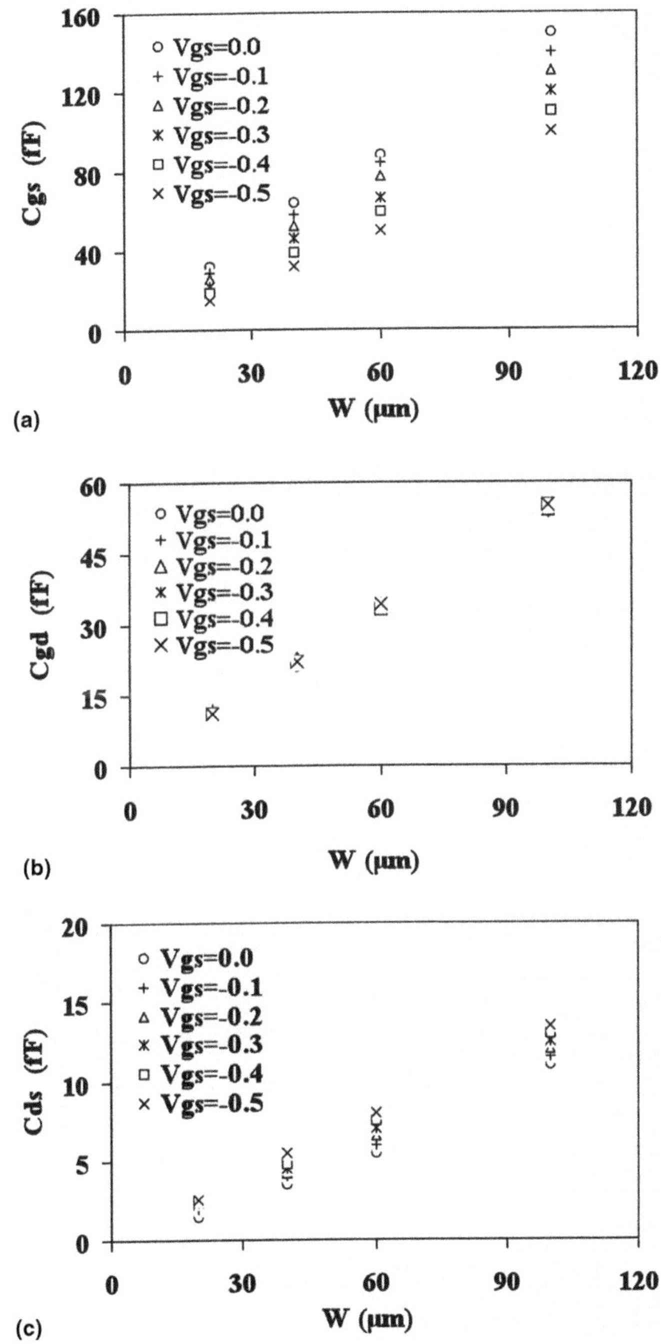

Figure 4.26 Extracted results of intrinsic elements C_{gs}, C_{gd}, C_{ds}, g_m, and g_{ds} (bias condition: $V_{gs} = 0$ V, $V_{ds} = 2$ V): (a) C_{gs}, C_{gd}, C_{ds}; (b) g_m, g_{ds}, G_{fD}; (c) R_1, τ *(continues)*

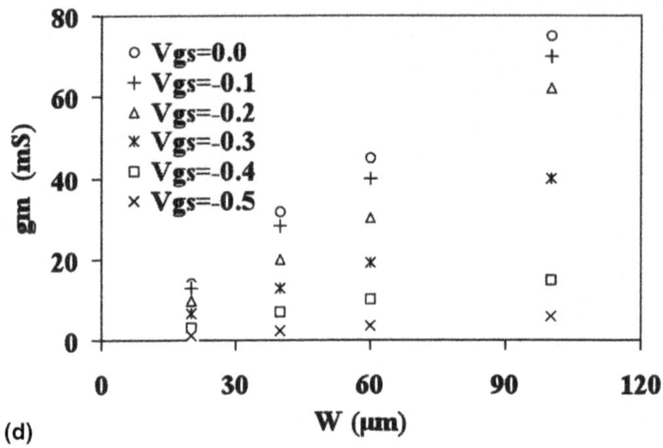

(d)

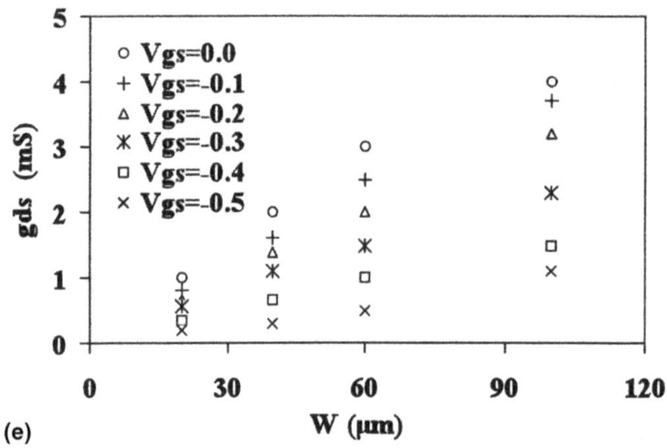

(e)

Figure 4.26 *(continued)*

The scaling formulas for intrinsic capacitances have the following form:

$$C_{gs}(W) = C_{gso}W \tag{4.75}$$

$$C_{gd}(W) = C_{gdo}W \tag{4.76}$$

$$C_{ds}(W) = C_{dso}W \tag{4.77}$$

where C_{gso}, C_{gdo}, and C_{dso} are the scaling factors.

The conductances are also proportional to the gatewidth of the device. The scaling formulas are as follows:

$$g_m(W) = g_{mo}W \tag{4.78}$$

$$g_{ds}(W) = g_{dso}W \tag{4.79}$$

where g_{mo} and g_{dso} are the scaling factors.

The intrinsic resistance R_i is inversely proportional to the gatewidth of device. The scaling formula is determined to be as follows:

$$R_i(W) = R_{io} / W \tag{4.80}$$

where R_{io} is the scaling factor for intrinsic resistance R_i.

Figure 4.27 compares the measured and modeled S parameters for the 2×20 µm, 2×40 µm, 2×60 µm and 2×100 µm PHEMTs in the frequency range of 50 MHz to 40 GHz under the bias condition of $V_{gs} = 0$ V, $V_{ds} = 2$ V. It can be seen that the modeled S parameters are in excellent agreement with the measured S parameters.

4.9 Semi-Analysis Method

It is well known that the intrinsic elements can be determined directly, after the extrinsic elements are obtained. However, sometimes only S parameters of the device are available, so it is difficult to extract the extrinsic elements according to the methods previously mentioned. Under such cases, a semi-analysis method is very useful [22, 23], the intrinsic elements determined are expressed as functions of the extrinsic elements. Assuming that the equivalent circuit composed of lumped elements is valid over the whole frequency range of the measurements, the extrinsic elements are iteratively determined using the variance of the intrinsic elements as an optimization criterion. The more detailed procedure is as follows:

1. *Set up the initial valued of extrinsic elements.*
2. *Calculate the intrinsic elements, which can be expressed as the functions of the extrinsic elements as well as frequency:*

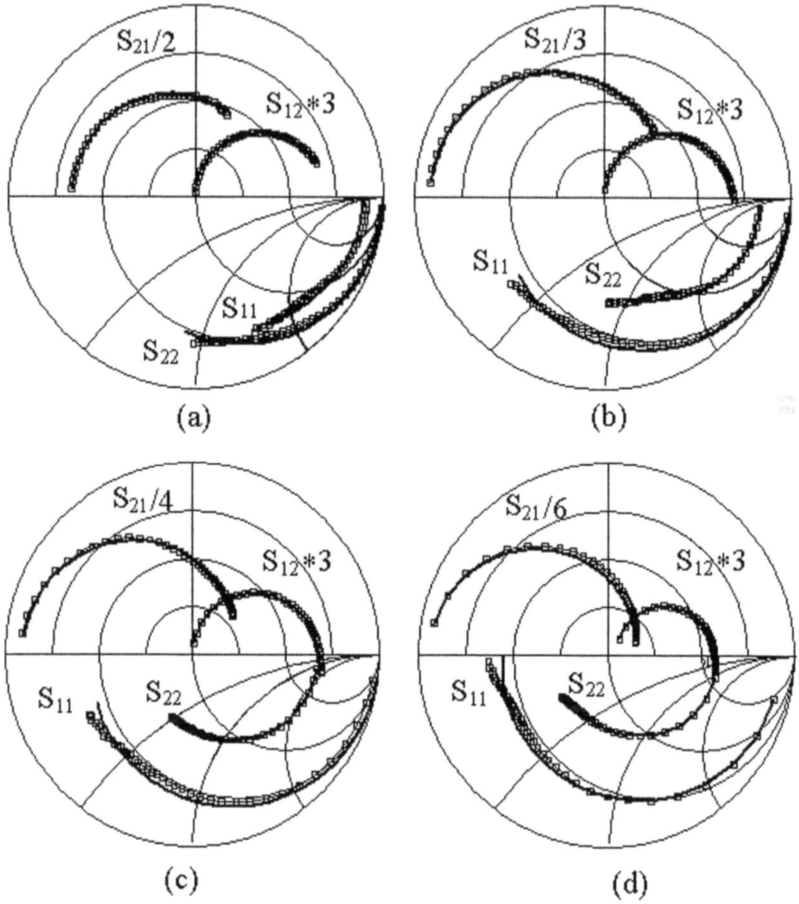

Figure 4.27 Comparison of modeled and measured S parameters for the PHEMTs, (bias condition: $V_{gs} = 0$ V, $V_{ds} = 2$ V). Squares indicate measured values, and lines indicated modeled ones: (a) 2 × 20 μm; (b) 2 × 40 μm; (c) 2 × 60 μm; (d) 2 × 100 μm

$$C_{gs} = f_1(C_{pg}, C_{pd}, C_{pgd}, L_g, L_d, L_s, R_g, R_d, R_s) \tag{4.81}$$

$$C_{gd} = f_2(C_{pg}, C_{pd}, C_{pgd}, L_g, L_d, L_s, R_g, R_d, R_s) \tag{4.82}$$

$$C_{ds} = f_3(C_{pg}, C_{pd}, C_{pgd}, L_g, L_d, L_s, R_g, R_d, R_s) \tag{4.83}$$

$$g_m = f_4(C_{pg}, C_{pd}, C_{pgd}, L_g, L_d, L_s, R_g, R_d, R_s) \tag{4.84}$$

$$\tau = f_5(C_{pg}, C_{pd}, C_{pgd}, L_g, L_d, L_s, R_g, R_d, R_s) \tag{4.85}$$

$$R_i = f_6(C_{pg}, C_{pd}, C_{pgd}, L_g, L_d, L_s, R_g, R_d, R_s) \tag{4.86}$$

$$g_{ds} = f_7(C_{pg}, C_{pd}, C_{pgd}, L_g, L_d, L_s, R_g, R_d, R_s) \tag{4.87}$$

For convenience, the function f can be expressed as follows:

$$f_k = f_k(\omega_i, Z_{ext}) \quad (k = 0,1\ldots7) \tag{4.88}$$

where Z_{ext} represents the extrinsic elements, and ω_i is the angular frequency.

3. Setup error criteria as follows:

$$\varepsilon_1^k(Z_{ext}) = \frac{1}{N-1}\sum_{i=0}^{N-1}\left| f_k(\omega_i, Z_{ext}) - \overline{\sum_{i=0}^{N-1} f_k(\omega_i, Z_{ext})}\right|^2 \quad (p,q=1,2) \tag{4.89}$$

$$\varepsilon_2(Z_{ext}) = \sum\sum\sum W_{pq}\left|S_{pq}^c(\omega_i, Z_{ext}) - S_{pq}^m(\omega_i)\right|^2 \quad (p,q=1,2) \tag{4.90}$$

where $S_{pq}^c(\omega_i, Z_{ext})$ represents the calculated S parameters, and $S_{pq}^m(\omega_i)$ represents the measured S parameters.

4. If error criteria are small enough, the iterative process will be over.

Although all the extrinsic and intrinsic elements can be determined from full-analysis methods, the extracted elements still may have a small variation with respect to frequency due to measurement and numerical calculation errors. To obtain the optimum values, the constrained optimization is needed. First, initial extrinsic elements are extracted from pinch-off condition S-parameters, and then the values of intrinsic elements are determined from Y-parameters after de-embedding the extrinsic elements. The previous element values are regarded as the initial values for optimization, and the variation ranges of elements are the dispersions of the initial values.

The fit to the measured data is obtained by defining an error criteria between the measured and modeled data. The total error is defined as:

$$E = \sum W_{ij} E_{ij} \quad i, j = 1, 2 \tag{4.91}$$

where W_{ij} are the weights placed on the relative rms errors, and the S-parameter errors, E_{ij} $(i, j = 1, 2)$ are defined as criteria:

$$E_{ij} = \frac{| s_{ij}^{m} - s_{ij}^{s} |}{| s_{ij}^{m} |} \tag{4.92}$$

Tables 4.7 and 4.8 show the extracted results for the intrinsic element values after optimization. Figure 4.28 shows the comparison of error percentage between direct extracted method and further optimization for PHEMT up to 40 GHz. After optimization, S_{11} and S_{12} have made an improvement of 5% above 30 GHz, whereas S_{22} has improved about 10% above 30 GHz. S_{21} has remained within an accuracy of 5%. This improvement is obtained at the cost of the accuracy of S_{11}, S_{12}, and S_{22} below 30 GHz.

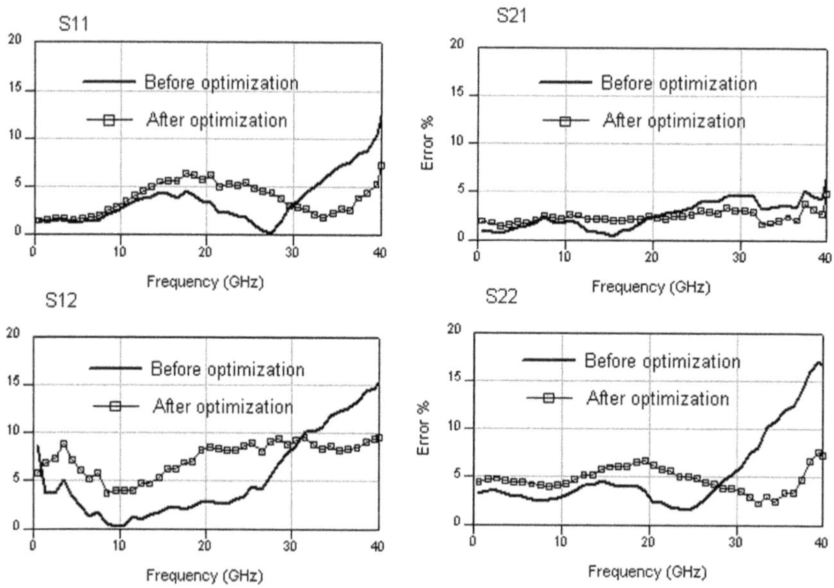

Figure 4.28 Comparison of error percentage between direct extracted method and optimization (bias condition: $V_{gs} = 0$ V, $V_{ds} = 2$ V). Squares indicate measured values, and lines indicate modeled values.

Table 4.7 Comparison between Direct Extracted Method and Further Optimization for Extrinsic Elements

Elements	Direct Extraction	After Optimization
R_d (Ω)	8	6
R_s (Ω)	1.5	2.0
R_g (Ω)	8	6.58
L_d (pH)	50	47
L_s (pH)	5	1
L_g (pH)	60	61.3

Table 4.8 Comparison between Direct Extracted Method and Further Optimization for Intrinsic Elements

Elements	Direct Extraction	After Optimization
C_{gs} (fF)	66	67.8
C_{gd} (fF)	23	23.5
C_{ds} (fF)	11	13
g_m (mS)	31	31.6
g_{ds} (mS)	1.6	1.54
τ (pS)	0.8	0.45
R_i (Ω)	0.1	0.1

4.10 Modeling up to 110 GHz

It is known that PHEMT application can go far higher than 50 GHz; in this section, we demonstrate this by extracting data of a 2 × 100 μm double heterojunction δ-doped PHEMTs from on-wafer S-parameter measurements up to 110 GHz [24].

Figure 4.29 shows the improved small-signal equivalent circuit model for PHEMT, which takes into account the conductance between gate and source and between gate and drain G_{fs} and G_{fd}. The corresponding Y parameters for intrinsic parts can be expressed as follows:

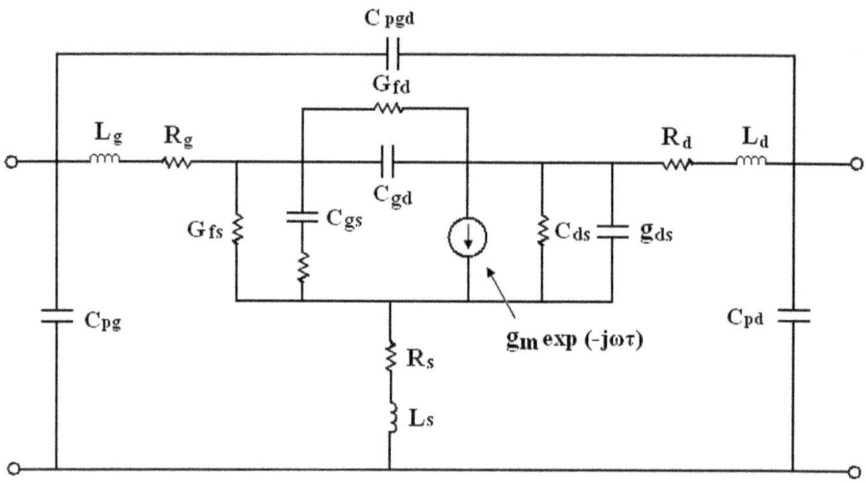

Figure 4.29 Improved small-signal equivalent circuit model for PHEMT

$$Y_{11} = G_{fs} + G_{fd} + \frac{j\omega C_{gs}}{1 + j\omega R_i C_{gs}} + j\omega C_{gd} \tag{4.93}$$

$$Y_{12} = -G_{fd} - j\omega C_{gd} \tag{4.94}$$

$$Y_{21} = \frac{g_m e^{j\omega\tau}}{1 + j\omega C_{gs} R_i} - G_{fd} - j\omega C_{gd} \tag{4.95}$$

$$Y_{22} = G_{fd} + g_{ds} + j\omega(C_{ds} + C_{gd}) \tag{4.96}$$

The differential conductances of the gate diodes, G_{fd} and G_{fs} can be determined at each operating point at low frequency:

$$G_{fs} = \mathrm{Re}(Y_{11} + Y_{12}) \tag{4.97}$$

$$G_{fd} = -\mathrm{Re}(Y_{12}) \tag{4.98}$$

Figures 4.30 and 4.31 show the frequency dependence of extrinsic pad capacitances and inductances up to 110 GHz, constant values are observed from 2 GHz to 110 GHz with the deviations from the mean

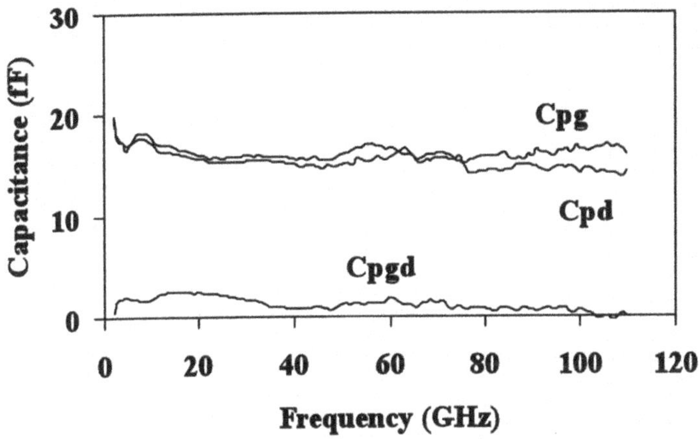

Figure 4.30 Frequency dependence of pad capacitances up to 110 GHz

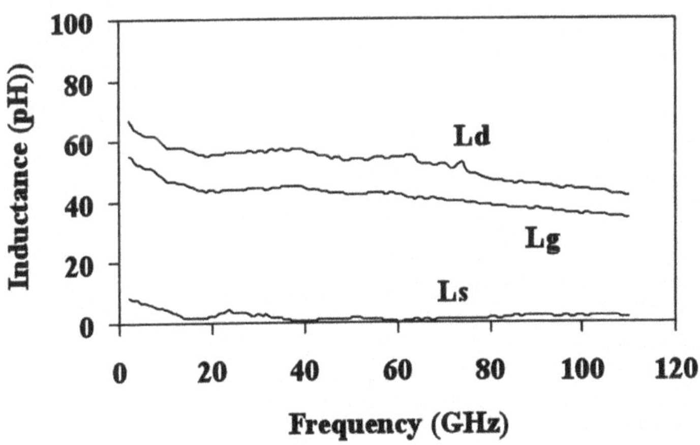

Figure 4.31 Frequency dependence of extrinsic inductances up to 110 GHz

values being less than 5%. Figure 4.32 compares the measured and modeled S parameters for the 2 × 100 μm PHEMT in the frequency range of 2 GHz to 110 GHz after optimization under the bias condition of V_{gs} = 0 V, Vds = 2 V. The modeled S parameters agree well with the measured S parameters in the entire frequency ranges.

Table 4.9 shows the extracted results for the parasitic and intrinsic element values after optimization. Figure 4.33 shows error percentage for PHEMT up to 110 GHz; error between measured and modeled S parameters are 0.8% for S_{11}, 8% for S_{12}, 7% for S_{21}, and 1.2% for S_{22}.

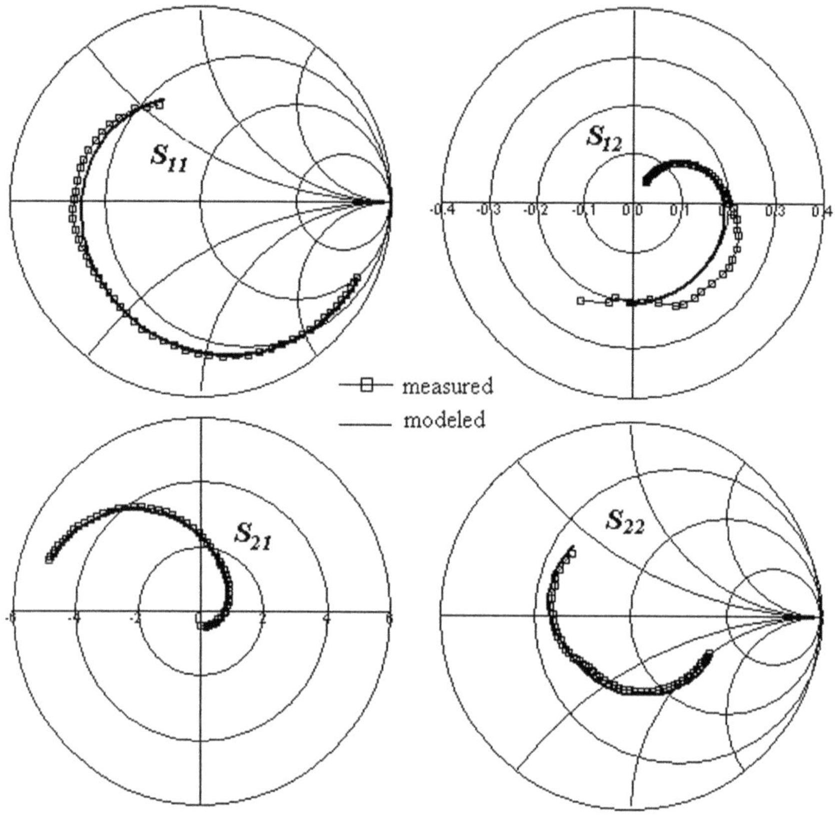

Figure 4.32 Comparison of modeled and measured S parameters for the 2 × 100 μm PHEMT up to 110 GHz (bias condition: V_{gs} = 0 V, V_{ds} = 2 V). Squares indicate measured values, and lines indicate modeled values.

4.11 Summary

In this chapter, we introduced the physical structure and operation concept of the HEMT device. The small-signal modeling and parameter extraction method were described, specifically determination methods for pad capacitances, feedline inductances, extrinsic resistances, and intrinsic elements. The scaling rules for intrinsic elements were discussed in more detail.

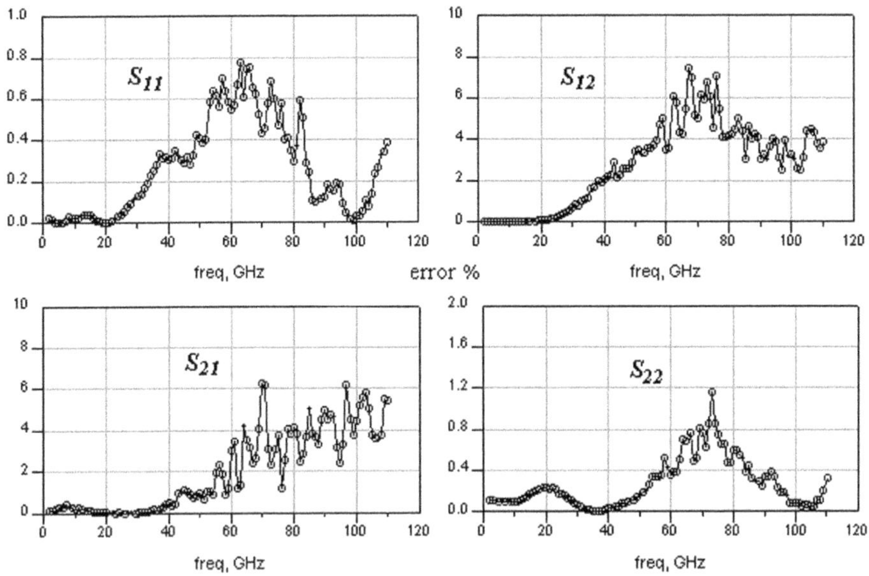

Figure 4.33 Error percentage for 2 × 100 μm PHEMT up to 110 GHz

Table 4.9 The Whole Elements Values after Optimization

Elements	Optimum Values	Elements	Optimum Values
C_{gs} (fF)	140	G_{fs} (mS)	0.15
C_{gd} (fF)	55	R_d (Ω)	2.0
C_{ds} (fF)	40	R_s (Ω)	1.0
g_m (mS)	78	R_g (Ω)	3.0
g_{ds} (mS)	7.0	L_d (pH)	50
τ (pS)	1.0	L_s (pH)	1.0
R_i (Ω)	5.0	L_g (pH)	42
g_{fd} (mS)	0.12		

Chapter 5

FET Nonlinear Modeling and Parameter Extraction

5.1 Introduction

State-of-the-art computer-aide design (CAD) methods for active microwave circuits rely heavily on models of real devices. The equivalent-circuit device models must be based on accurate parameter extraction from experimental data. The model permits the RF performance of a device or integrated circuit to be determined as a function of process and device design information or bias and RF operating conditions. Figure 5.1 shows the flowchart for an ideal microwave and RF circuit simulator. Such an integrated simulator allows both the active devices and passive elements to be optimized, based upon the parameters accessible in the fabrication process. That is, factors such as device geometry, ion-implant species, dose and energy that result in optimized RF output power, power-added efficiency (PAE), gain, and impedance can be determined. However, the simulators mention above are time consuming, to save the time, state-of-the-art CAD simulators such as SPICE are all based on the device empirical equivalent model, and the model parameters can be determined from various microwave and RF performance measurements.

It is well known, that nonlinear simulation of an active circuit is an important tool when designing frequency transistors (i.e., mixers, harmonic generators), attenuators, oscillators, and power amplifiers. The quality of the model can be measured in different ways: comparison of the measured and simulated I-V characteristics, two dimensional simulations, RF load-pull measurements; or comparison of measured and simulated bias-dependent S-parameters. The commonly used nonlinear

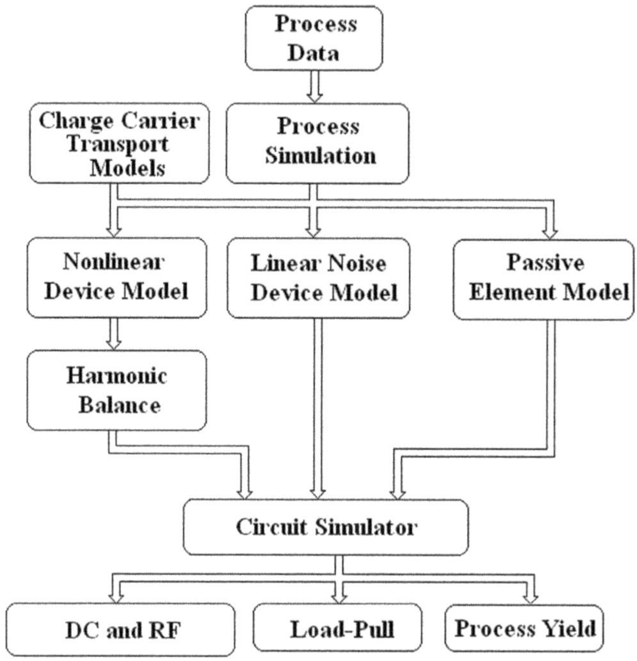

Figure 5.1 A flowchart for ideal microwave and RF circuit simulator

FET models are the physics-based nonlinear model [1-4], table-based nonlinear model [5], and empirical equivalent-circuit-based model [6].

In this chapter, we will introduce the nonlinear model for FETs previously mentioned, and focus on the empirical equivalent-circuit-based model, which consists of the drain-current DC model and nonlinear intrinsic capacitances models.

5.2 Physics-Based Model

The physics-based models are derived from solutions of the basic semiconductor device equations. The physical simulation of the FET device is based on four semi-classical semiconductor equations (i.e., Poisson equation, current continuity equation, and energy and momentum equations) with analytical expressions for the FET channel. The model accounts for process-related parameters (e.g., geometry, recess depth, material parameters, doping profile), surface-depletion effects, substrate conduction, contact resistivities, avalanche breakdown, and forward gate conduction. The difficulty in applying physical device models to microwave CAD simulators is the large execution times required.

The physical models solve the semiconductor device equations using some form of numerical technique such as finite differences or finite elements. Due to lengthy execution times, however, the operation of the models usually is limited to DC solutions, and parasitics can not be taken into account.

Equivalent-circuit-based models are commonly used to investigate large-signal device performance. However, these models require extensive characterization of the device after fabrication, as well as some knowledge of process variation statistics. For the well-proven physical model used here, there is no need for an extensive series of measurements since all the data are provided from the process parameters and physical structure of the device. Equivalent-circuit models also suffer from problems associated with curve-fitting errors and discontinuous or inaccurate high-order derivatives in current and voltage expressions, leading to inaccurate predictions of inter-modulation distortion (IMD). Physical models solve the semiconductor equations explicitly for the device and avoid the problem of ill-defined relationships between equivalent-circuit elements and physical behavior.

5.3 Table-Based Model

In recent years, table-based models have become important tools for rapid accurate device modeling. Several groups in the United State and Europe [6, 7] developed and put these models into practical use. The models are highly accurate and device independent, and they expedite simulations because they are based directly on the measured data. Performance between measured points is interpolated with spline functions, and these functions should be differentiable to a high order of derivatives in order to ensure a correct description of harmonics and convergence within harmonic balance (HB) simulations. Specific software and equipment is needed to extract and use these types of models, and technology-related changes in the device characteristics require repeating the extraction procedure. Care should be taken to extract data carefully because the model can give unexpected results when operated outside the specified range.

5.4 Empirical Nonlinear Model

Although the nonlinear performance for FET devices can be predicted by using the two-dimensional nonlinear differential equations which describing the electron transport in the channel. However, this method has several disadvantages: It requires specialized software; it is not suited to the inclusion of package parasitics; and device-circuit interactions are not easily taken into account. An alternative approach is to use a circuit analysis based on a circuit model of the device. Equivalent-circuit models of active device (e.g., FET, HBT) are the backbone of the nonlinear microwave CAD simulator software. All commercially available software includes one or more of each type device models. The modeling concept is that the complex active device is represented by using the basic resistances, capacitances, and control sources. The difference in the various models is the expression used to characterize the drain-current generator. The popular FET models in the commercial software will are introduced in the next section.

The basic nonlinear equivalent circuit model for microwave and RF FETs is shown in Figure 5.2. The model contains only elements considered to be of first order importance to device operation. There are seven nonlinear elements: three current generators, three charges, and a nonlinear input resistance.

The various components in the model are defined in the following list:

$I_{ds}(V_{in}, V_{out})$ Gate-to-source and drain-to-source voltages-controlled drain-current source

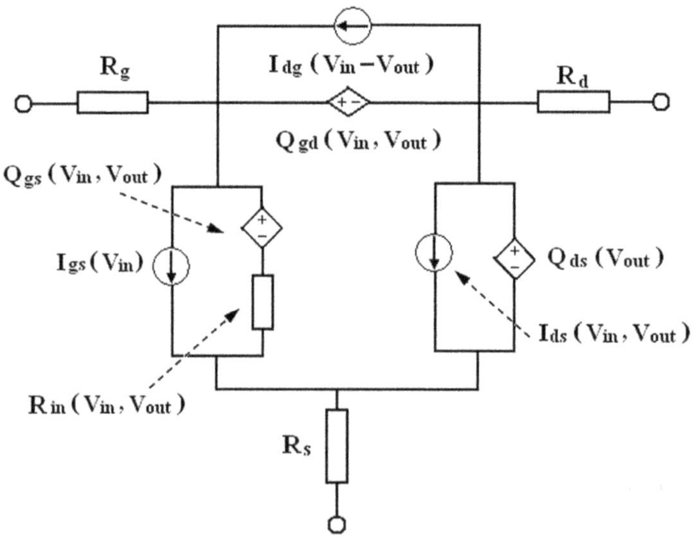

Figure 5.2 Basic nonlinear-equivalent circuit model for microwave and RF FETs

$I_{dg}(V_{out}-V_{in})$	Drain-to-gate voltage-controlled current source
$I_{gs}(V_{in})$	Gate-to-source voltage-controlled current source
$Q_{gs}(V_{in},V_{out})$	Gate-to-source and drain-to-source voltages-controlled gate-to-source charge storage
$Q_{gd}(V_{in},V_{out})$	Gate-to-source and drain-to-source voltages-controlled gate-to-drain charge storage
$Q_{ds}(V_{out})$	Drain-to-source voltages-controlled charge storage
$R_{in}(V_{in},V_{out})$	Gate-to-source and drain-to-source voltages-controlled intrinsic resistance

It can be observed that the main nonlinearity is the drain-source current I_{ds}. The nonlinear current source I_{gs} and I_{gd} are used to represent forward conduction of the gate source and reverse breakdown of the gate-drain Schottky barrier diodes. Figure 5.3 shows the relationship between nonlinear and linear equivalent-circuit elements, where

$$g_m = \frac{\partial I_{ds}}{\partial V_{gs}} \tag{5.1}$$

$$g_{ds} = \frac{\partial I_{ds}}{\partial V_{ds}} \tag{5.2}$$

$$C_{gs} = \frac{\partial Q_{gs}}{\partial V_{gs}} + \frac{\partial Q_{gd}}{\partial V_{gs}} \tag{5.3}$$

$$C_{gd} = \frac{\partial Q_{gs}}{\partial V_{gd}} + \frac{\partial Q_{gd}}{\partial V_{gd}} \tag{5.4}$$

$$C_{ds} = \frac{\partial Q_{ds}}{\partial V_{ds}} \tag{5.5}$$

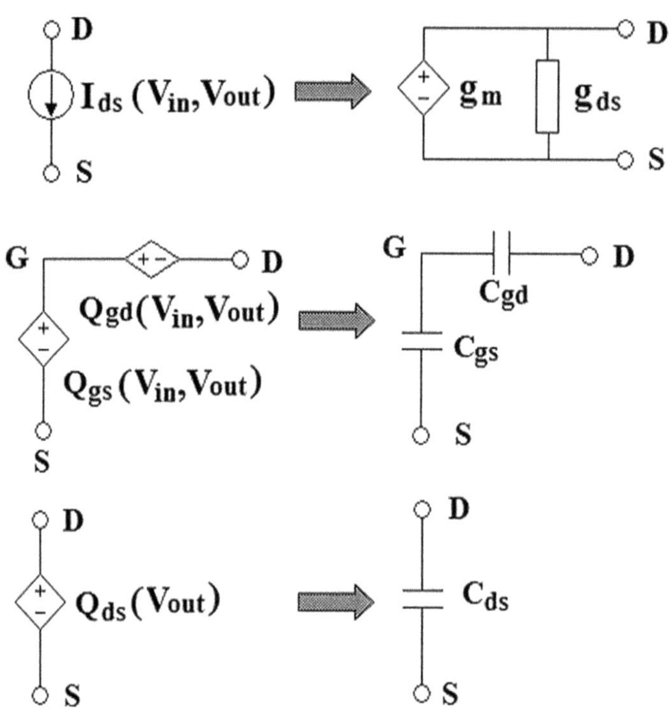

Figure 5.3 Relationship between nonlinear and linear equivalent-circuit elements

The advantage of the equivalent-circuit model is that it can be implemented in the commercial circuit simulators; however, the model can not be used for guiding the device design, and accuracy is limited by the microwave and RF measurement. Table 5.1 shows the comparison of the different three models.

Table 5.1 Comparison of Different Models

Model	Physics-Based	Table-Based	Equivalent-Circuit
Device design	Yes	No	No
Accuracy	High	Middle	Low
Compatibility	Low	Fair	Good

5.5 Commercial FET Nonlinear Model

The commonly used FET nonlinear equivalent circuit models are as follows:

1. Statz model [7–9]
2. TriQuint model [10, 11]
3. Curtice model [12–14]
4. Tajiama model [15]
5. Materka model [16]
6. Angelov model [17–19]

We will introduce the nonlinear models that are available in the commercial microwave simulators (e.g., simulation program with integrated circuit emphasis [SPICE], advance design system [ADS]) in more detail in the following sections.

5.5.1 Statz Model

The Statz model is a popular nonlinear FET model that is available in most large-signal circuit simulation packages used by microwave engineers. The model can be divided into two shells: (1) inner shell represents the intrinsic FET device, and (2) an outer shell represents the device parasitics.

The drain current can be expressed as follows:

$$I_{ds} = \frac{\beta(V_{gs} - V_{TO})^2}{1 + b(V_{gs} - V_{TO})}(1 + \lambda V_{ds})\tanh(\alpha V_{ds}) \tag{5.6}$$

In this equation, the model parameters are defined in the following list:

I_{ds}: drain current (unit A)

V_{TO}: threshold voltage (unit V)

β: transconductance parameter (A/V^2)

α: saturation voltage parameter (determine the voltage that I_{ds} saturates) (unit V^{-1})

λ: channel length modulation parameter (unit V^{-1})

b: the fitting parameter that controls $I_{ds} - V_{gs}$ characteristic transition from quadratic to linear behavior (unit V^{-1})

The tanh function in (5.6) consumes considerable computer time, and can be further approximated to the tanh function below saturation by a simple polynomial K_t of the form:

$$K_t = \tanh(\alpha V_{ds}) = \begin{cases} 1-(1-\alpha V_{ds}/3)^3 & 0 < V_{ds} < 3/\alpha \\ 1 & V_{ds} \geq 3/\alpha \end{cases} \tag{5.7}$$

It can be found that in the saturation region ($V_{ds} \leq 3/\alpha$), the tanh function is replaced by unity.

The corresponding transconductance and output conductance can be expressed as follows:

$$g_m = \frac{\beta\left(V_{gs} - V_{TO}\right)\cdot[2 + b(V_{gs} - V_{TO})]}{[1 + b(V_{gs} - V_{TO})]^2}(1 + \lambda V_{ds})K_t \tag{5.8}$$

$$g_{ds} = \frac{\beta\left(V_{gs} - V_{TO}\right)^2}{[1 + b(V_{gs} - V_{TO})]}\left[\lambda K_t + \left(1 + \lambda V_{ds}\right)\alpha\left(1 - \alpha V_{ds}/3\right)^2\right] \tag{5.9}$$

The extraction of the five DC model parameters can be carried out using the following procedure:

1. The threshold voltage can be approximately estimated from the gate-to-source voltage, when drain-current tends to zero.

$$V_{TO} = V_{gs}\Big|_{I_{ds}\to 0} \tag{5.10}$$

2. When gate-to-source voltage satisfies $1 + b\left(V_{gs} - V_{TO}\right) \approx 1$ in the saturation region, the DC expression becomes:

$$I_{ds} = \beta(1 + \lambda V_{ds})(V_{gs} - V_{TO})^2 \tag{5.11}$$

If gate-to-source voltage V_{gs} remains invariant, the channel length modulation parameter λ can be extracted by using two different drain currents I_{ds} under two different drain-to-source voltages V_{ds}:

$$\lambda = \frac{I_{ds}^2 - I_{ds}^1}{V_{ds}^2 I_{ds}^1 - V_{ds}^1 I_{ds}^2} \tag{5.12}$$

Substituting (5.12) into (5.11), we have:

$$\beta = \frac{I_{ds}}{\left(V_{gs} - V_{TO}\right)^2 \left(1 + \lambda V_{ds}\right)} \tag{5.13}$$

3. The saturation voltage parameter α can be determined in the linear region:

$$\alpha = \frac{3}{V_{ds}} \left(1 - \sqrt[3]{1 - \frac{I_{ds}}{\beta(1 + \lambda V_{ds})(V_{gs} - V_{TO})^2}} \right) \tag{5.14}$$

4. When $V_{gs} \gg V_{TO}$ and $V_{ds} \geq 3/\alpha$, the drain current becomes:

$$I_{ds} = \frac{\beta(V_{gs} - V_{TO})^2}{1 + b(V_{gs} - V_{TO})} \cdot (1 + \lambda V_{ds}) \tag{5.15}$$

Therefore, the fitting parameter b is given by

$$b = \frac{\beta(V_{gs} - V_{TO})}{I_{ds}} \cdot (1 + \lambda V_{ds}) - \frac{1}{(V_{gs} - V_{TO})} \tag{5.16}$$

In the Statz MESFET model, the dependence of intrinsic capacitances C_{gs} and C_{gd} on the intrinsic terminal voltages are given by (5.17) and (5.18)

$$C_{gs} = \frac{C_{gso}}{\sqrt{1 - \frac{V_{new}}{V_{bi}}}} \frac{1}{4}(1 + k_1)(1 + k_2) + \frac{1}{2} C_{gdo}(1 - k_2) \tag{5.17}$$

$$C_{gd} = \frac{C_{gso}}{\sqrt{1 - \frac{V_{new}}{V_{bi}}}} \frac{1}{4}(1 + k_1)(1 - k_2) + \frac{1}{2} C_{gdo}(1 + k_2) \tag{5.18}$$

where

$$V_{new} = \frac{1}{2}[V_{eff1} + V_{TO} + \sqrt{(V_{eff1} - V_{TO})^2 + \delta^2}\,] \tag{5.19}$$

$$V_{eff1} = \frac{1}{2}[V_{gs} + V_{gd} + \sqrt{(V_{gs} - V_{gd})^2 + \alpha^{-2}}\,] \tag{5.20}$$

$$k_1 = \frac{V_{eff1} - V_{TO}}{\sqrt{(V_{eff1} - V_{TO})^2 + \delta^2}} \tag{5.21}$$

$$k_2 = \frac{V_{gs} - V_{gd}}{\sqrt{(V_{gs} - V_{gd})^2 + \alpha^{-2}}} \tag{5.22}$$

where C_{gs} and C_{gd} are the gate-to-source and gate-to-drain capacitances, respectively, and C_{gso} and C_{gdo} are the corresponding zero-bias gate-source and gate-drain capacitances. The δ– parameter models the behavior of C_{gs} and C_{gd} around and below pinchoff.

The capacitance model parameters (C_{gso}, C_{gdo}, and δ) can be extracted from capacitance values at multi-bias points. Setting V_{gs} equal to V_{gd} ($k_2 = 0$) and neglecting δ in (5.19)–(5.22), simplifies (5.17) to

$$C_{gs} = \frac{1}{2}\frac{C_{gso}}{\sqrt{1 - \dfrac{V_{gs} + \dfrac{1}{2\alpha}}{V_{bi}}}} + \frac{1}{2}C_{gdo} \tag{5.23}$$

By performing the Y-parameters at several bias points, (5.23) yield a number of values for C_{gs}. When the build in voltage V_{bi} is known, plotting these values of C_{gs} versus

$$\left[\sqrt{1 - (V_{gs} + \frac{1}{2\alpha})/V_{bi}}\right]^{-1}$$

allows for linear regression, yielding C_{gso} as slope and C_{gdo} as intercept.

An improved method including a reduction of the number of parameters to be simultaneously extracted and a further simplification of the capacitance equations is proposed in [9]. The simplification of the procedure is achieved by eliminating V_{bi} from the equations. From (5.17), it is clear that this can be done only by making V_{new} equal to zero. Substituting $V_{new} = 0$ in (5.18), one obtains (5.24) for V_{eff1}:

$$V_{new} = 0$$

$$\Leftrightarrow \quad \left(V_{eff1} + V_{TO}\right)^2 = \left(V_{eff1} - V_{TO}\right)^2 + \delta^2$$

$$\Leftrightarrow \quad V_{eff1} = \frac{\delta^2}{4V_{TO}} \tag{5.24}$$

Using (5.18), with α large, and (5.24), one finally obtains the desired gate–bias voltage V_{gs} (δ is 0.2):

$$V_{gs} \approx V_{eff1} = \frac{0.01}{V_{TO}} \tag{5.25}$$

Setting a large gate-drain voltage guarantees that the term $(V_{gs} - V_{gd})^2$ in (5.20) is large compared with α^{-2}, and that the assumption of a large α is valid. Under these conditions, V_{new} is negligible in (5.17) and (5.18) and $k_1 = 1$ and $k_2 = 1$. The capacitance equations then reduce to (5.26) and (5.27):

$$C_{gso} = C_{gs} \tag{5.26}$$

$$C_{gdo} = C_{gd} \tag{5.27}$$

Evidence for (5.26) and (5.27) can also be found on simulated characteristics (Figures 5.4 and 5.5). It has been demonstrated that the parameters to be extracted can be calculated easily from measured S-parameters for one particular bias point. The value of V_{bi} can be easily determined from measurements at one additional bias point.

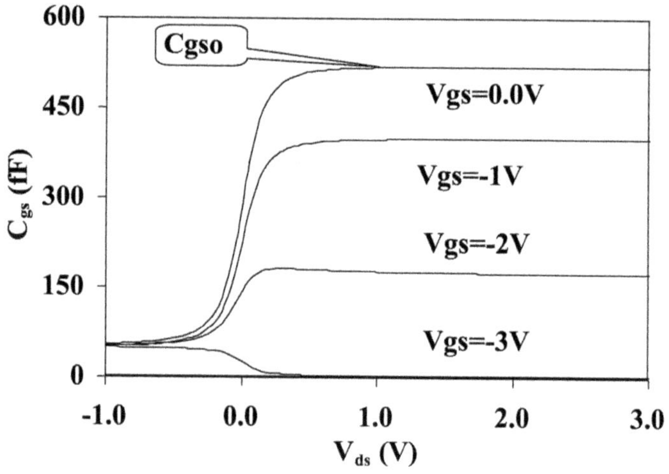

Figure 5.4 C_{gs} versus V_{ds} for different values of V_{gs}

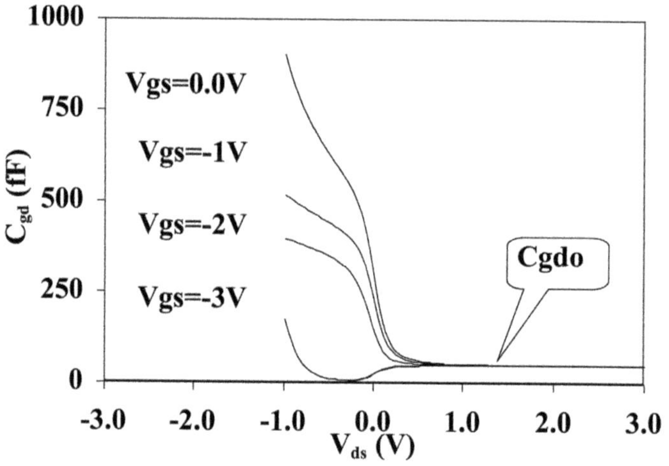

Figure 5.5 C_{gd} versus V_{ds} for different values of V_{gs}

To avoid charge conservation problems in harmonic balance (HB), it is preferable to work with the charges; thus, we use a charge data set calculated from measured capacitance data:

$$Q = 2C_{gso}V_{bi}\left(1 + \sqrt{1 - \frac{V_{eff1}}{V_{bi}}}\right) + C_{gdo}V_{eff2} \tag{5.28}$$

5.5.2 TriQuint Nonlinear Model

TriQuint models are developed by TriQuint company [10, 11]. There are two models and implemented in microwave & RF CAD simulator: (1) TOM model (TriQuint's own model); and (2) the TOM III model (capacitance-based model).

5.5.2.1 TOM Model

The TOM model (also called TriQuint scalable nonlinear FET model) is actually an improved Statz model. As with most proposed MESFET models, the drain conductance is modeled by multiplying the expression for I_{ds} by $(1+\lambda V_{ds})$. Although this term can represent drain conductance at a particular bias point, it does not model variations with bias correctly, predicting a conductance that increases at higher values of I_{ds} while the observed conductance actually decreases. The Statz model also fails to provide an accurate model at low currents where V_{gs} is near cutoff. An improved expression for I_{ds} is developed:

$$I_{ds} = \frac{I_{dso}}{1+\delta V_{ds} I_{dso}} \tag{5.29}$$

where δ is the fitting factor, and I_{dso} is given by the expression:

$$I_{dso} = \begin{cases} \beta\left(V_{gs}-V_T\right)^Q\left[1-\left(1-\dfrac{\alpha V_{ds}}{3}\right)^3\right] & 0 < V_{ds} < \dfrac{3}{\alpha} \\ \beta\left(V_{gs}-V_T\right)^Q & V_{ds} > \dfrac{3}{\alpha} \end{cases} \tag{5.30}$$

The parameter Q is necessary to model the non square law, dependence of I_{ds} that is observed for device with small or positive pinch-off voltage. To address the poor fit near pinch-off values V_{gs}, V_T can be made a function of drain voltage:

$$V_T = V_{TO} - \gamma V_{ds} \tag{5.31}$$

From expressions (5.29)–(5.31), it can be observed that in the TriQuint model, two new fitting factors δ and γ have been used instead of b and λ in Statz model. The frequency dispersion has been taken into account in the RF simulator for DC and RF measured drain currents.

Due to the frequency dispersion, the extracted transconductance and output drain conductance from RF measurements are different from the extracted values from DC measurement, that is,

$$g_m^{RF} = \sqrt{\{\mathrm{Re}^2(Y_{21}) + [\mathrm{Im}(Y_{21}) + \omega C_{gd}]^2\}\left(1 + \omega^2 C_{gs}^2 R_i^2\right)} \neq \frac{\partial I_{ds}}{\partial V_{gs}} \tag{5.32}$$

$$g_{ds}^{RF} = \mathrm{Re}(Y_{22}) \neq \frac{\partial I_{ds}}{\partial V_{ds}} \tag{5.33}$$

where Y_{21} and Y_{22} are the intrinsic Y- parameters.

To consider the frequency dispersion, a shunt circuit is necessary in parallel with drain current at the output port. Figure 5.6 shows the TriQuint equivalent-circuit model in commercial simulators.

The circuit elements R_{db} and C_{bs} are both linear elements that are used to control the frequency at which the current source I_{db} becomes a factor. Note that at DC the source I_{db} has no impact on the response and the drain-source current is just the DC value. At very high fre-

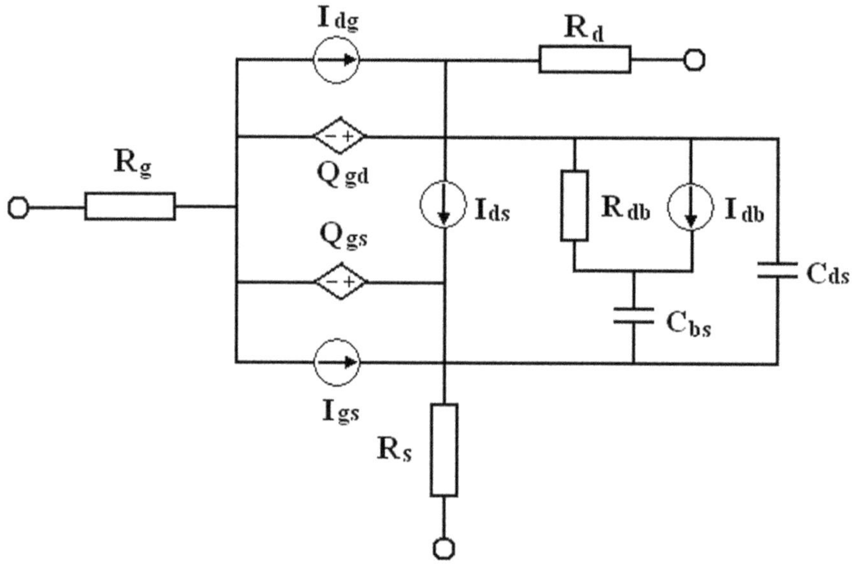

Figure 5.6 TriQuint equivalent-circuit model in commercial simulators

quency and with R_{db} set to a very large quantity, the sources I_{ds} and I_{db} add, giving the AC value for the drain-source current.

The frequency dispersion current I_{db} is given by:

$$I_{db}\left(V_{gs}, V_{ds}\right) = I_{ds}^{AC} - I_{ds}^{DC} \tag{5.34}$$

$$I_{ds}^{AC} = I_{ds}\left(V_{gs}, V_{ds}, \gamma_{AC}\right) \tag{5.35}$$

$$I_{ds}^{DC} = I_{ds}\left(V_{gs}, V_{ds}, \gamma_{DC}\right) \tag{5.36}$$

where γ_{AC} and γ_{dc} represent the changes of threshold voltage with V_{ds}.

5.5.2.2 TOM III Model

In the TOM III model, the drain current I_{ds} is given by [11]:

$$I_{ds} = \beta V_G^Q \frac{\alpha V_{ds}}{\left(1+\left(\alpha V_{ds}\right)^k\right)^{1/k}} \tag{5.37}$$

where

$$V_G = QV_{ST} \ln\left(1+e^u\right) \tag{5.38}$$

$$u = \frac{V_{gs} - V_{TO} + \gamma V_{ds}}{QV_{ST}} \tag{5.39}$$

$$V_{ST} = V_{STO}\left(1 + M_{STO} V_{ds}\right) \tag{5.40}$$

The additional dc model parameters are defined as follows:

k: knee function power law coefficient

V_{STO}: subthreshold slope voltage

M_{STO}: parameter for subthreshold slope voltage dependence on V_{ds}

The empirical equations for the gate charge and capacitances are:

$$Q_{GG} = Q_{GGL} f_p + Q_{GGH}\left(1 - f_p\right) \tag{5.41}$$

$$C_{GS} = C_{GSL} f_p + Q_{GSH}\left(1 - f_p\right) + \left(Q_{GGL} - Q_{GGH}\right)\frac{\partial f_P}{\partial V_{gs}} \tag{5.42}$$

$$C_{GD} = C_{GDL} f_p + Q_{GDH}\left(1 - f_p\right) + \left(Q_{GGL} - Q_{GGH}\right)\frac{\partial f_P}{\partial V_{gd}} \tag{5.43}$$

with

$$f_P = e^{-\delta_c I_{ds} V_{ds}}$$

where Q_{GGH} represents the high-power gate charge, C_{GSH} and C_{GDH} are the high-power gate-to-source and gate-to-drain capacitances, respectively, Q_{GGL} represents the low-power gate charge, C_{GSL} and C_{GDL} are the low-power gate-to-source and gate-to-drain capacitances, respectively.

5.5.3 Curtice Nonlinear Model

The most commonly used Curtice models are the Curtice-Quadratic model [12] and the Curtice-Cubic model [13, 14].

5.5.3.1 Curtice-Quadratic Model

The quadratic dependence of the drain current with respect to the gate voltage is calculated with the following expression in the region $V_{ds} \geq 0$:

$$I_{ds} = \beta(1 + \lambda V_{ds})(V_{gs} - V_{TO})^2 \cdot \tanh(\alpha V_{ds}) \tag{5.44}$$

Assuming symmetry, in the reverse region, the drain and source swap roles, and the expression becomes:

$$I_{ds} = \beta(1+\lambda V_{ds})(V_{gs} - V_{TO})^2 \cdot \tanh(\alpha V_{ds}) \tag{5.45}$$

The drain current is set to zero in case the gate-to-source junction voltage V_{gs} drops below the threshold voltage V_{TO}.

To model the non-square-law performance for I_{ds}, an advanced Curtice-Quadratic model is available in commercial software; the corresponding drain current is expressed as follows:

$$I_{ds} = \beta_n(1+\lambda V_{ds})(V_{gs} - V_{ton})^Q \cdot \tanh(\alpha V_{ds}) \tag{5.46}$$

with

$$\beta_n = \frac{\beta}{1+U(V_{gs} - V_{TO})} \tag{5.47}$$

$$V_{ton} = V_{TO} + \gamma V_{ds} \tag{5.48}$$

where Q, γ, and U are the fitting factors.

5.5.3.2 Curtice-Cubic model

By using a cubic approximation, drain current in Curtice-Cubic model is calculated with the following expression:

$$I_{ds} = (A_0 + A_1 V_1 + A_2 V_1^2 + A_3 V_1^3)\tanh(\gamma V_{ds}(t)) \tag{5.49}$$

where V_1 is the input voltage:

$$V_1 = V_{gs}(t-\tau)\cdot[1+\beta(V_{ds}^o - V_{ds}(t))] \tag{5.50}$$

where

β = coefficient for pinch-off change

V_{ds}^o = output voltage at which A_0, A_1, A_2, and A_3 are evaluated

τ = internal time delay of FET

The coefficients A_0, A_1, A_2 and A_3 can be evaluated from data in the saturation region. One disadvantage of the cubic relationship is that unlike the quadratic, a pinchoff voltage may result in zero current or zero transconductance, but not both.

Figure 5.7 shows the Curtice nonlinear equivalent-circuit model, which is available in the commercial software. A shunt network (C_f and R_c) in parallel with the output port is used to model frequency-dependent output conductance.

In this model, the drain-gate avalanche current I_{dg} is taken to be

$$I_{dg} = \begin{cases} \dfrac{V_{dg}(t) - V_B}{R_1} & V_{dg} > V_B \\ 0 & V_{dg} < V_B \end{cases} \tag{5.51}$$

where $V_B = V_{BO} + R_2 I_{ds}$, R_1 is the approximate break down resistance, R_2 is the resistance related break-down voltage to channel currents, and V_{BO} is the gate-drain junction reverse bias breakdown voltage.

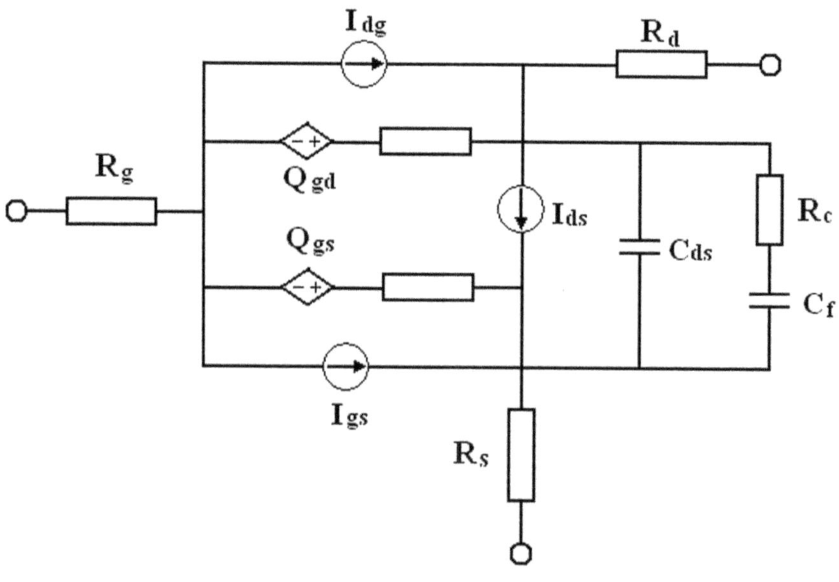

Figure 5.7 Curtice nonlinear equivalent-circuit model

The forward-biased gate-current I_{gs} is taken to be

$$I_{gs} = \begin{cases} \dfrac{V_{gs}(t) - V_{bi}}{R_F} & V_{gs} > V_{bi} \\ 0 & V_{gs} < V_{bi} \end{cases} \tag{5.52}$$

where V_{bi} is the built-in voltage, and R_F is the effective value of forward bias resistance.

5.5.4 Tajima Nonlinear Model

The Tajima nonlinear model was proposed in 1981 [15], and is the most earliest model for FET device. The drain current formula is given by

$$I_{ds} = I_{d1}(V_{gs}^{'})I_{d2}(V_{ds}) \tag{5.53}$$

with

$$I_{d1}(V_{gs}^{'}) = \frac{1}{k}\left\{1 + \frac{V_{gs}^{'}}{V_P} - \frac{1}{m} + \frac{1}{m}\exp\left[-m(1 + \frac{V_{gs}^{'}}{V_P})\right]\right\} \tag{5.54}$$

$$I_{d2}(V_{ds}) = I_{dsp}\left\{1 - \exp\left[\frac{-V_{ds}}{V_{dss}}\right] - a\left(\frac{V_{ds}}{V_{dss}}\right)^2 - b\left(\frac{V_{ds}}{V_{dss}}\right)^3\right\} \tag{5.55}$$

$$k = 1 - \frac{1}{m}[1 - \exp(-m)] \tag{5.56}$$

$$V_P = V_{po} + pV_{ds} + V_{bi} \tag{5.57}$$

$$V_{gs}^{'} = V_{gs} - V_{bi} \tag{5.58}$$

where m, a, b, and P are the fitting factors.

5.5.5 Materka Nonlinear Model

The basic drain-current formula for the Materka model is given by [16]:

$$I_{ds} = I_{DSS}\left(1-\frac{V_{gs}}{V_P}\right)^2 \tanh\left(\frac{\alpha V_{ds}}{V_{gs}-V_P}\right) \tag{5.59}$$

$$V_P = V_{to} + \gamma V_{ds}$$

The advanced drain-current formula is as follows:

$$I_{ds} = I_{D1}(V_{gs}) I_{D2}(V_{ds}) \tag{5.60}$$

$$I_{D1}(V_{gs}) = \left(1-\frac{V_{gs}}{V_P}\right)^{(E_e + K_e V_{gs})} \tag{5.61}$$

$$I_{D2}(V_{ds}) = I_{DSS}\tanh\left(\frac{\alpha V_{ds}}{I_{DSS}(1-k_g V_{gs})}\right)(1+\frac{SV_{ds}}{I_{DSS}}) \tag{5.62}$$

where k_g, E_e, K_e, and S are the DC model parameters.

5.5.6 Angelov Nonlinear Model

The previously mentioned nonlinear models are intended mainly to describe the performance of MESFETs; a general FET models, of course, which model both HEMTs and MESFETs is needed. In particular, the characteristic peak in the transconductance versus gate-voltage dependence found in most HEMTs must be correctly modeled. The Angelov model is suitable for both HEMTs and MESFETs.

The drain current in the Angelov model is calculated with the following expression:

$$I_{ds} = I_{pk}(1+\tanh(\varphi))\tanh(\alpha V_{ds})(1+\lambda V_{ds}) \tag{5.63}$$

or

$$I_{ds} = I_{pk}(1+\tanh(\varphi))\tanh(\alpha V_{ds})\exp(\lambda V_{ds}) \tag{5.64}$$

where I_{pk} is the drain current at which we have maximum transconductance, with the contribution from the output conductance subtracted. λ is the channel length modulation parameter. The parameter λ is the same as those in the Statz and Curtice models.

The linear approach in the drain part of (5.73) $1 + \lambda V_{ds}$ gives a good approximation of I_{ds}, and in most cases, the linear approach gives satisfactory results. This term can also be considered as a first term of a power series of exponential dependence:

$$\exp(\lambda V_{ds}) = 1 + \lambda V_{ds} + \frac{1}{2}(\lambda V_{ds})^2 + \cdots\cdots \tag{5.65}$$

The exponential function describes the device behavior better when the drain current is small and generally converges well in HB simulation. Equation (5.74) is also well suited for devices with complex $I_{ds} = f(V_{ds})$ dependence and with such devices as kink effect, soft breakdown, and high-voltage FET.

ϕ is in general a power series function centered at V_{pk} with V_{gs} as a variable, i.e.,

$$\varphi = P_1(V_{gs} - V_{pk}) + P_2(V_{gs} - V_{pk})^2 + P_3(V_{gs} - V_{pk})^3 + \cdots\cdots \tag{5.66}$$

Substituting (5.76) into (5.74), the drain-current formula can be simplified as follows:

$$I_{ds} = I_{pk}(1+\tanh(P_1(V_{gs} - V_{pk})))\tanh(\alpha V_{ds})\exp(\lambda V_{ds}) \tag{5.67}$$

where V_{pk} is the gate voltage at which the maximum transconductance occurs.

The same type of modeling functions are chosen to model the dependencies on gate and drain voltage of capacitances C_{gs} and C_{gd}:

$$C_{gs} = C_{gso}[1+\tanh(\varphi_1)][1+\tanh(\varphi_2)] \tag{5.68}$$

$$C_{gd} = C_{gdo}[1+\tanh(\varphi_3)][1-\tanh(\varphi_4)] \tag{5.69}$$

where

$$\varphi_1 = P_{0gsg} + P_{1gsg}V_{gs} + P_{2gsg}V_{gs}^2 + P_{3gsg}V_{gs}^3 + \cdots\cdots \tag{5.70}$$

$$\varphi_2 = P_{0gsd} + P_{1gsd}V_{ds} + P_{2gsd}V_{ds}^2 + P_{3gsd}V_{ds}^3 + \cdots\cdots \tag{5.71}$$

$$\varphi_3 = P_{0gdg} + P_{1gdg}V_{gs} + P_{2gdg}V_{gs}^2 + P_{3gdg}V_{gs}^3 + \cdots\cdots \tag{5.72}$$

$$\varphi_4 = P_{0gdd} + (P_{1gdd} + P_{1cc}V_{gs})V_{ds} + P_{2gdd}V_{ds}^2 + P_{3gdd}V_{ds}^3 + \cdots\cdots \tag{5.73}$$

When accuracy on the order of 5% to 10% of C_{gs} and C_{gd} is sufficient, (5.68)–(5.73) can be simplified to

$$C_{gs} = C_{gsp} + C_{gso}[1 + \tanh(P_{10} + P_{11}V_{gs})][1 + \tanh(P_{20} + P_{21}V_{ds})] \tag{5.74}$$

$$C_{gd} = C_{gdp} + C_{gdo}[1 + \tanh(P_{30} + P_{31}V_{ds})][1 - \tanh(P_{40} + P_{41}V_{gd})] \tag{5.75}$$

The necessary condition for the charge conservation is

$$\frac{\partial C_{gs}}{\partial V_{dg}} = \frac{\partial C_{gd}}{\partial V_{gs}} \tag{5.76}$$

By integrating C_{gs} and C_{gd} with the terminal voltage to obtain Q_{gs} and Q_{gd}, we have

$$Q_{gs} = C_{gsp}V_{gs} + C_{gso}[V_{gs} + L_{c1} + V_{gs}T_{ch2} + L_{c1}T_{ch2}) \tag{5.77}$$

$$Q_{gd} = C_{gdp}V_{gd} + C_{gdo}[V_{gd} + L_{c4} + V_{gd}T_{ch3} + L_{c4}T_{ch3}) \tag{5.78}$$

where

$$L_{c1} = \frac{\log[\cosh(P_{10} + P_{11}V_{gs})]}{P_{11}} \tag{5.79}$$

$$T_{h2} = \tanh[P_{20} + P_{21}V_{ds}] \tag{5.80}$$

$$L_{c4} = \frac{\log[\cosh(P_{40} + P_{41}V_{gd})]}{P_{41}} \tag{5.81}$$

$$T_{h3} = \tanh[P_{30} + P_{31}V_{ds}] \tag{5.82}$$

5.6 Example of Compact Modeling Technique

To explain the modeling procedure in more detail for an actual device, we introduce two examples for compact modeling technique, that were developed by [20, 21]: (1) for DC modeling, and (2) for RF modeling.

5.6.1 DC Modeling

An all-region current-based empirical dc model is developed based on the Statz model in this section. The improvement lies in allowing the Statz model parameters to vary with gate-source voltage. Good agreement is obtained between simulation results and measured results for 0.25 μm gatelength and 40 μm gatewidth double heterostructure PHEMT.

Table 5.2 shows the extracted dc parameters for a PHEMT device using the existing Statz model. Notice that the extracted parameters have quite different values in the high-current ($I_{ds} > 1$ mA) and low-current regions ($I_{ds} \leq 1$ mA). This is because the model parameters are actually dependent on the gate-to-source voltage V_{gs}. Therefore, an all-region current-based model can be developed by making β, α, and λ as a function of gate-source voltage.

Table 5.2 Extracted Statz Parameters for High and Low Current Application

Parameters	V_{to}	β	α	λ	b
High-current	–0.67	2.19E–2	6.51	0.265	0.44
Low-current	–0.686	8.16E–3	11.87	1.39	0

Figure 5.8 shows extracted model parameters β, α, and λ as a function of gate-source voltage V_{gs} using the Statz model for the PHEMT device. It is observed that α and λ decrease rapidly with increasing V_{gs} for low values of V_{gs}, and decrease slowly for higher values. So α and λ can be modeled using the conventional decline formula:

$$\alpha(V_{gs}) = \begin{cases} \dfrac{\alpha_0}{(1+\dfrac{V_{gs}}{\alpha_1})^{\alpha_2}} & V_{gs} > V_{TO} \\ 0 & V_{gs} \le V_{TO} \end{cases} \tag{5.83}$$

$$\lambda(V_{gs}) = \begin{cases} \dfrac{\lambda_0}{(1+\dfrac{V_{gs}}{\lambda_1})^{\lambda_2}} & V_{gs} > V_{TO} \\ 0 & V_{gs} \le V_{TO} \end{cases} \tag{5.84}$$

The transconductance parameter β has a more complicated relationship with V_{gs}. It can be modeled using the exponential function and hyperbolic tangent function:

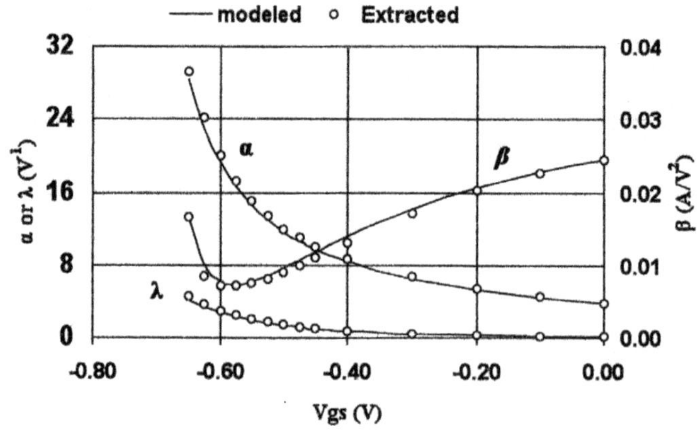

Figure 5.8 A comparison between the modeled and extracted model parameters β, α, and λ versus gate-to-source voltage V_{gs}

$$\beta(V_{gs}) = \beta_0 \exp(\beta_1(V_{gs} - V_{TO})) + \beta_2 \tanh(\beta_3(V_{gs} - V_{TO})) \quad (5.85)$$

As an example to verify the improved model just proposed, the extracted results for a AlGaAs/InGaAs/GaAs PHEMT with 0.25 μm mushroom gate (1 × 40 μm) was fabricated using Nanyang Technical University in-house process technology. The extracted model parameters are presented in Table 5.3. Figure 5.9 depicts the measured and calculated data for the characteristics of drain-source current I_{ds} versus V_{ds} and V_{gs} under low-current ($V_{gs} < -0.4$ V) conditions. Figure 5.10 shows these characteristics for high-current ($V_{gs} \geq -0.4$ V) condition.

To estimate the fitting error of the modeled quantities, the absolute error and relative error are computed and listed in Table 5.4. It can be seen that the model exhibits a very high accuracy for all current regions. Under low-current conditions ($I_{ds} \leq 1$ mA), absolute error is less than 15 μA, and the relative error is less than 6%. Under high-current conditions ($I_{ds} > 1$ mA), absolute error is less than 150 μA and the relative error is less than 5%. The fit of measured and modeled G_m and G_{ds} as a function of V_{gs} are shown in Figures 5.11 and 5.12. It can be seen that the modeled data agree well with the measured data.

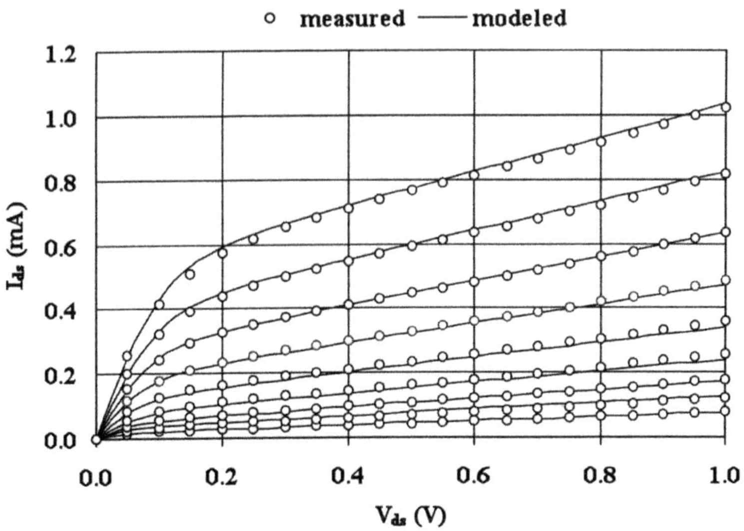

Figure 5.9 A comparison of dc measured and modeled I_{ds} versus V_{ds} (V_{gs} = −0.65 → −0.45 V, step 0.025 V)

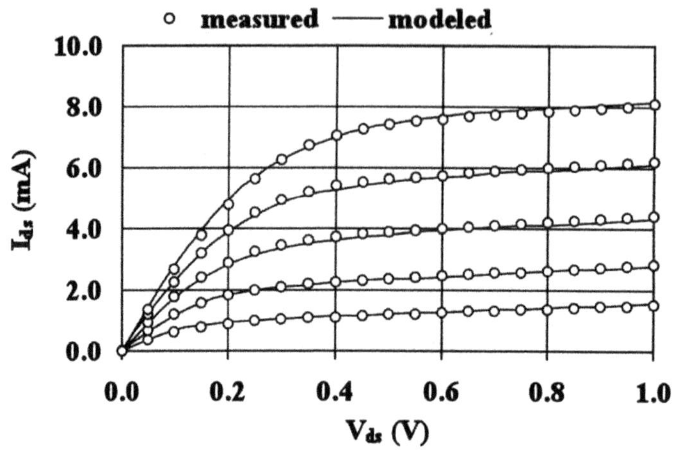

Figure 5.10 A comparison of DC measured and modeled I_{ds} versus V_{ds} (V_{gs} = −0.4 → −0.0 V, step 0.1 V)

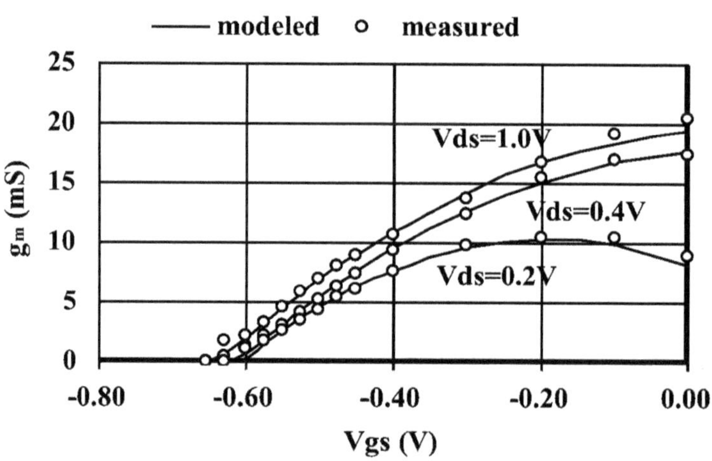

Figure 5.11 A comparison of measured and modeled g_m versus V_{gs}

5.6.2 RF Modeling

In this section, a compact nonlinear PHEMT model up to 40 GHz suitable for low-current applications is developed [21]. This model is suitable to be implemented into Agilent ADS to perform various analyses of PHEMT. The single-stage 2.45 GHz and 5.8 GHz amplifiers for RF identification (RFID) was designed using this model, Good agreement is

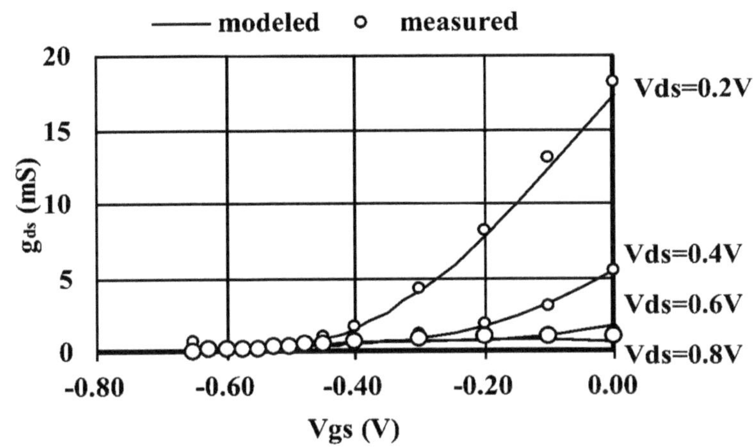

Figure 5.12 A comparison of measured and modeled g_{ds} versus V_{gs}

Table 5.3 Extracted Model Parameters of The Improved Model

Parameter	Value	Unit	Parameter	Value	Unit
V_{TO}	–0.68	V	α_0	3.84	V^{-1}
b	0.75	V^{-1}	α_1	0.763	V
β_0	3.568 E–2	A/V^2	α_2	1.046	
β_1	–29.54	V^{-1}	λ_0	0.085	V^{-1}
β_2	2.83 E–2	A/V^2	λ_1	1.454	V
β_3	1.93	V^{-1}	λ_2	6.55	

Table 5.4 Accuracy of DC Characteristics of Improved Model

	Absolute Error (μA)	Relative Error
High-current	150	5%
Low-current	15	6%

obtained between simulation results and measured results for PHEMT device and amplifier performances.

A modified RF drain-current I_{ds} formula based on the Curtice model is given as follows:

$$I_{ds} = \beta(V_{gs} - V_t)^Q \tanh(\alpha V_{ds})\exp(\lambda V_{ds})$$
$$V_t = V_{to} - \gamma V_{ds} \tag{5.86}$$

where β is the transconductance coefficient, α is the saturation voltage parameter, λ is the channel-length modulation coefficient, and V_{to} is the threshold voltage. Q is necessary to model non-square-law dependence of I_{ds}, and γ is used to model threshold voltage change with V_{ds}.

The effect of a frequency dispersion of transconductance g_m and drain-output conductance g_{ds} causes a significant difference between the DC I-V characteristic and a RF I-V characteristic with fit with g_m and g_{ds} at RF operation. To overcome this problem, the RF drain-current model (g_m and g_{ds} model) is used instead of the DC I-V model.

For a PHEMT operated in small-signal conditions, the RF drain-current I_{ds} can be modeled by using transconductance and output drain conductance:

$$g_m = \frac{\partial I_{ds}}{\partial V_{gs}} = Q(V_{gs} - V_t)^{Q-1} \tanh(\alpha V_{ds})\exp(\lambda V_{ds}) \tag{5.87}$$

$$g_{ds} = \frac{\partial I_{ds}}{\partial V_{ds}} = (V_{gs} - V_t)^Q (\lambda \tanh(\alpha V_{ds})\exp(\lambda V_{ds}) + \alpha \exp(\lambda V_{ds})\text{sech}^2(\alpha V_{ds})) \tag{5.88}$$

At each bias point g_m and g_{ds} are obtained from the bias-dependent S parameters.

The new empirical capacitance equations for PHEMTs are described by

$$\begin{aligned} C_{gs} &= \frac{\partial Q_{gs}}{\partial V_{gs}} + \frac{\partial Q_{gd}}{\partial V_{gs}} \\ &= (C_{gs0} + C_{gs1}V_{gs} + C_{gs2}V_{gs}^2)(1 + \lambda_{gs}V_{ds}) \\ &\quad + \lambda_{gd}V_{gd}(C_{gd0} + \frac{C_{gd1}V_{gd}}{2} + \frac{C_{gd2}V_{gd}^2}{3}) \end{aligned} \tag{5.89}$$

$$
\begin{aligned}
C_{gd} &= \frac{\partial Q_{gd}}{\partial V_{gd}} + \frac{\partial Q_{gs}}{\partial V_{gd}} \\
&= (C_{gd0} + C_{gd1}V_{gd} + C_{gd2}V_{gd}^2)(1 + \lambda_{gd}V_{ds}) \\
&\quad -\lambda_{gs}V_{gs}(C_{gs0} + \frac{C_{gs1}V_{gs}}{2} + \frac{C_{gs2}V_{gs}^2}{3})
\end{aligned}
\tag{5.90}
$$

where Q_{gs} and Q_{gd} are the gate-source and gate-drain charges, C_{gso} and C_{gdo} are the gate-source and gate-drain capacitances under zero-bias condition ($V_{gs} = 0$, $V_{ds} = 0$).C_{gs1}, C_{gs2}, C_{gd1}, and C_{gd2} are fitting parameters. The terms $1 + \lambda_{gs}V_{ds}$ and $1 + \lambda_{gd}V_{ds}$ are used to model the small variation of the capacitances with respect to drain-source voltage V_{ds}.

By integrating C_{gs} and C_{gd} with the terminal voltage, Q_{gs} and Q_{gd} are given as

$$
\begin{aligned}
Q_{gs} &= C_{gs0}V_{gs}(1 - \lambda_{gs}V_{gd}) + \frac{V_{gs}^2}{2}(C_{gs1} + C_{gs0}\lambda_{gs} - C_{gs1}\lambda_{gs}V_{gd}) \\
&\quad + \frac{V_{gs}^3}{3}(C_{gs2} + C_{gs1}\lambda_{gs} - C_{gs1}\lambda_{gs}V_{gd}) + \frac{V_{gs}^4}{4}C_{gs2}\lambda_{gs}
\end{aligned}
\tag{5.91}
$$

$$
\begin{aligned}
Q_{gd} &= C_{gs0}V_{gd}(1 - \lambda_{gd}V_{gs}) + \frac{V_{gd}^2}{2}(C_{gd1} - C_{gd0}\lambda_{gd} + C_{gd1}\lambda_{gd}V_{gs}) \\
&\quad + \frac{V_{gd}^3}{3}(C_{gd2} - C_{gd1}\lambda_{gd} + C_{gd2}\lambda_{gd}V_{gs}) + \frac{V_{gd}^4}{4}C_{gd2}\lambda_{gd}
\end{aligned}
\tag{5.92}
$$

To avoid the frequency dispersion effect of the drain conductance g_{ds} and transconductance g_m, the drain-current I_{ds} model parameters will not be extracted from DC I-V characteristics, and all model parameters will be extracted based on S parameters measurement. The procedure includes several phases:

- The pad capacitances C_{pg}, C_{pd}, and C_{pgd} are determined by measuring the S parameter of open test structure, so the pad capacitances can be separated from other parasitic elements.
- The parasitic resistances and inductances are extracted by using the pinch-off bias condition. The parasitic elements except for pad capacitances are extracted by using pinch-off bias condition only.

- Once extrinsic elements are obtained, intrinsic elements can be determined analytically for each bias point.
- The RF drain-current model parameters are obtained by fitting bias-dependent g_m and g_{ds}
- Capacitance model parameters can be determined by fitting C_{gs} and C_{gd} with respect to V_{gs} and V_{ds}

The S parameter measurements for model extraction and verification were carried out using an Agilent 8510C network analyzer from 50 MHz to 40 GHz, DC bias was supplied by an Agilent 4156A. All measurements were carried out on wafer using Cascade Microtech's Air-Coplanar Probes ACP50-GSG-150, with all instruments under IC CAP software control. The wafer probes were calibrated using Line-Reflect-Match (LRM) calibration method.

The fit of measured and modeled g_m and g_{ds} as a function of V_{ds} and V_{gs} are shown in Figures 5.13 and 5.14. It can be seen that the modeled data agree well with the measured data. The model parameters are extracted from S parameters of 90 bias points. Figure 5.15 presents the comparison between measured DC I-V characteristic and RF drain current calculated from RF g_m and g_{ds}. Note that this comparison shows the discrepancies between DC measurement and RF drain current modeled due to the frequency dispersion behavior of the g_m and g_{ds}. Figures 5.16 and 5.17 show the fitting results for C_{gs} and C_{gd} as a function of V_{ds} with varying V_{gs}, respectively, and reveal good agreement

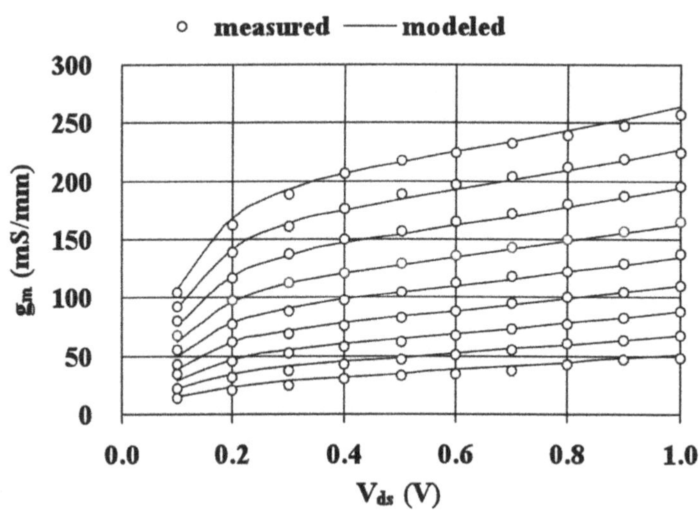

Figure 5.13 Measured and modeled transconductance versus V_{gs} and V_{ds} (V_{gs} = −0.65 → −0.45 V, step 0.025 V)

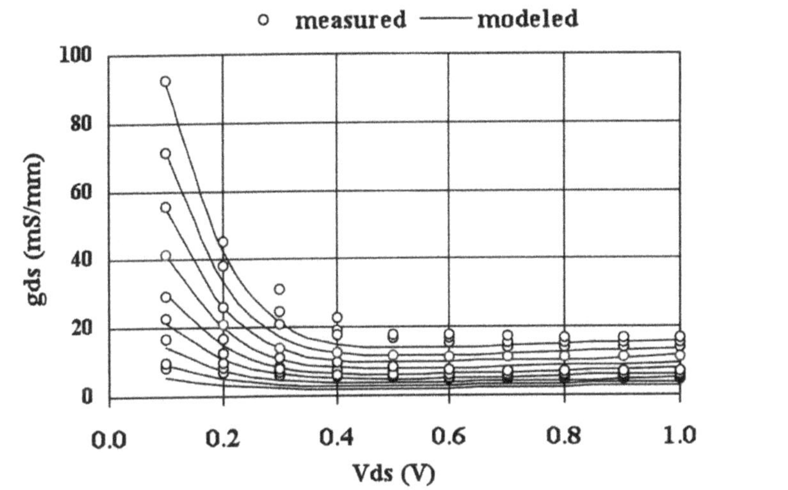

Figure 5.14 Measured and modeled output conductance versus V_{gs} and V_{ds} (V_{gs} = −0.65 → −0.45 V, step 0.025 V)

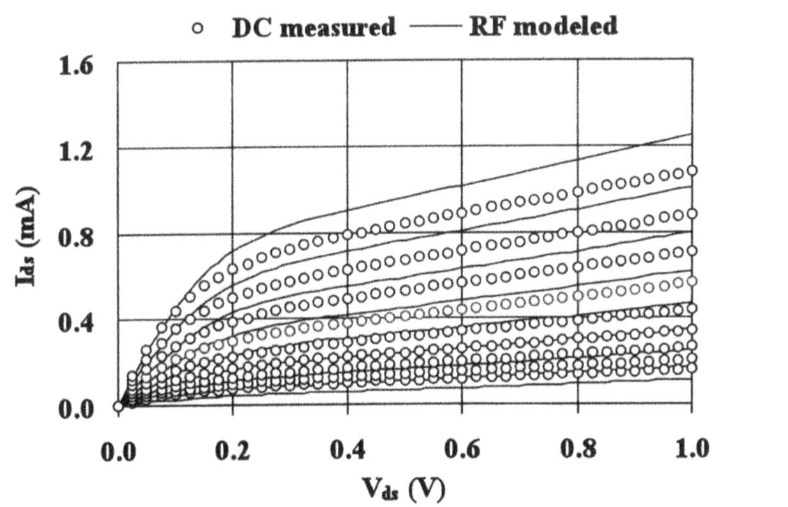

Figure 5.15 Comparison between measured DC I-V characteristic and RF drain current calculated from RF g_m and g_{ds} (V_{gs} = −0.55 → −0.35 V, step 0.025 V)

between the modeled and measured data. The extracted model parameters are summarized in Table 5.5.

The S parameters of the 40 μm PHEMT in the frequency range up to 40 GHz are simulated by using an improved model in ADS software. Figure 5.18 shows the simulated and measured S-parameters of the

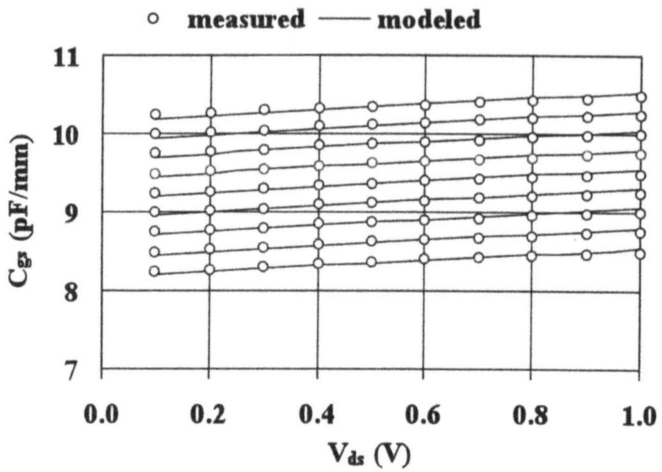

Figure 5.16 Measured and modeled gate-source capacitance versus V_{gs} and V_{ds} (V_{gs} = −0.55 → −0.35 V, step 0.025 V)

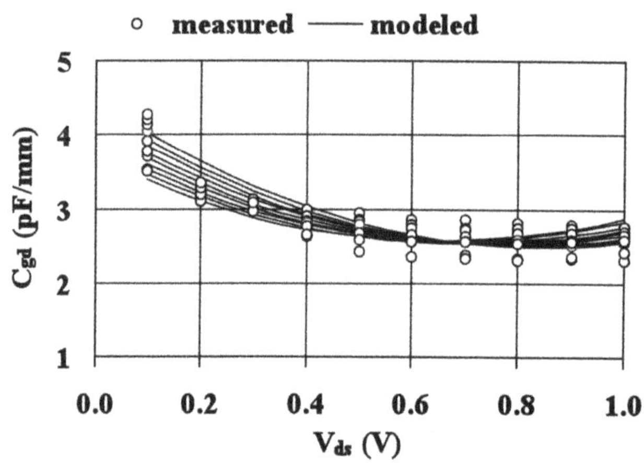

Figure 5.17 Measured and modeled gate-drain capacitance versus V_{gs} and V_{ds} (V_{gs} = −0.55 → −0.35 V, step 0.025 V)

PHEMT at different bias points. Figures 5.19 and 5.20 show the simulated and measured f_t and f_{max} at different bias points. The simulated results agree well with the measurements at all bias points.

Figure 5.21 shows the comparison between the measured and simulated output power levels for the fundamental (2.45 GHz) and second harmonic at 1 mA bias point. The measured and simulated harmonics for the different bias points and the fundamental at 2.45 GHz and 5.8 GHz are shown in Figures 5.22 and 5.23. The simulations run fast,

Table 5.5 Extracted Parameters of Nonlinear PHEMT Model

Pad capacitance	Parameters	Value	Unit
	C_{pg}	16	fF
	C_{pd}	18	fF
	C_{pdg}	2.5	fF
Parasitic elements	R_d	9	Ω
	R_s	6.5	Ω
	R_g	11	Ω
	L_d	78	pH
	L_s	0.09	pH
	L_g	56.35	pH
Intrinsic elements	C_{ds}	12.8	fF
	C_{rf}	12.0	fF
	R_C	968	W
RF drain-current model parameters	V_{TO}	–0.68	V
	β	0.025	A/V^2
	α	6.758	V^{-1}
	Q	3.208	
	λ	0.074	V^{-1}
	γ	0.0529	V^{-1}
Capacitance model parameters	C_{gs0}	52.583	fF
	C_{gs1}	34.77	fF/V
	C_{gs2}	–7.36	fF/V^2
	λ_{gs}	7.2E–6	V^{-1}
	C_{gd0}	25.611	fF
	C_{gd1}	25.698	fF/V
	C_{gs2}	11.397	fF/V^2
	λ_{gd}	0.1277	V^{-1}

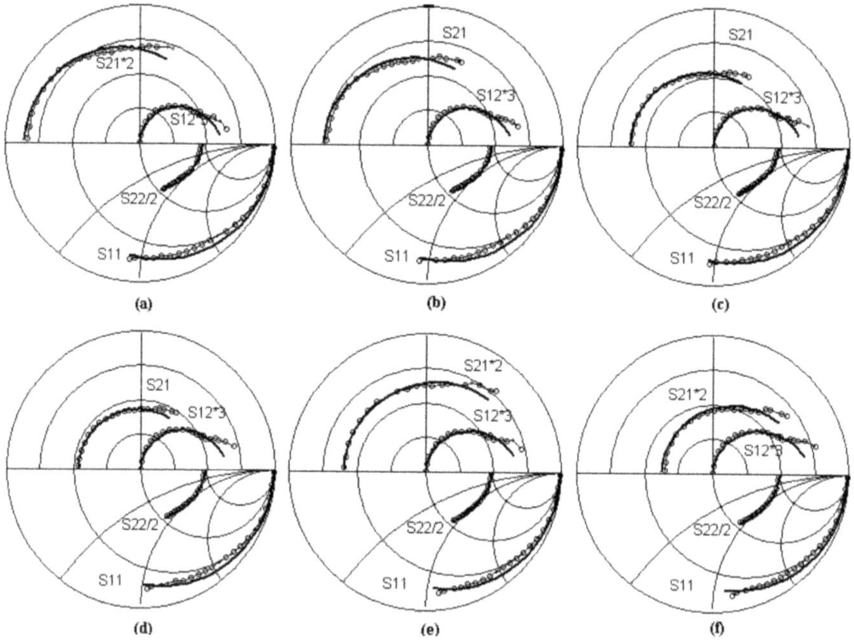

Figure 5.18 Comparison of modeled and measured S parameters (50 MHz to 40 GHz) for different bias point: (a) $I_{ds} = 1$ mA; (b) $I_{ds} = 0.8$ mA; (c) $I_{ds} = 0.6$ mA; (d) $I_{ds} = 0.4$ mA; (e) $I_{ds} = 0.2$ mA; (f) $I_{ds} = 0.1$ mA.

show good correspondence between measured and simulated data, and have no convergence problems. These indicate that our model simulates these figures of merit with high accuracy.

The 2.45 GHz and 5.8 GHz single-stage amplifiers for RFID have been designed using this model in Agilent ADS software. Figure 5.24 shows the schematic circuit of low-current amplifier. The amplifier consists of input and output microstrip matching networks, and PHEMT device. The performances of amplifiers can be measured by using a microwave tuner instead of microstrip matching networks. The simulated results for a 2.45 GHz amplifier and a 5.8 GHz amplifiers with respect to supply current are shown in Figures 5.25 and 5.26, and agree well with the measured results, which verify the model accuracy.

5.7 Summary

The nonlinear models for FETs, which include the physics-based nonlinear model, table-based nonlinear model, and empirical equivalent-

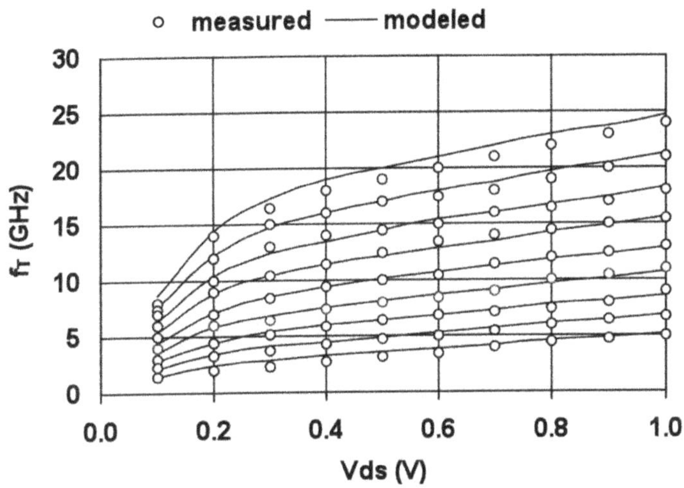

Figure 5.19 Measured and modeled cut off frequency at different bias points (V_{gs} = –0.55 → –0.35 V, step 0.025 V)

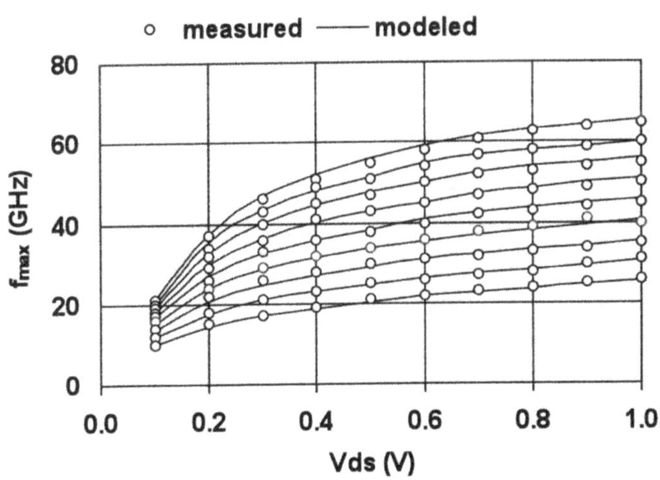

Figure 5.20 Measured and modeled maximum frequency of oscillation at different bias points (V_{gs} = –0.55 → –0.35 V, step 0.025 V)

circuit-based model, were introduced in this chapter. The compact modeling technique for FET is also described.

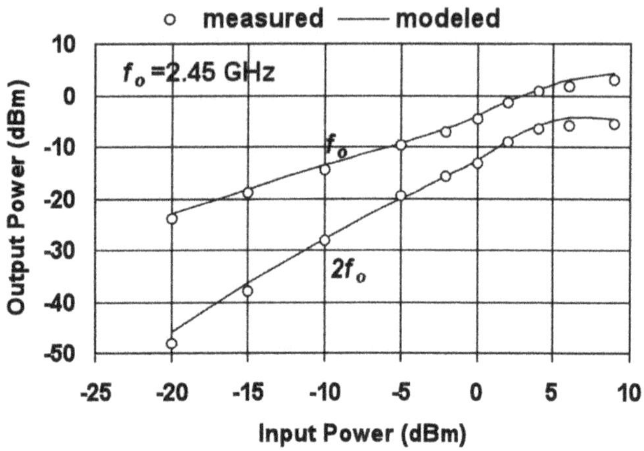

Figure 5.21 Comparison between measured and modeled output power levels (f_o = 2.45 GHz, V_{gs} = −0.35 V, V_{ds} = 1 V)

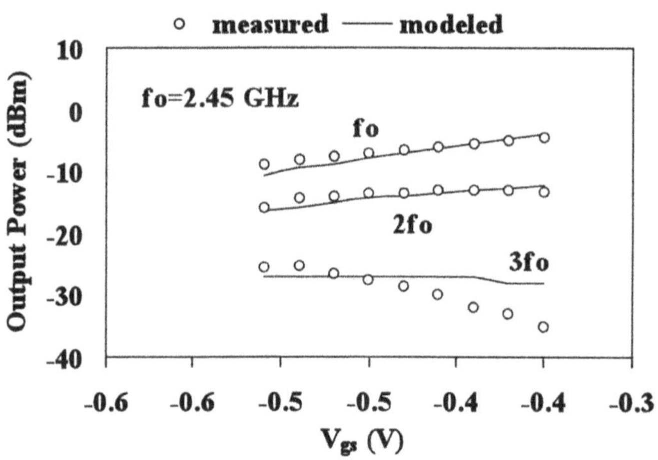

Figure 5.22 Comparison between measured and modeled output power levels for different V_{gs} (f_o = 2.45 GHz, V_{ds} = 1 V)

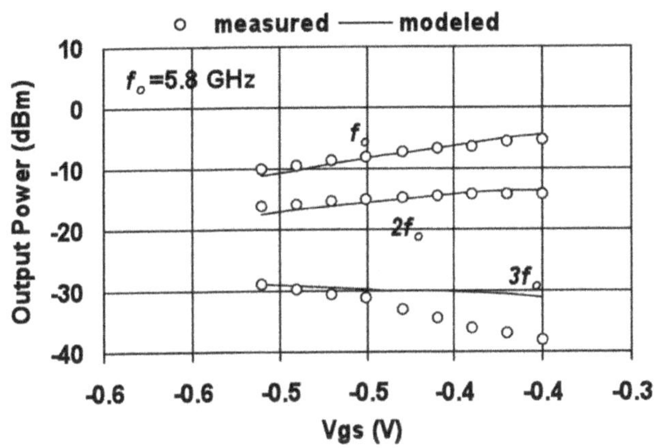

Figure 5.23 Comparison between measured and modeled output power levels for different V_{gs} (f_o = 5.8 GHz, V_{ds} = 1 V)

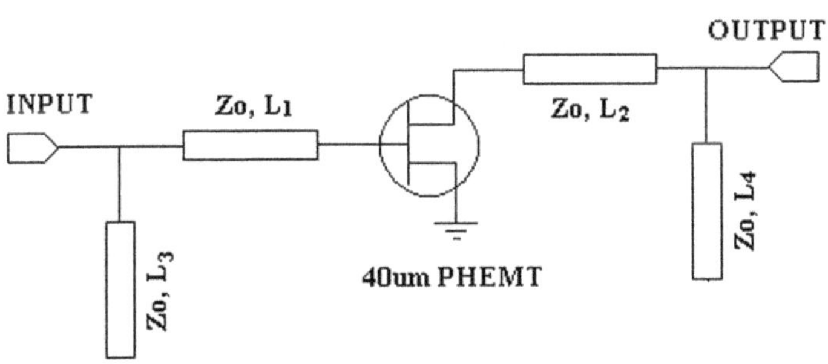

Figure 5.24 2.45 GHz and 5.8 GHz RFID low-current amplifier topology

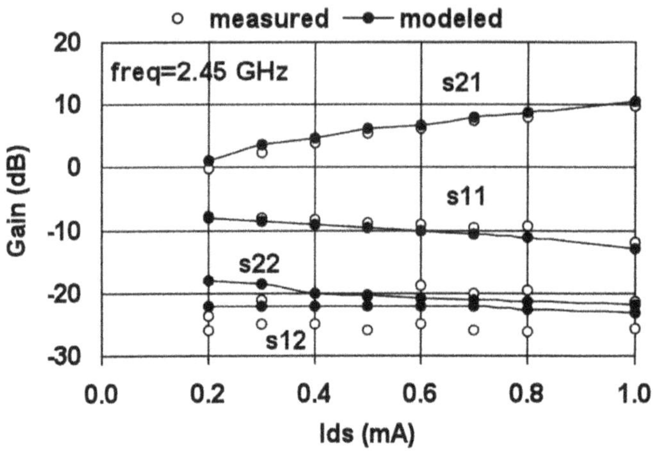

Figure 5.25 Measured and modeled *S*-parameters for 2.45 GHz single-stage amplifier

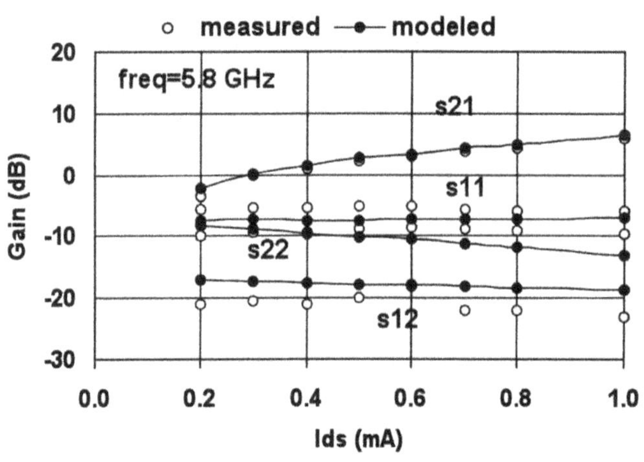

Figure 5.26 Measured and modeled *S*-parameters for 2.45 GHz single-stage amplifier

Chapter 6

Microwave Noise Modeling and Parameter Extraction Technique for FETs

6.1 Overview of Noise Model

Low-noise design is one of the key issues in most of the radio frequency (RF) receiver circuits. The complete characterization of the transistor in terms of noise and scattering parameters is necessary for the computer-aided design (CAD) of low-noise amplifiers. Therefore, accurate modeling of high-frequency noise is indispensable in developing low-noise amplifiers (LNAs) with short development time. Figure 6.1 shows the typical flowchart for field-effect transistor (FET) noise modeling procedure.

From Figure 6.1, it can be observed that small-signal equivalent circuit model is the basis of the noise equivalent-circuit model. The intrinsic noise sources include gate-induced noise, channel thermal noise, and low-frequency noise (1/f noise). The parasitic noise sources are the thermal noise generated by the extrinsic resistances.

The noise performance of FETs has been a subject of study for more than 50 years. It remains a subject of active research, as metal-oxide semiconductor field-effect transistors (MOSFETs) continue to set records for noise performance at both room and cryogentic temperatures. In spite of a considerable effort in this field, a noise model that could be helpful in understanding noise sources within a MESFET (HEMT, PHEMT) and at the same time could be useful in a circuit design still need to be designed.

The published studies of the noise properties of FETs may be divided into two distinctive groups. The first group, as a starting point of analy-

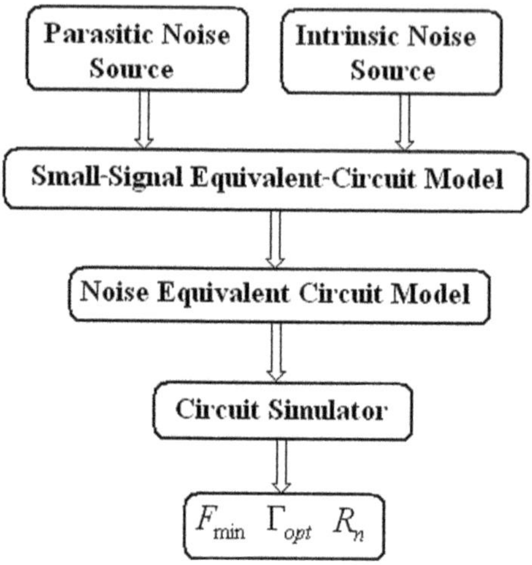

Figure 6.1 A flowchart for FET noise modeling

sis, considers fundamental equations of transport in semiconductors [1–8]. Most papers in this category over the years may be viewed as giving progressively more sophisticated treatment to the problem originally tackled by Van der Ziel [1]. The second group of published studies [9–17] addresses the issue of what needs to be known about the device in addition to its equivalent circuit to predict the noise performance.

From the circuit point of view, the FET device can be treated as a black box of a noisy two-port network. As is well known, the noise behavior of a linear noisy two-port network can be characterized by three noise parameters: minimum noise figure F_{min}, noise conductance g_n (or noise resistance R_n), and optimum source impedance Z_{opt} (or optimum source admittance Y_{opt}), with

$$F = F_{\min} + \frac{g_n}{R_s}\left|Z_s - Z_{opt}\right|^2 \tag{6.1}$$

$$F = F_{\min} + \frac{R_n}{G_s}[(G_s - G_{opt})^2 + (B_s - B_{opt})^2] \tag{6.2}$$

where F is the noise figure, $Z_s = R_s + jX_s$ is the source admittance, and $Y_s = G_s + jB_s$ is the optimum source admittance. These noise parameters can easily be calculated from the equivalent circuit of the noisy FET presented in Figure 6.2(a) [17]. The circuit comprises the well-known small-signal equivalent and the four noise sources

$$\overline{e_g^2}\,,\ \overline{e_s^2}\,,\ \overline{i_g^2}\,,\ \text{and}\ \overline{i_d^2}$$

The two current noise sources $\overline{i_g^2}$ and $\overline{i_d^2}$ represent the internal noises of the intrinsic FET device; these noise sources are correlated. By means of a simple circuit manipulation, these four noise sources are transformed into two correlated noise sources $\overline{v^2}$ and $\overline{i^2}$ preceding the

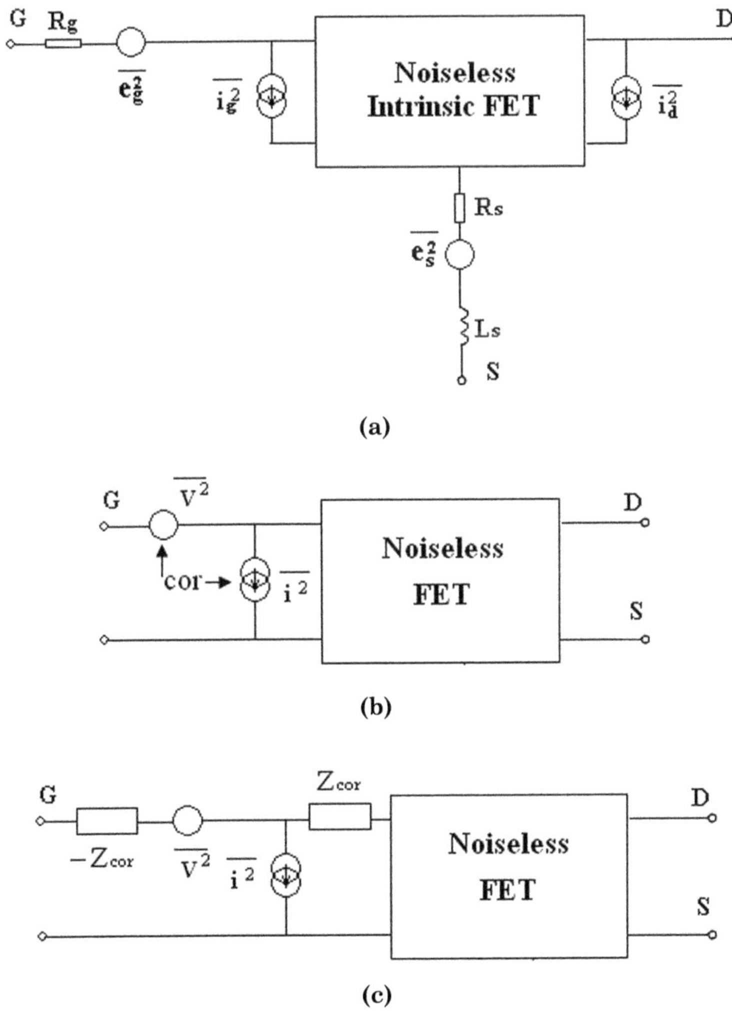

Figure 6.2 Circuit transformations for the noise figure determination

extrinsic FET, which is now considered noiseless (Figure 6.2(b)). Lastly, $\overline{v^2}$ and $\overline{i^2}$ are de-correlated by introduction of two (noiseless) correlation impedances (Figure 6.2(c)). At this step, the calculation of F_{min} and Z_{opt} is straightforward and gives:

$$F_{\min} = 1 + 2g_n(R_{cor} + R_{opt}) \tag{6.3}$$

$$Z_{opt} = R_{opt} + jX_{opt} = \sqrt{R_{cor}^2 + \frac{r_n}{g_n}} - jX_{cor} \tag{6.4}$$

where

$$Z_{cor} = R_{cor} + jX_{cor}$$

$$g_n = \overline{i^2} / 4kT\Delta f$$

$$r_n = \overline{v^2} / 4kT\Delta f$$

The commonly used noise models for FET include the Fukui noise model, the Pucel noise model (or PRC model), and the Pospieszalski noise model. These noise models are discussed in more detail in the next sections.

6.1.1 Fukui Noise Model

The four noise parameters could be expressed in terms of equivalent-circuit elements as follows [9, 10]:

$$F_{\min} = 1 + k_1 f C_{gs} \sqrt{\frac{R_g + R_s}{g_m}} \tag{6.5}$$

$$R_n = \frac{k_2}{g_m^2} \tag{6.6}$$

$$R_{opt} = k_3 \left[\frac{1}{4g_m} + R_g + R_s \right] \tag{6.7}$$

$$X_{opt} = k_4 / fC_{gs} \tag{6.8}$$

where K_1, K_2, K_3, and K_4 are fitting factors, and f is frequency. It can be observed clearly that minimum noise figure is a linear function of frequency, optimum source reactance X_{opt} is inversely proportional to frequency, and optimum source resistance R_{opt} and noise resistance R_n are frequency independent.

Comparing these expressions with the experimental data, all noise parameters can be determined by using the following formulas:

$$k_1 = \frac{1}{C_{gs}\sqrt{(R_s + R_g)/g_m}} \cdot \frac{\partial[(F_{\min} - 1)]}{\partial f} \tag{6.9}$$

$$k_2 = R_n g_m^2 \tag{6.10}$$

$$k_3 = \frac{R_{opt}}{R_g + R_s + 1/4g_m} \tag{6.11}$$

$$k_4 = C_{gs} \frac{\partial X_{opt}}{\partial(1/f)} \tag{6.12}$$

Based on the relationship between small-signal model parameters (g_m, C_{gs}, R_g, and R_s) and the geometrical and material parameters of FET device:

$$g_m = k_5 W \left[\frac{N}{aL}\right]^{1/3} \tag{6.13}$$

$$C_{gs} = k_6 W \left[\frac{NL^2}{a}\right]^{1/3} \tag{6.14}$$

$$R_g = \frac{17z^2}{hLW} \tag{6.15}$$

$$R_s = \frac{1}{W}\left[\frac{2.1}{a^{0.5}N^{0.66}} + \frac{1.1L_{sg}}{(a-a_s)N^{0.82}}\right] \tag{6.16}$$

where N is the free carrier concentration in the active channel, L is the gate length, W is the total device gatewidth, a is the active channel thickness, z is the unit gatewidth, h is the gate metallization height, L_{sg} is the distance between the source and gate electrodes, and a_s is the depletion layer thickness at the surface in the source-gate space.

Substituting (6.13)–(6.16) into (6.5)–(6.8), the relationship between noise parameters and geometrical and material parameters can be obtained.

6.1.2 Pucel Noise Model

The Pucel noise model is also called "PRC" noise model [1–3], the corresponding gate-induced noise current $\overline{i_g^2}$ and drain channel noise current $\overline{i_d^2}$ are expressed as follows:

$$\overline{i_g^2} = 4kT\Delta f\omega^2 C_{gs}^2 R / g_m \tag{6.17a}$$

$$\overline{i_d^2} = 4kT\Delta f g_m P \tag{6.17b}$$

The cross-correlation between $\overline{i_g^2}$ and $\overline{i_d^2}$ can be expressed as:

$$\overline{i_g^* i_d} = C\sqrt{\overline{i_g^2}\,\overline{i_d^2}} = 4kT\Delta f\omega C_{gs} C\sqrt{PR} \tag{6.18}$$

where R and P are the gate- and drain-current noise model parameters, respectively, and C is the correlation coefficient. k is the Boltzmann constant, T is the absolute temperature (normally is 290 K), and Δf is the bandwidth.

The corresponding noise parameters can be expressed as follows [17]

$$F_{\min} = 1 + 2\sqrt{P + R - 2C\sqrt{PR}}\,\frac{f}{f_c}\sqrt{g_m(R_s + R_g) + \frac{PR(1-C^2)}{P + R - 2C\sqrt{PR}}} \tag{6.19}$$

$$g_n = g_m \left(\frac{f}{f_c} \right)^2 (P + R - 2C\sqrt{PR}) \tag{6.20}$$

$$R_{opt} = \sqrt{\frac{g_m(R_s + R_g) + \dfrac{PR(1 - C^2)}{P + R - 2C\sqrt{PR}}}{P + R - 2C\sqrt{PR}}} \frac{1}{\omega C_{gs}} \tag{6.21}$$

$$X_{opt} = \frac{1}{\omega C_{gs}} \frac{P - C\sqrt{PR}}{P + R - 2C\sqrt{PR}} \tag{6.22}$$

In several commercial software applications, such as the SPICE circuit simulator, gate-induced noise current $\overline{i_g^2}$ has been neglected, that is $R = 0$. Equations (6.19)–(6.22) can be simplified as follows:

$$F_{\min} = 1 + 2\sqrt{P} \frac{f}{f_c} \sqrt{g_m(R_s + R_g)} \tag{6.23}$$

$$g_n = P g_m \left(\frac{f}{f_c} \right)^2 \tag{6.24}$$

$$R_{opt} = \sqrt{\frac{g_m(R_s + R_g)}{P}} \frac{1}{\omega C_{gs}} \tag{6.25}$$

$$X_{opt} = \frac{1}{\omega C_{gs}} \tag{6.26}$$

Here f_c is the cutoff frequency, and is expressed as follows:

$$f_c = \frac{g_m}{2\pi C_{gs}} \tag{6.27}$$

According to [12], parameter P can be approximated for operating points below the onset of saturation by:

$$P = \frac{I_{ds}}{E_c L g_m} \tag{6.28}$$

In this expression, I_{ds} is the dc drain current, E_c the critical field, L is the gate length, and g_m is the transconductance.

6.1.3 Pospieszalski Noise Model

The Pospieszalski noise model is shown in Figure 6.3(a), the intrinsic part is shown in Figure 6.3(b) [18]. The thermal noise voltage sources generated by extrinsic resistances R_g, R_d, and R_s are as follows:

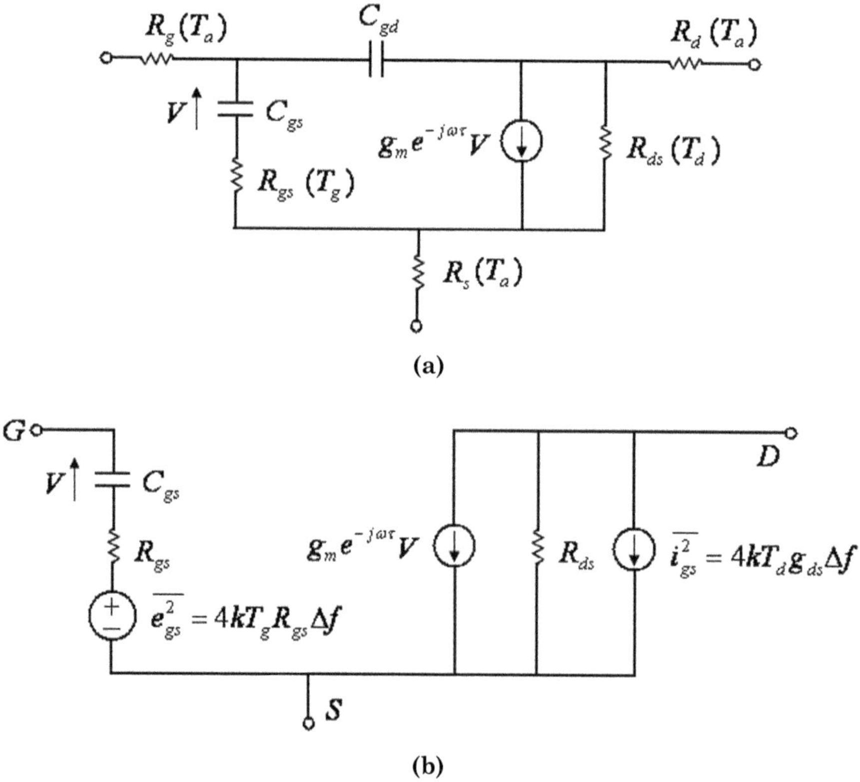

Figure 6.3 (a) Pospieszalski noise model (b) intrinsic part

$$\overline{e_g^2} = 4kT_a R_g \Delta f \tag{6.29}$$

$$\overline{e_d^2} = 4kT_a R_d \Delta f \tag{6.30}$$

$$\overline{e_s^2} = 4kT_a R_s \Delta f \tag{6.31}$$

where T_a is the ambient temperature. The two uncorrelated current noise sources $\overline{e_{gs}^2}$ and $\overline{i_{ds}^2}$ represent the internal noise sources of the intrinsic FET; these two noise sources are characterized by their mean quadratic value in a bandwidth Δf centered on the frequency f, and can be given by the following expressions:

$$\overline{e_{gs}^2} = 4kT_g R_{gs} \Delta f \tag{6.32}$$

$$\overline{i_{ds}^2} = 4kT_d g_{ds} \Delta f \tag{6.33}$$

$$\overline{e_{gs}^* i_{ds}} = 0 \tag{6.34}$$

where T_g and T_d are the equivalent noise temperature of the intrinsic resistance R_{gs} and output conductance g_{ds}, respectively.

In the Pospieszalski noise model, the elements of the noise correlation matrix are

$$\overline{i_1^2} = 4kT_g \Delta f R_{gs} \left| \frac{j\omega C_{gs}}{1 + j\omega C_{gs} R_{gs}} \right|^2 \tag{6.35}$$

$$\overline{i_2^2} = 4k\Delta f \left(\frac{T_d}{R_{ds}} + T_g R_{gs} \left| \frac{g_m}{1 + j\omega C_{gs} R_{gs}} \right|^2 \right) \tag{6.36}$$

$$\overline{i_1 i_2^*} = 4kT_g \Delta f \frac{g_m^* \omega C_{gs} R_g}{| 1 + j\omega C_{gs} R_{gs} |^2} \tag{6.37}$$

Figure 6.4 presents the noise temperatures as a function of drain current for commercially available low-noise GaAs PHEMT (NEC NE32500) [19]. The results show that T_g equals the ambient temperature in the lower drain-current ranges, and that T_d increase with drain current.

6.1.4 Gate Leakage Current Model

It is well established that a short gate on a thin, heavily doped layer is necessary to obtain the excellent microwave characteristics in FETs. As a result, the influence of gate-leakage currents occurring in the high-drain and gate-bias condition on their performance becomes significant, especially on the noise performance of short-gate FET's at low frequencies of about a few gigahertz [20–22]. Therefore, an additional noise current source $\overline{i_{gL}^2}$ is needed in parallel with noise current source $\overline{i_g^2}$ at the input port for the "PRC" noise model (as shown in Figure 6.5) [20]. Alternatively, an additional resistor R_{pgs} is needed in parallel with the noise source $\overline{i_g^2}$ for the Pospieszalski noise model (as shown in Figure 6.6) [21].

$$\overline{i_{gL}^2} = 2qI_{gL}\Delta f \tag{6.38}$$

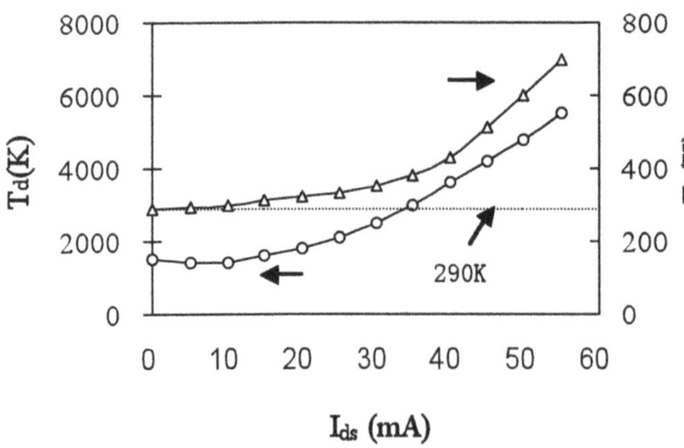

Figure 6.4 T_d and T_g versus drain current

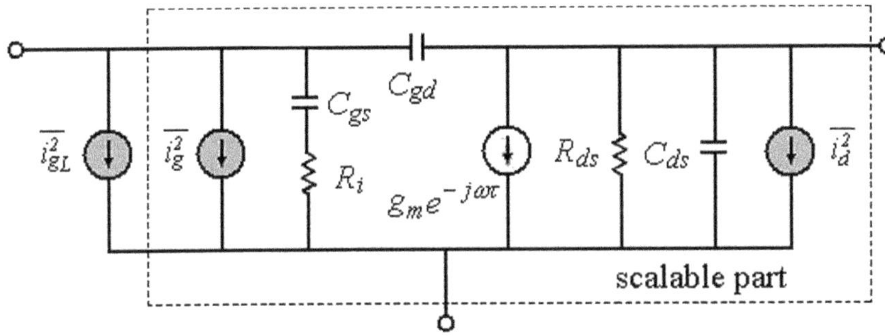

Figure 6.5 Improved noise model with gate-leakage current for PRC noise model

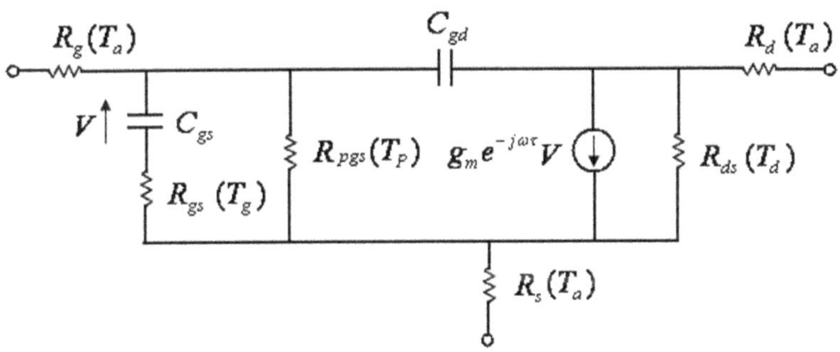

Figure 6.6 Improved noise model with gate leakage current for Pospieszalski noise model

$$\overline{i_{pgs}^2} = 4kT_p R_{pgs} \Delta f \tag{6.39}$$

where T_P is the equivalent noise temperature of the resistance R_{pgs}.

Figure 6.7 shows the noise parameters F_{min}, g_n, and Z_{opt} versus frequency. To demonstrate the strong influence of the gate-leakage at low-frequency range, the data are shown in the frequency range from 0.1 GHz up to 18 GHz. The plots show that a gate-current has a significant impact on the noise parameters and leads to an increase of F_{min} and g_n at low frequencies. The optimum generator impedance Z_{opt} shows a shift to higher values of G_{opt} [21]. For gate-leakage current higher than 1 μA, it is necessary to use the improved model for an exact modeling of all noise parameters. At very low gate-leakage current, the additional noise currents are negligible.

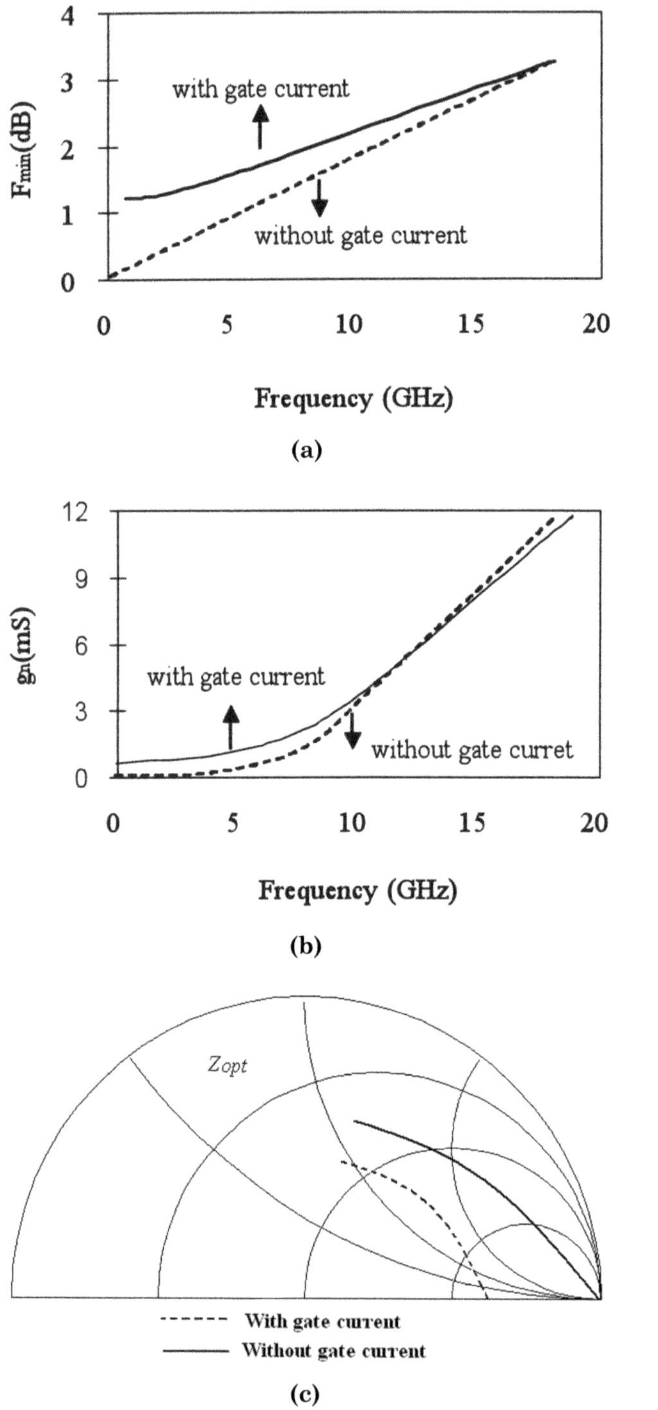

Figure 6.7 The influence of gate current on noise parameters

6.2 Scalable Noise Model

To optimize the noise performance of FETs for low noise application, full-scalable, analytical expressions for the noise parameters that provide good accuracy and can be scalable based on the layout structure, as well as gatewidth, are very attractive for monolithic microwave integrated circuit (MMIC) design and have a number of advantages including the following:

The noise parameters of differently sized device under the same process conditions can be readily obtained using scalable normalization model parameters, so we can achieve low cost and save a lot of time.

It is useful for understanding the physical mechanisms and for evaluating the influence of the different parameters of the small-signal equivalent circuit model.

A set of expressions for the four noise parameters of AlGaAs/InGaAs/GaAs PHEMT is derived from an accurate noise equivalent-circuit model without any assumptions and approximations. The effects of all intrinsic elements and gate-leakage current are taken into account. The scaling rules for the noise parameters of intrinsic part are determined based on these analytical expressions.

The corresponding short-circuit Y parameters of the intrinsic small-signal equivalent-circuit can be expressed as:

$$Y = \begin{bmatrix} \dfrac{j\omega C_{gs}}{1+j\omega R_i C_{gs}} + j\omega C_{gd} & -j\omega C_{gd} \\ \dfrac{g_m e^{-j\omega\tau}}{1+j\omega C_{gs} R_i} - j\omega C_{gd} & g_{ds} + j\omega(C_{ds} + C_{gd}) \end{bmatrix} \tag{6.40}$$

The corresponding admittance noise correlation matrix can be expressed as:

$$C_Y = 4kT \begin{bmatrix} \alpha I_{gL} / 2V_T + (\omega C_{gs})^2 R / g_m & j\omega C_{gs} C\sqrt{PR} \\ -j\omega C_{gs} C\sqrt{PR} & g_m P \end{bmatrix} \tag{6.41}$$

where V_T represents the thermal voltage ($V_T = kT/q$).

Based on the noise correlation matrix technique [23], the four noise parameters can be directly calculated from the chain noise correlation matrix:

$$R_n = C_{A11} \tag{6.42}$$

$$G_{opt} = \sqrt{\frac{C_{A22}}{C_{A11}} - \left(\frac{\mathrm{Im}(C_{A12})}{C_{A11}}\right)^2} \tag{6.43}$$

$$B_{opt} = \frac{\mathrm{Im}(C_{A12})}{C_{A11}} \tag{6.44}$$

$$F_{\min} = 1 + 2[\mathrm{Re}(C_{A12}) + G_{opt} C_{A11}] \tag{6.45}$$

where

$$C_{A11} = \frac{C_{Y22}}{|Y_{21}|^2} \tag{6.46}$$

$$C_{A12} = \frac{(Y_{11})^* C_{Y22} - (Y_{21})^* C_{Y21}}{|Y_{21}|^2} \tag{6.47}$$

$$C_{A22} = C_{Y11} + \frac{|Y_{11}|^2 C_{Y22}}{|Y_{21}|^2} - 2\,\mathrm{Re}\left(\frac{Y_{11}}{Y_{21}} C_{Y21}\right) \tag{6.48}$$

Substituting (6.40) and (6.41) into (6.42)–(6.45), we have:

$$F_{\min} = 1 + 2k_3 + \frac{2}{g_m k_1}\sqrt{\omega^2 C_{gs}^2 PR(1 - C^2) + g_m k_1 k_2} \tag{6.49}$$

$$R_n = \frac{P}{g_m k_1} \tag{6.50}$$

$$G_{opt} = \sqrt{\omega^2 C_{gs}^2 \frac{R}{P}(1 - C^2) + \frac{g_m k_1 k_2}{P}} \tag{6.51}$$

$$B_{opt} = -\omega[(C_{gs} + C_{gd}) - C\sqrt{\frac{R}{P}} C_{gs}] \tag{6.52}$$

where

$$k_1 = \frac{g_m + 2\omega^2 C_{gd}(\tau + C_{gs} R_i)}{g_m(1 + + C_{gs}^2 R_i^2)}$$

$$k_2 = \frac{\alpha I_{gL}}{2V_T}$$

$$k_3 = \frac{\omega^2 C_{gs}}{g_m k_1}[PC_{gs}R_i - C\sqrt{PR}(\frac{C_{gd}}{g_m} + C_{gs}R_i + \tau)]$$

According to the scaling rules of FETs [24], the intrinsic small-signal model parameters have standard scaling: The intrinsic capacitances, transconductance, and output conductance are proportional to gate-width, and the intrinsic resistance is inversely proportional to gate-width, that is,

$$\begin{bmatrix} C_{gs} \\ C_{gd} \\ C_{ds} \\ g_m \\ g_{ds} \\ R_i \end{bmatrix} = \begin{bmatrix} n & 0 & 0 & 0 & 0 & 0 \\ 0 & n & 0 & 0 & 0 & 0 \\ 0 & 0 & n & 0 & 0 & 0 \\ 0 & 0 & 0 & n & 0 & 0 \\ 0 & 0 & 0 & 0 & n & 0 \\ 0 & 0 & 0 & 0 & 0 & \frac{1}{n} \end{bmatrix} \begin{bmatrix} C_{gs}^c \\ C_{gd}^c \\ C_{ds}^c \\ g_m^c \\ g_{ds}^c \\ R_i^c \end{bmatrix} \tag{6.53}$$

where n represents the number of the elementary cells for larger gate-width size device, defined by

$$n = \frac{W}{W^c} \tag{6.54}$$

and where superscript c denotes the elementary cell.

Assuming the time delay remains invariant at the same bias condition for the same device process, then $\tau = \tau^c$. Substituting (6.54) into (6.53), we have

$$Y = nY^c \tag{6.55}$$

From (6.55), it can be found that a large-size gatewidth FET can be viewed as consisting of n elementary cells with the same gatewidth connected in parallel. Based on the noise correlation matrix technique, the admittance noise correlation matrix for large- size gatewidth FET can be expressed as:

$$C_Y = nC_Y^c \tag{6.56}$$

Because the gate-leakage current is not scalable, all the scaling rules for the noise parameters in this book not include the effect of gate-leakage current. Substituting (6.55) and (6.56) into (6.41), we have

$$P = P^c \tag{6.57}$$

$$R = R^c \tag{6.58}$$

$$C = C^c \tag{6.59}$$

Substituting (6.57) to (6.59) into (6.49)–(6.52), it can be found that the scaling rules of the noise parameters for intrinsic part (dashed box in Figure.6.5) are as follows: minimum noise figure F_{min} remains invariant, noise resistance R_n is inversely proportional to number of the elementary cells, and optimum source conductance G_{opt} and optimum source susceptance B_{opt} are proportional to number of the elementary cells, that is,

$$F_{\min} = F_{\min}^{c} \tag{6.60}$$

$$R_n = \frac{1}{n} R_n^c \tag{6.61}$$

$$G_{opt} = nG_{opt}^{c} \tag{6.62}$$

$$B_{opt} = nB_{opt}^{c} \tag{6.63}$$

Once the small-signal elements are extracted from the S-parameter measurements, the extraction of the four unknown noise model parameters can be carried out using the procedure based on the noise correlation matrix technique. The noise model parameters can be determined as follows:

$$P = \frac{C_{Y22}}{4kTg_m \Delta f} \tag{6.64}$$

$$R = \frac{C_{Y11} - 2\alpha q I_{gL}}{4kT(\omega C_{gs})^2 \Delta f} g_m \tag{6.65}$$

$$C = \frac{\mathrm{Im}(C_{Y12})}{4kT\omega C_{gs}\sqrt{PR}} \tag{6.66}$$

$$\alpha = \frac{1}{2qI_{gL}} C_{Y11}\Big|_{\omega \to 0} \tag{6.67}$$

The extracted noise model parameters P, R, C, and α are summarized in Table 6.1. It can be found that P, R, and C remain invariant for four different gatewidth size PHEMTs under the same bias conditions. However, the gate-leakage current cannot be scalable.

Table 6.1 The Noise Model Parameters for Differently Sized PHEMTs

Elements	2 × 20 μm	2 × 40 μm	2 × 60 μm
P	1.50	1.50	1.55
R	1.10	1.10	1.15
C	0.30	0.32	0.35
αI_{gL}	5×10^{-6}	4×10^{-6}	2×10^{-6}

In Figure 6.8, we compare the measured and computed noise parameters versus frequency for 2 × 20 μm PHEMT under single-bias conditions (V_{gs} = 0 V, V_{ds} = 2 V). Good agreements are observed. The new expressions are also compared with the model proposed by [17]. Solid lines correspond to the values calculated from (6.49) to (6.52), whereas dashed lines correspond to the values calculated from [17]. It clearly can be found that the new expressions are more accurate than [17]. Because of the gate-leakage current effect, the minimum noise figure does not cross the zero point.

To illustrate the efficiency of the scaling rules for noise parameters, in Figures 6.9 and 6.10 we compare the measured and predicted results for larger-size PHEMTs (2 × 40 μm and 2 × 60 μm). The 2 × 40 μm and 2 × 60 μm PHEMTs can be viewed as consisting of two and three elementary cells (2 × 20 μm PHMT) connected in parallel, respectively. The measured data for the intrinsic part of 2 × 40 μm and 2 × 60 μm PHEMTs are obtained after de-embedding the extrinsic elements from measured noise parameters, and the predicted data are obtained by using scaling process (6.60)–(6.63) from elementary cell (2 × 20 μm PHEMT). Good agreement is obtained between measured and predicted data to validate the scaling approach for noise parameters.

6.3 Noise Parameters Extraction Method

The commonly used methods for determining noise parameters for semiconductor devices are the tuner-based method [25–29] and noise figure measurement system based method [30–33].

6.3.1 Tuner-Based Extraction Method

Determining the noise parameters is typically performed by analyzing the variation of measured noise figure as a function of the source

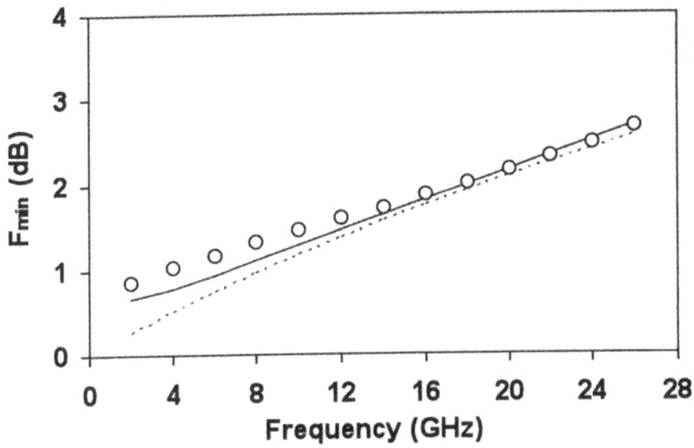

(a)

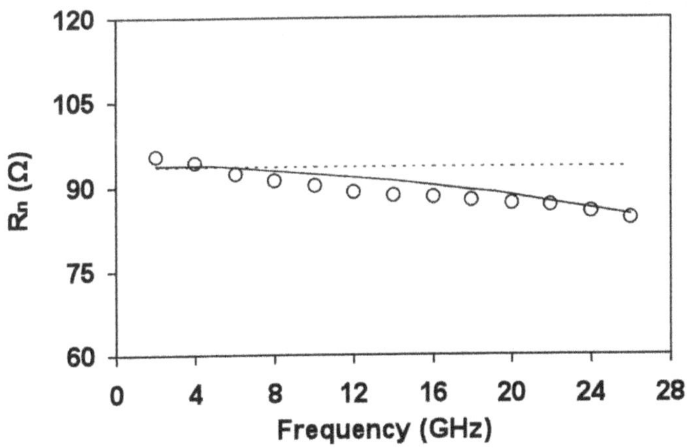

(b)

Figure 6.8 Comparison of measured and calculated noise parameters for 2 × 20 μm PHEMT based on the new expressions. oo = Experimental data, — = calculated data from (6.60)–(6.63), --- = calculated data from [17] (bias condition: V_{gs} = 0 V, V_{ds} = 2 V) *(continues)*

impedance. A minimum of four independent measurements is required. However, for increasing accuracy, more than four measurements are usually performed, and curve-fitting techniques are then used to determine the noise parameters. Figure 6.11 shows the tuner-based noise parameters measurement system, and a typical source reflection coefficient distributed Smith chart is shown in Figure 6.12.

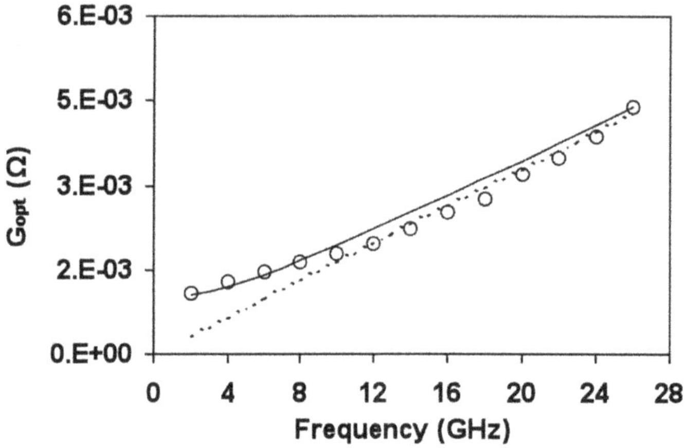

(c)

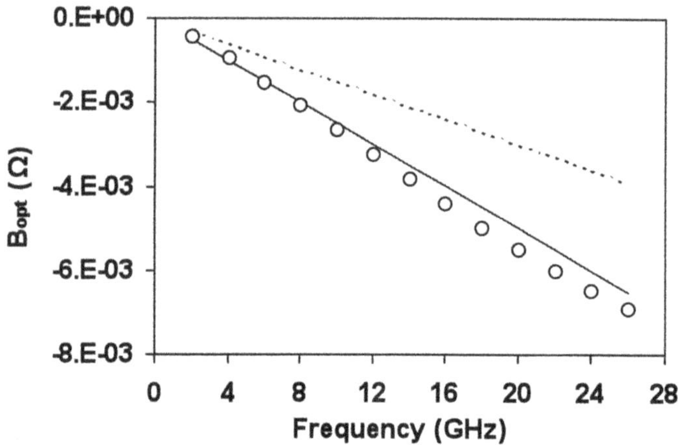

(d)

Figure 6.8 Comparison of measured and calculated noise parameters for 2× 20 μm PHEMT based on the new expressions. oo = Experimental data, — = calculated data from (6.60)–(6.63), --- = calculated data from [17] (bias condition: V_{gs} = 0 V, V_{ds} = 2 V) *(continued)*

The noise figure can be regards as a nonlinear function of the source admittance ($Y_s = G_s + jB_s$):

$$F = A + BG_s + \frac{C + BB_s^2 + DB_s}{G_s} \tag{6.68}$$

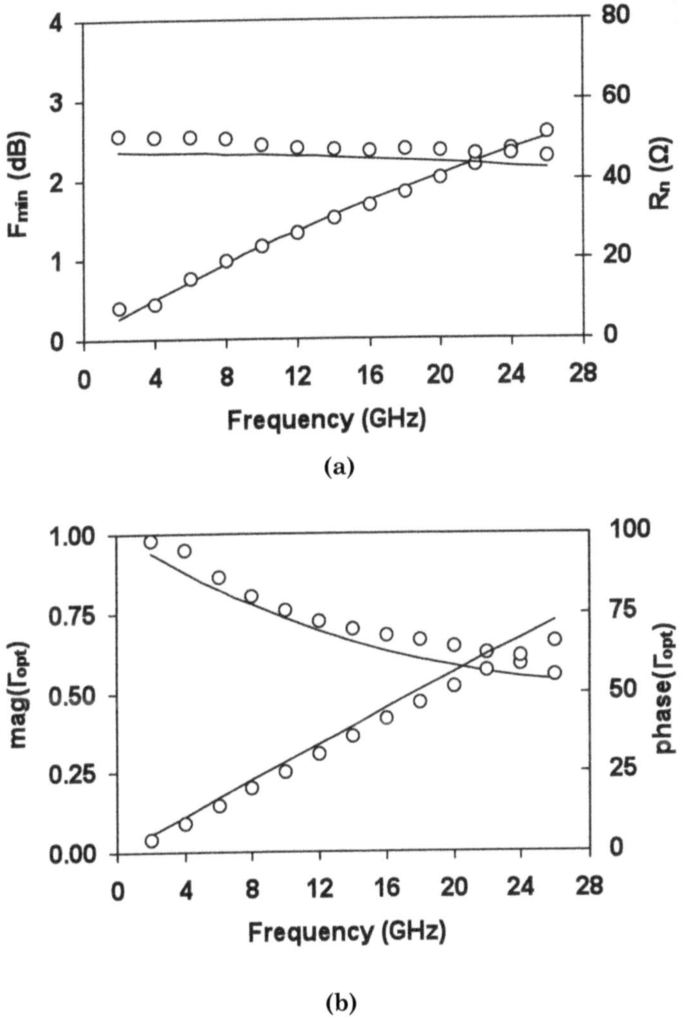

Figure 6.9 Comparison of measured and predicted noise parameters for 2 × 40μm PHEMT. oo = Experimental data, — = predicted data from 2 × 20μm PHEMT

where

$$A = F_{\min} - 2R_n G_{opt} \tag{6.69}$$

$$B = R_n \tag{6.70}$$

$$C = R_n(G_{opt}^2 + B_{opt}^2) \tag{6.71}$$

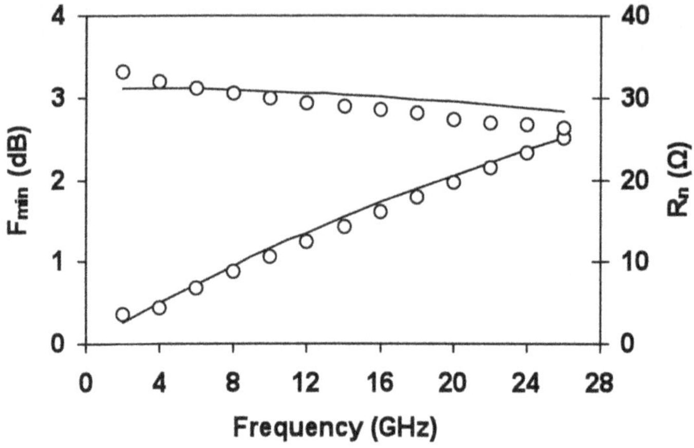

(a)

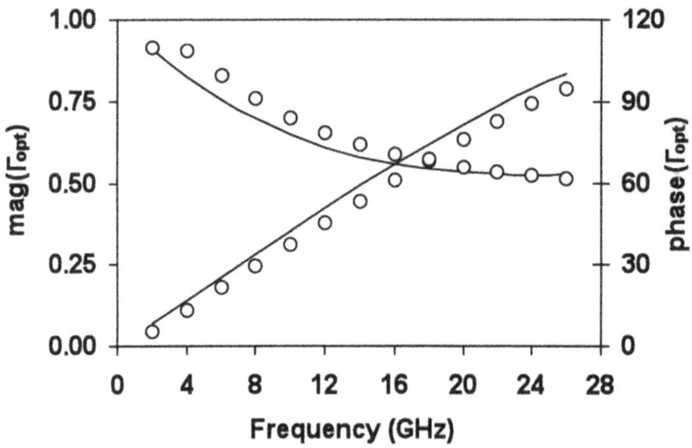

(b)

Figure 6.10 Comparison of measured and predicted noise parameters for 2 × 60 μm PHEMT. oo = experimental data, — = predicted data from 2 × 20 μm PHEMT

$$D = -2R_n B_{opt} \tag{6.72}$$

A least-squares fit of n observed noise figures to the plane of (6.67) is sought; therefore, the following error criterion is established [25]:

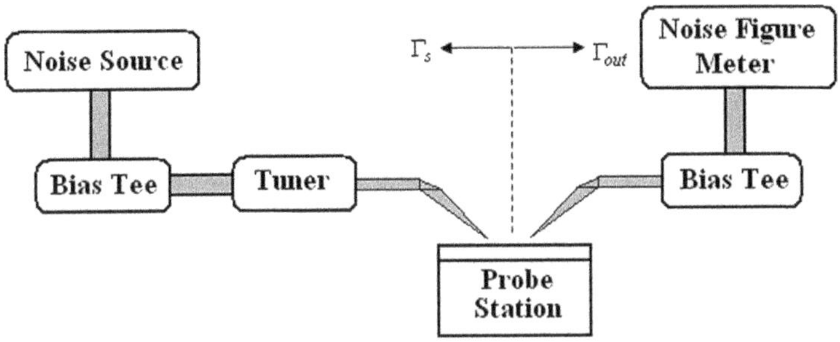

Figure 6.11 Tuner based noise parameters measurement system

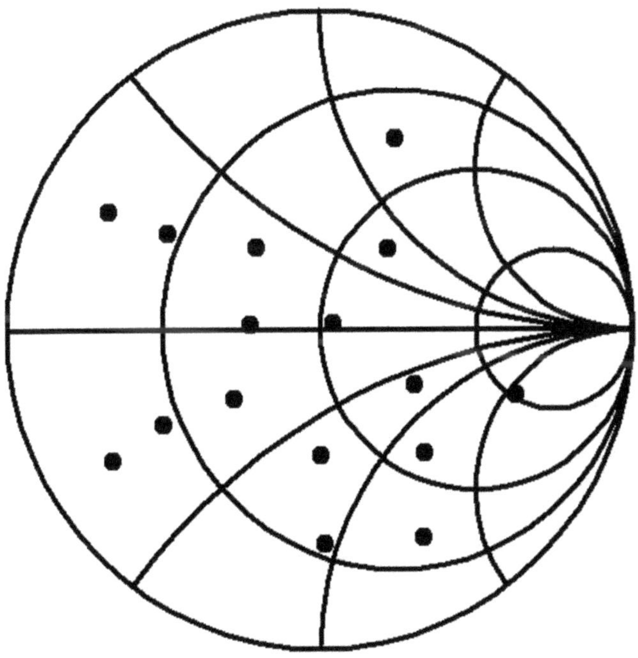

Figure 6.12 Typical source reflection coefficient distributed Smith chart

$$\varepsilon = \frac{1}{2}\sum_{i=1}^{n}\left[A + B\left(G_{si} + \frac{B_{si}^2}{G_{si}}\right) + \frac{C}{G_{si}} + \frac{DB_{si}}{G_{si}} - F_{si}\right]^2 \tag{6.73}$$

where G_{si} and B_{si} $(i = 1, 2 \ldots n)$ are the source conductances and susceptances, F_i is the corresponding noise figures, and W_i is the weighting factors to be used if certain data known to be less reliable than the average. Then

$$\frac{\partial \varepsilon}{\partial A} = \sum_{i=1}^{n} P = 0 \tag{6.74}$$

$$\frac{\partial \varepsilon}{\partial B} = \sum_{i=1}^{n} \left(G_i + \frac{B_i^2}{G_i} \right) P = 0 \tag{6.75}$$

$$\frac{\partial \varepsilon}{\partial C} = \sum_{i=1}^{n} \frac{1}{G_i} P = 0 \tag{6.76}$$

$$\frac{\partial \varepsilon}{\partial D} = \sum_{i=1}^{n} \frac{B_i}{G_i} P = 0 \tag{6.77}$$

$$P = A + B\left(G_i + \frac{B_i^2}{G_i} \right) + \frac{C}{G_i} + \frac{DB_i}{G_i} - F \tag{6.78}$$

By solving equations (6.73)–(6.76), the four unknown noise parameters can be obtained as follows:

$$F_{\min} = A + \sqrt{4BC - D^2} \tag{6.79}$$

$$R_n = B \tag{6.80}$$

$$G_{opt} = \frac{\sqrt{4BC - D^2}}{2B} \tag{6.81}$$

$$B_{opt} = -\frac{D}{2B} \tag{6.82}$$

Figure 6.13 shows a typical experimental setup for determining of the noise parameters. It is composed of a wafer-probe station, an automatic network analyzer (ANA) HP 8510C up to 40 GHz, a noise measurement system (NMS) up to 26.5 GHz, and an electronic broadband

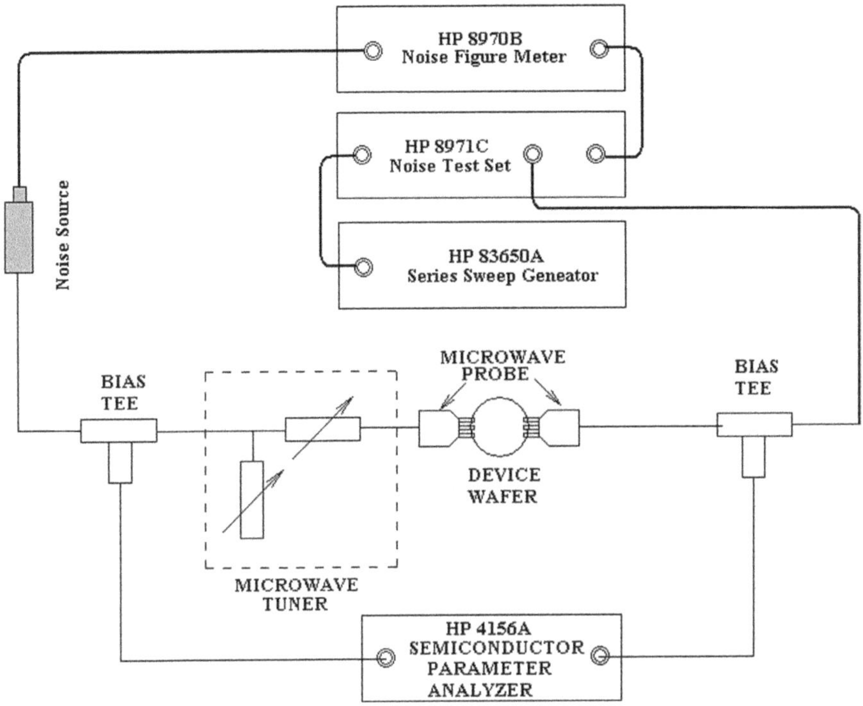

Figure 6.13 Commercial noise parameters measurement setup

noise source HP 346C up to 50 GHz. The NMS consists of a noise figure test set HP 8971C and the noise figure meter HP 8970B. The local oscillator is an HP 83650B synthesized sweeper up to 50 GHz. HP 8971C consists of a low noise preamplifier, a mixer and a Yttrium-Iron-Garnet (YIG) filter. The second stage (HP 8970B) is centered at the intermediate frequency IF = 450 MHz. DC bias was supplied by an Agilent 4156A.

6.3.2 Noise Figure Measurement Based Extraction Method

Although the tuner-based method gives accurate results, it is time-consuming and requires expensive automatic broadband microwave tuners that require complex calibration procedures. Some authors have proposed improved methods which are using the equivalent transistor noise model to provide additional information to reduce complexity in the measurement procedure. Other successful techniques are based on match source reflection 50 Ω measurements system (F_{50}) without an automatic tuner [30–33].

6.3.2.1 Analytical Expressions of Noise Parameters

The equivalent-circuit model of the noisy FET device is shown in Figure 6.14. By neglecting the influence of gate-drain feedback capacitance C_g of the intrinsic equivalent-circuit model of PHEMT, the calculation of the four parameters can be carried out analytically as follows [9–10]:

$$F_{\min}^{INT'} = 1 + K_B \omega \tag{6.83}$$

$$G_{opt}^{INT'} = K_C \omega \tag{6.84}$$

$$B_{opt}^{INT'} = K_D \omega \tag{6.85}$$

$$R_n^{INT'} = K_E \tag{6.86}$$

where $F_{\min}^{INT'}$, $G_{opt}^{INT'}$, $B_{opt}^{INT'}$, and $R_n^{INT'}$ represent intrinsic noise parameters without C_{gd}. K_B, K_C, K_D, and K_E are fitting factors, and ω is the angular frequency. From (6.82)–(6.85), it can be seen that the equivalent noise resistance $R_n^{INT'}$ is frequency independent, $G_{opt}^{INT'}$ and $B_{opt}^{INT'}$ are proportional to ω, and $F_{\min}^{INT'}$ is a linear function of ω.

However, the gate-drain feedback capacitance C_{gd} is very important because the noise correlation coefficient will be decreased when C_{gd} is neglected. The influence of gate-drain feedback capacitance C_{gd} can be

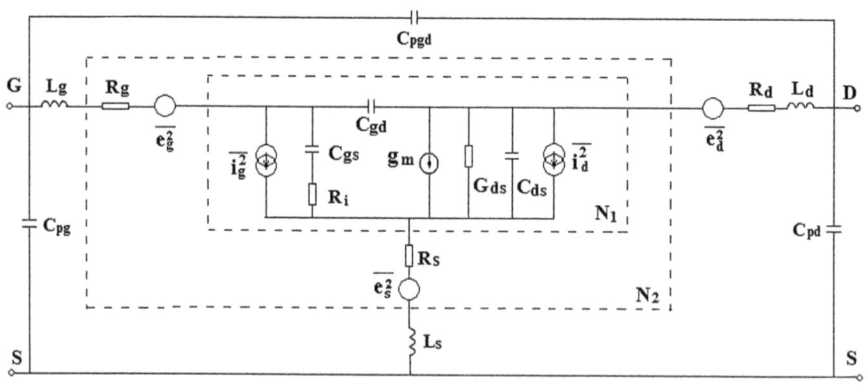

Figure 6.14 Noisy small-signal equivalent-circuit model of FET

considered using noise correlation matrix technique. The Y parameter noise matrix of the network N_{GD} consists only of gate-drain feedback capacitance C_{gd}; $C_Y^{GD} = 4kT\,\mathrm{Re}(Y^{GD})$ is a zero matrix, so the intrinsic matrix can be calculated as

$$C_Y^{INT} = C_Y^{GD} + C_Y^{INT'} = C_Y^{INT'} \tag{6.87}$$

where $C_Y^{INT'}$ is the Y noise matrix of network N_1 (except for C_{gd}). $C_Y^{INT'}$ can be determined by translating the chain noise matrix as follows

$$C_{Y11}^{INT'} = C_{A22}^{INT'} + C_{A11}^{INT'}\left|Y_{11}^{INT'}\right|^2 - (Y_{11}^{INT'})^* C_{A21}^{INT'} - Y_{11}^{INT'} C_{A12}^{INT'} \tag{6.88}$$

$$C_{Y12}^{INT'} = C_{A11}^{INT'}(Y_{21}^{INT'})^* Y_{11}^{INT'} - C_{A21}^{INT'}(Y_{21}^{INT'})^* \tag{6.89}$$

$$C_{Y21}^{INT'} = C_{A11}^{INT'}(Y_{11}^{INT'})^* Y_{21}^{INT'} - C_{A12}^{INT'} Y_{21}^{INT'} \tag{6.90}$$

$$C_{Y22}^{INT'} = C_{A11}^{INT'} \left| Y_{21}^{INT'} \right|^2 \tag{6.91}$$

where

$$C_{A11}^{INT'} = R_n^{INT'} \tag{6.92}$$

$$C_{A22}^{INT'} = R_n^{INT'} \left| Y_{opt}^{INT'} \right|^2 \tag{6.93}$$

$$C_{A21}^{INT'} = \frac{F_{\min}^{INT'} - 1}{2} - R_n^{INT'}(Y_{opt}^{INT'})^* \tag{6.94}$$

$$C_{A12}^{INT'} = \frac{F_{\min}^{INT'} - 1}{2} - R_n^{INT'} Y_{opt}^{INT'} \tag{6.95}$$

The intrinsic chain noise matrix is obtained by translating C_Y^{INT} as

$$C_{A11}^{INT} = \frac{C_{Y22}^{INT'}}{|Y_{21}^{INT}|^2} \tag{6.96}$$

$$C_{A21}^{INT} = \frac{Y_{11}^{INT}}{|Y_{21}^{INT}|^2} C_{Y22}^{INT'} - \frac{C_{Y12}^{INT'}}{(Y_{21}^{INT})^*} \tag{6.97}$$

$$C_{A12}^{INT} = \frac{(Y_{11}^{INT})^*}{|Y_{21}^{INT}|^2} C_{Y22}^{INT'} - \frac{C_{Y12}^{INT'}}{Y_{21}^{INT}} \tag{6.98}$$

$$C_{A22}^{INT} = C_{Y11}^{INT'} + \frac{|Y_{11}^{INT}|^2}{|Y_{21}^{INT}|^2} C_{Y22}^{INT'} - \frac{Y_{11}^{INT}}{Y_{21}^{INT}} C_{Y21}^{INT'} - \frac{(Y_{11}^{INT})^*}{(Y_{21}^{INT})^*} C_{Y12}^{INT'} \tag{6.99}$$

Because $j\omega C_{gd} << Y_{21}^{INT'} \approx g_m$, we have

$$Y_{21}^{INT} = Y_{21}^{INT'} - j\omega C_{gd} \approx Y_{21}^{INT'} \tag{6.100}$$

$$Y_{11}^{INT} = Y_{11}^{INT'} + j\omega C_{gd} \propto \omega \tag{6.101}$$

The four noise parameters of the intrinsic part are obtained from:

$$G_{opt}^{INT} = \sqrt{\frac{C_{A22}^{INT}}{C_{A11}^{INT}} - [\frac{\text{Im}(C_{A12}^{INT})}{C_{A11}^{INT}}]^2} = G_{opt}^{INT'} \tag{6.102}$$

$$B_{opt}^{INT} = \frac{\text{Im}(C_{A12}^{INT})}{C_{A11}^{INT}} = B_{opt}^{INT'} - \omega C_{gd} \propto \omega \tag{6.103}$$

$$F_{\min}^{INT} = 1 + 2(C_{A12}^{INT} + C_{A11}^{INT} Y_{opt}^{INT}) = F_{\min}^{INT'} \tag{6.104}$$

$$R_n^{INT} = C_{A11}^{INT} = R_n^{INT'} \tag{6.105}$$

where $F_{\min}^{INT}$, G_{opt}^{INT}, B_{opt}^{INT}, and R_n^{INT} represent intrinsic noise parameters.

From (6.102)–(6.105), it is shown that $F_{\min}^{INT}$, G_{opt}^{INT}, and R_n^{INT} remain invariant and B_{opt}^{INT} remains proportional to ω. Equations (6.83)–(6.86) are valid for the intrinsic device where the parasitic resistances due to R_g and R_s affect only the fitting factors. Therefore, the four frequency-dependent noise parameters become four frequency-independent constants; hence these can be obtained directly from the 50 Ω measurement system.

6.3.2.2 Noise Parameter Extraction Procedure

Once the parasitic elements (C_{pg}, C_{pd}, C_{pdg}, L_g, L_d, L_s, R_d) are known, extracting of the four unknown noise parameters can be carried out using the the following procedure:

1. Measurement of the S parameters of the PHEMT.
2. Transformation of the S parameters to admittance parameters and subtraction of pad capacitances (C_{pg}, C_{pd}, C_{pdg}).
3. Transformation of the Y parameters to impedance parameters and subtraction of the series inductances (L_g, L_d, L_s) and series resistance R_d that correspond to the sub-network N_2.
4. Measurement of the noise figure (F_m) of the system that consists of an input bias network, an output bias network, and the active device.
5. Measurement of the S parameters of the input and output bias network. The bias networks include coaxial switches, bias tees, probe tips, and cables between them (Figure 6.15). Because the two ports of the bias network are different types, one-port is coaxial and another is coplanar, it is difficult to measure S-parameters using vector network analyzer (VNA) on wafer measurement directly.

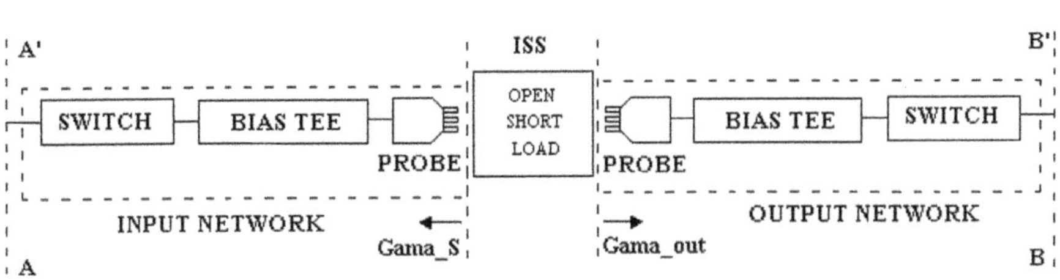

Figure 6.15 Block diagram of input and output network measurement method

Therefore we use a one-port measurement method to determine the S-parameters of the bias networks.

For a two-port network, the input reflection coefficient can be expressed as a function of the load reflection coefficient:

$$S_{in} = S_{11} + \frac{S_{12}S_{21}\Gamma_L}{1 - S_{22}\Gamma_L} \tag{6.106}$$

where S_{in} and Γ_L are the input and load reflection coefficient, respectively.

First, the one-port coaxial open short load (OSL) calibration is performed at plane A-A' (input port measurement) or B-B' (output port measurement); then, the S parameters are measured when the probe tip is connected to the open, short, load standards on the impedance standard substrate ISS (Cascade-Microtech), corresponding to Γ_L equals 1, –1, and 0 respectively. The S parameters of the bias network can be calculated directly as follows:

$$S_{11} = S_{11}^{LOAD} \tag{6.107}$$

$$S_{22} = \frac{S_{11}^{OPEN} + S_{11}^{SHORT} - 2S_{11}}{S_{11}^{OPEN} - S_{11}^{SHORT}} \tag{6.108}$$

$$S_{12} = S_{21} = \sqrt{(S_{11}^{OPEN} - S_{11})(1 - S_{22})} \tag{6.109}$$

where S_{11}^{OPEN}, S_{11}^{SHORT}, and S_{11}^{LOAD} are measured input reflection coefficients of the probes if the probe tip is terminated by the open, short, and load standard, respectively.

The source reflection coefficient and the corresponding admittance are shown in Figure 6.16. The output reflection coefficient and admittance are shown in Figure 6.17. It can be noticed that the system is not an accurate 50 Ω system, the real parts of Y_s and Y_{out} (G_s, G_{out}) have small deviations from the 50 Ω system ($G_s = G_{out} =$ 20 mS), the imaginary parts of Y_s and Y_{out} (B_s, B_{out}) with small deviations from the 50 Ω system ($B_s = B_{out} = 0$). It also can be observed that Y_s is close to a 50 Ω system as Y_{out}, because a high performance bias tee has been used at the input. Actually, the perfect 50 Ω system is not available because of the losses of input and output network cannot be neglected.

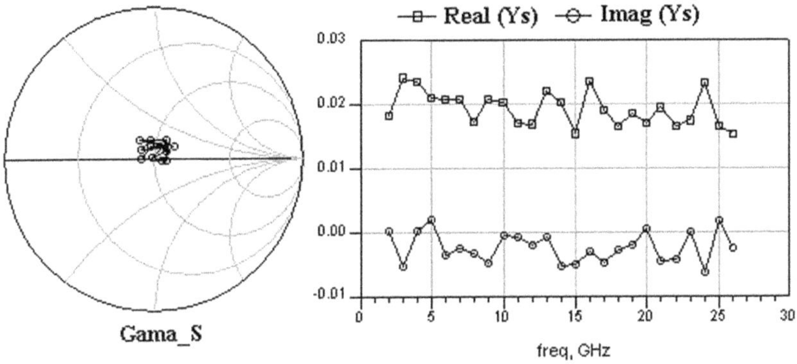

Figure 6.16 Source reflection coefficient and admittance

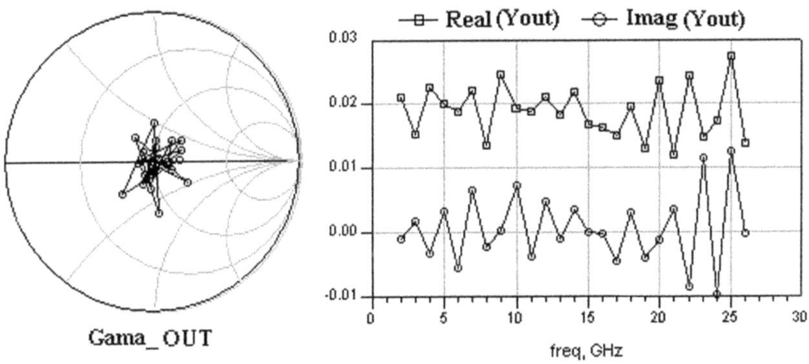

Figure 6.17 Output reflection coefficient and admittance

The source admittance and the output admittance are:

$Y_s = (20 \pm 5) - j(0 \pm 5)$ mS

$Y_{out} = (20 \pm 8) - j(0 \pm 11)$ mS

6. Calculate noise figure of device under test (DUT)

It is known that the noise figure of a cascade of noisy two-port networks is given by

$$F_m = F_{IN} + \frac{F_D - 1}{G_{IN}} + \frac{F_{OUT}}{G_{IN} G_D} \tag{6.110}$$

In this case, F_{in} and G_{in} are noise figure and available gain of the input bias network, respectively, F_{out} is the noise figure of the output bias network, and F_d and G_d are noise figure and available gain of the DUT, respectively.

Because input and output bias networks are passive networks, we get

$$F_{IN} = 1/G_{IN}\,,\quad F_{OUT} = 1/G_{OUT} \tag{6.111}$$

$$F_D = G_{IN}F_m - \frac{1-G_{OUT}}{G_{IN}G_D} \tag{6.112}$$

The available gains G_{in}, G_{out}, and G_d are determined by S parameters that correspond to the input bias network, output bias network, and active device:

$$G_{IN} = \frac{|S_{21}^{IN}|^2}{1-|S_{22}^{IN}|^2} \tag{6.113}$$

$$G_D = \frac{|S_{21}|^2 (1-|\Gamma_S|^2)}{|1-S_{11}\Gamma_S|^2 (1-|S_{22}^{'}|^2)} \tag{6.114}$$

where

$$S_{22}^{'} = S_{22} + \frac{S_{12}S_{21}\Gamma_S}{1-S_{11}\Gamma_{in}}$$

The determination of G_{out} can be carried out similar to G_{in}.

7. Setting initial values of four noise parameters K_B, K_C, K_D, and K_E ($F_{\min}^{INT}$, G_{opt}^{INT}, B_{opt}^{INT}, and R_n^{INT}) and calculating of chain noise matrix C_A^D using (6.83)–(6.86).

As we know, the accuracy of the numerical optimization methods that minimize the difference between measured and modeled data versus frequency can vary depending on the optimization method and the starting values. Therefore, the method of setting initial values of K_B, K_C, K_D, and K_E is very important.

A direct extraction method for determining the equivalent noise resistance R_n and the magnitude of the optimum generator admittance $|Y_{opt}|$ based on 50 Ω measurement system is proposed by [31]. Based on this method, an improved method for determining the initial values of all four noise parameters can be used in this chapter.

In the case of a 50 Ω generator impedance ($Y_s = G_o = 20$ mS), the noise figure can be written as

$$F_{50} = 1 + R_n G_0 + \frac{R_n}{G_0}(2G_0 G_{cor} + |Y_{opt}|^2) \tag{6.115}$$

where G_{cor} is the real part of the correlation admittance. G_{cor} can be approximated by G_{11} (real part of Y_{11}). Because G_{11} is close to zero, G_{cor} can be neglected, and we get

$$F_{50} \approx 1 + R_n G_0 + \frac{R_n}{G_0}|Y_{opt}|^2 \tag{6.116}$$

Since R_n is nearly frequency independent whereas $|Y_{opt}|$ varies proportional to ω^2, the plot of F_{50} versus ω^2 is linear, and the value at $\omega = 0$ is $(1 + R_n G_0)$. Thus, R_n can be easily deduced from the F_{50} extrapolation for $\omega = 0$:

$$R_n \approx \frac{(F_{50}^{\omega=0} - 1)}{G_0} \tag{6.117}$$

The slope of F_{50} versus ω^2 provides the magnitude of the optimum generator admittance $|Y_{opt}|$:

$$|Y_{opt}| \approx \frac{\omega G_0}{2\pi R_n}\sqrt{slope(F_{50})} \tag{6.118}$$

Because the real part G_{opt} and the imaginary part B_{opt} have the same order of magnitude, we can set $G_{opt} \approx -B_{opt}$. After R_n and G_{opt} are determined, the initial value of minimum noise figure F_{min} can be written as:

$$F_{\min} \approx 1 + 2R_n G_{opt} \tag{6.119}$$

The four noise-fitting parameters K_B, K_C, K_D and K_E are determined using the (6.83)–(6.86). Figure 6.18 shows the experimental evaluations of F_{50} versus the square of the frequency for three different PHEMTs of sizes (2 × 20 μm, 2 × 40 μm and 2 × 60 μm). The ripple in the graphs is caused by the non perfect 50 Ω source impedance. It grows up with increasing gatewidth due to a decreasing multiplication factor R_n/G_s.

The initial values for the four noise parameters K_B, K_C, K_D, and K_E are summarized in Table 6.2. They have been determined using the aforementioned method (frequency unit is GHz). Then the chain noise matrix C_A^D is determined using (6.92)~(6.95).

8. Transformation of the chain noise correlation matrix C_A^D to the impedance noise correlation matrix, and addition of series inductances (L_g, L_d, L_s) and series resistance R_d.

Table 6.2 Initial Values of Four Noise Parameters for Three Differently Sized PHEMTs

Device	K_B	K_C	K_D	K_E
2 × 20 μm	1.3E–2	5.6E–5	–5.6E–5	120
2 × 40 μm	1.0E–2	9.3E–5	–9.3E–5	55
2 × 60 μm	8E–3	1.33E–4	–1.33E–4	30

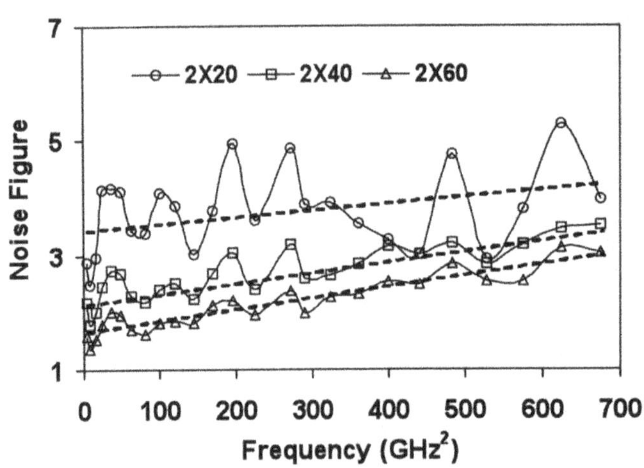

Figure 6.18 Evaluation of the noise figures F_{50} of PHEMTs (2 × 20 μm, × 40 μm, and 2 × 60 μm) versus the square of the frequency (bias condition: V_{ds} = 2 V, V_{gs} = 0 V; I_{ds} = 6 mA, 12 mA, 18 mA).

9. Transformation of the impedance noise correlation matrix to the admittance noise correlation matrix, and addition of pad capacitances (C_{pg}, C_{pd}, C_{pdg}).
10. Transformation of the admittance noise correlation matrix to the chain noise correlation matrix and calculation of the noise figure of DUT:

$$F_{MODEL} = 1 + 2[C_{A12} + C_{A11}(\sqrt{\frac{C_{A22}}{C_{A11}} - [\frac{\mathrm{Im}(C_{A12})}{C_{A11}}]^2} + j\frac{\mathrm{Im}(C_{A12})}{C_{A11}}) + (\frac{C_{A22}}{C_{A11}} + G_S^2 - 2G_S\sqrt{\frac{C_{A22}}{C_{A11}} - [\frac{\mathrm{Im}(C_{A12})}{C_{A11}}]^2})$$

(6.120)

11. Calculation of the error criteria as a function of noise figure of DUT:

$$\varepsilon = \frac{1}{N-1}\sum_{i=0}^{N-1} | F_{MODEL}(f_i) - F_{MEASURE}(f_i) |^2$$ (6.121)

where N is the number of considered frequency points, $F_{MEASURE}(f_i)$ is the measured noise figure at the frequency f_i and $F_{MODEL}(f_i)$ is the calculated corresponding noise figure derived from extracted values of the model parameters.

12. If $\varepsilon > \varepsilon_0$, the values of F_{min}^{INT}, G_{opt}^{INT}, B_{opt}^{INT}, and R_n^{INT} are updated to reduce ε using the least-squares method.

6.3.2.3 Measurement System Setup

Figure 6.19 shows the experimental setup. DC bias was supplied by an Agilent 4156A. All measurements were carried out on-wafer using Cascade Microtech's Air-Coplanar Probes ACP50-GSG-100. The wafer probes were calibrated using Line-Reflect-Match (LRM) calibration method for S parameter measurement. The noise parameter measurement method proposed here has been tested on-wafer up to 26 GHz using AlGaAs/InGaAs/GaAs PHEMTs with 0.25 μm mushroom gates, grown and fabricated using NTU's in-house developed process technology.

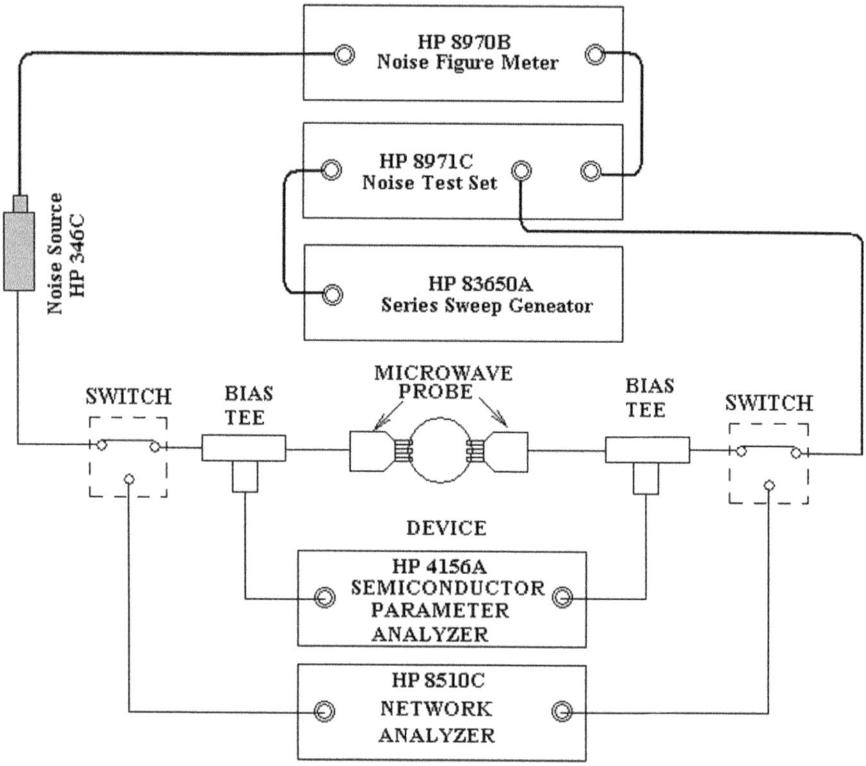

Figure 6.19 Microwave transistor S parameter and noise measurement setup

The three inductances L_g, L_d, and L_s are dependent on the length of the device feed lines and are not scalable. However, the drain resistance R_d is inversely proportional to the size of the device. Table 6.3 shows the extracted results of the four parasitic elements. When parasitic elements, pad capacitances, series inductances, and drain resistance are determined, the noise parameters can be obtained by using the aforementioned method.

Table 6.3 Extracted Parasitic Inductance and R_d Values

Elements	2×20 μm	2×40 μm	2×60μ
L_g (pH)	95	85	75
L_d (pH)	80	70	60
L_s (pH)	10	10	10
R_d Ω	10	5	3.5

Figure 6.20 compares the measured and modeled S parameters for the 2 × 20 μm, 2 × 40 μm, 2 × 60 μm and 2 × 100 μm PHEMT in the frequency range of 50 MHz to 40 GHz. The modeled S parameters agree very well with the measured ones. Figure 6.21 compares the measured and modeled noise figure F_{50} for the (a) 2 × 20 μm, (b) 2 × 40 μm, and (c) 2 × 60 μm PHEMT in the frequency range of 2 GHz to 26 GHz under the bias condition V_{gs} = 0 V, V_{ds} = 2 V (I_{ds} = 6 mA, 12 mA, 18 mA). An adequate comparison for the noise figure of the system is given in Figure 6.21 as well. The modeled noise figures agree well with the measured ones based on the 50 Ω measurement system.

The transistor noise parameters determined from F_{50} by using the new method are compared with the noise parameters measured with the commercial ATN system NP5 (commercial noise parameters measurement system) based on a broadband tuner. Figures 6.22 to 6.24 show these comparisons as a function of frequency. A good agreement between measured and modeled results can be indicated, and the validity of the method is confirmed.

Table 6.4 Optimum Values of Noise Parameters for Three Differently Sized PHEMTs

Elements	K_B	K_C	K_D	K_E
2 × 20 μm	7.5E–3	2.6E–5	–4.3E–5	110
2 × 40 μm	6.8E–3	4.8E–5	–7.0E–5	60
2 × 60 μm	5.4E–2	7.0E–5	–1.0E–4	33

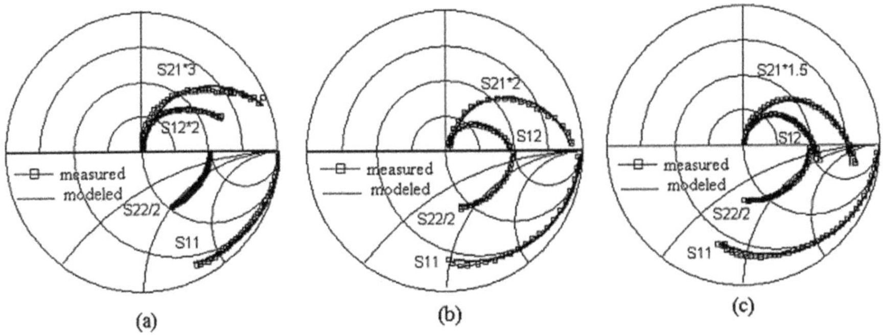

Figure 6.20 Comparison of modeled and measured S-parameters for (a) 2 × 20 μm, (b) 2 × 40 μm, (c) 2 × 60 μm, pinch-off PHEMT (bias condition: V_{gs} = –3 V, V_{ds} = 0 V). The squares indicate the measured values, and the lines the

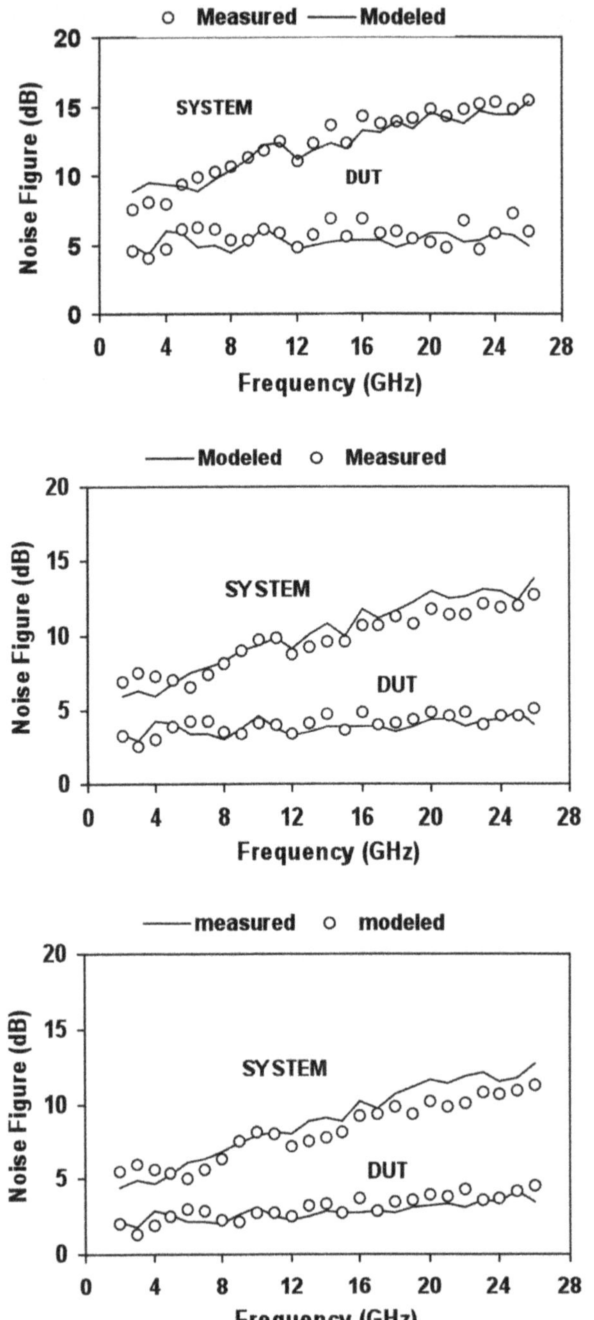

(a)

(b)

(c)

Figure 6.21 Comparison of measured and modeled noise figures F_{50} for (a) 2 × 20 μm, (b) 2 × 40 μm, (c) 2 × 60 μm PHEMTs and the system, respectively (bias condition: V_{gs} = 0 V, V_{ds} = 2 V; I_{ds} = 6 mA, 12 mA, 18 mA)

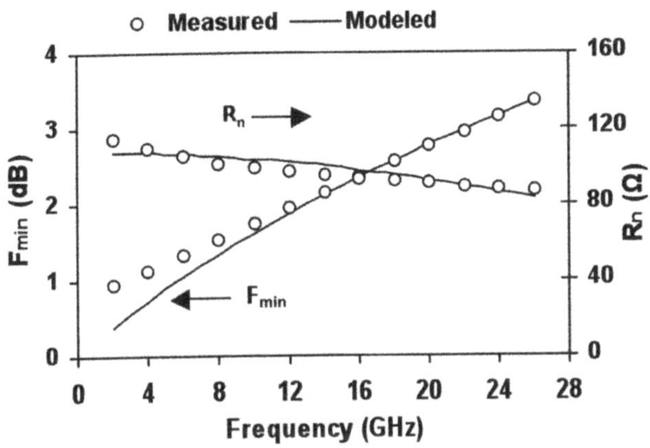

(a)

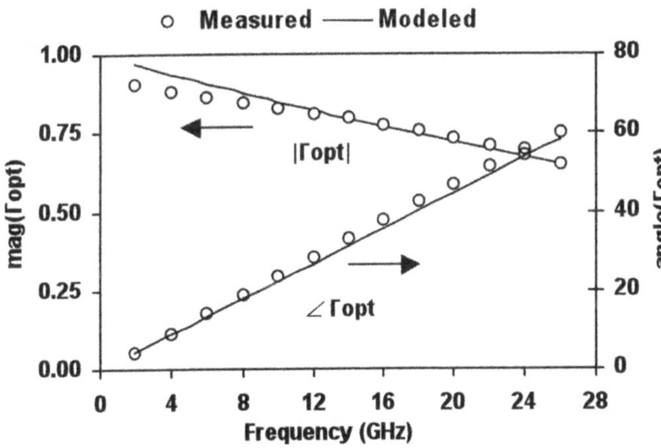

(b)

Figure 6.22 Comparison of noise parameters directly measured using commercial ATN system (o) determined with the new technique, (–) for the 2 × 20 μm PHEMT (bias condition: V_{gs} = 0 V, V_{ds} = 2 V, I_{ds} = 6 mA)

Table 6.4 summarizes the extracted optimum four fitting parameters that correspond to transistor noise parameters. A comparison with Table 6.3 proves that initial values and optimum values match very well for noise resistance K_E (R_n). The dispersions between the initial values of K_B, K_C, and K_D (F_{min}, G_{opt}, and B_{opt}) and optimum values of the large-sized devices (2 × 40 μm and 2 × 60 μm) are smaller than smaller-sized devices for the following reasons:

1. The system is not an accurate 50 Ω system for the device under test.
2. The large-sized device satisfies the assumption ($G_{opt} >> G_{cor}$) better.

After carefully examining the measured data, the scaling formulae are determined to be as follows:

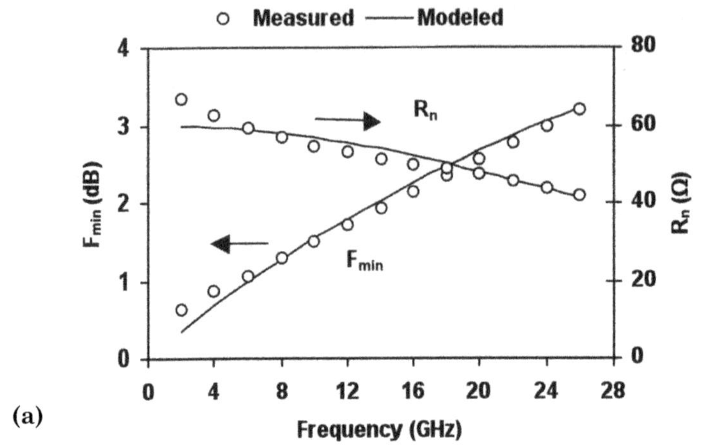

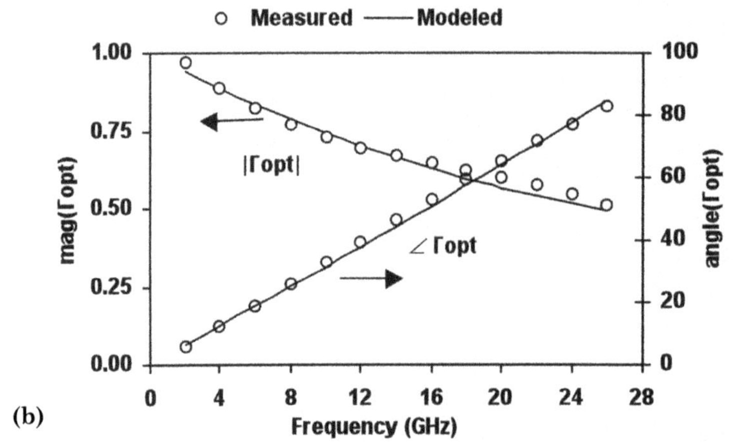

Figure 6.23 Comparison of noise parameters directly measured using commercial ATN system (o) and determined with the new technique, (-) for the 2 × 40 μm PHEMT (bias condition: V_{gs} = 0 V, V_{ds} = 2 V, I_{ds} = 12 mA)

$$\frac{K_C^I}{K_C^{II}} = \frac{W_g^I}{W_g^{II}} \tag{6.122}$$

$$\frac{K_D^I}{K_D^{II}} = \frac{W_g^I}{W_g^{II}} \tag{6.123}$$

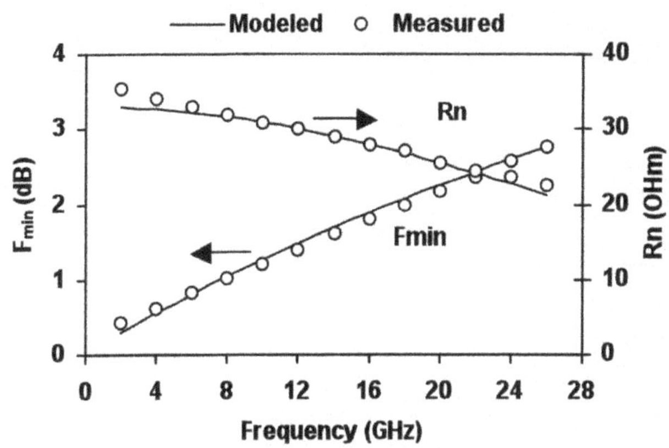

(a)

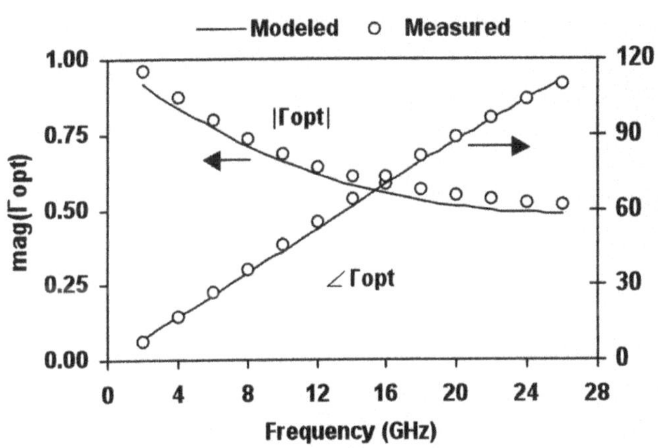

(b)

Figure 6.24 Comparison of noise parameters directly measured using commercial ATN system (o) and determined with the new technique, (-) for the 2 × 60 μm PHEMT. (bias condition: V_{gs} = 0 V, V_{ds} = 2 V, I_{ds} = 18 mA)

$$\frac{K_E^I}{K_E^{II}} = \frac{W^I}{W_g^{II}} \tag{6.124}$$

that is,

$$R_n \propto \frac{1}{W}$$

$$G_{opt} \propto W$$

$$B_{opt} \propto W$$

Notice that K_B is not scalable, which means that F_{min} is not dependent on the width. In Figure 6.22(a), Figure 6.23(a) and Figure 6.24(a), we can find that the modeled optimum noise figure F_{min} agrees well with the measured F_{min}, especially at high frequencies. However, there are small differences between measured and modeled data at low frequencies; the differences increase if gatewidth decreases. The reason for that lies in a gate-leakage current.

It is well-established that a short gate on a thin, heavily-doped layer is necessary to obtain an excellent microwave performance. As a result, problems with gate-leakage current and additional noise become significant. For our device, the leakage currents are in the same order for all devices (roughly 1 μA) independent on the gatewidth. This means that the influence of the gate current on noise behavior is larger for small devices than for large ones. Because the extracted noise parameters of our 50 Ω method are based primarily on S-parameters and a model with two noise sources, influence of a gate current is not taken into account in contrast to a complete noise parameter characterization. However, the influence of a gate current can be considered by introducing a new empirical formula:

$$F_{\min}^{INT} = K_A + K_B\omega$$

Table 6.5 Optimum Values of Noise Fitting Parameters for Three Differently Sized PHEMTs after Model Extension

Elements	K_A	K_B	K_D	K_D	K_E
2 × 20 μm	1.14	6.5E–3	2.6E–5	–4.3E–5	110
2 × 40 μm	1.05	6.3E–3	4.8E–4	–7.0E–5	60
2 × 60 μm	1.0	5.4E–2	7.0E–4	–1.0E–4	33

The extracted optimum five fitting parameters after model extension are summarized in Table 6.5. We can find that the parameters have small changes for the 2 × 20 μm and 2 × 40 μm devices, while the parameters remain unvaried for the 2 × 60 μm device, Fig. 6.25 shows that the application of the proposed formula leads to a good agreement

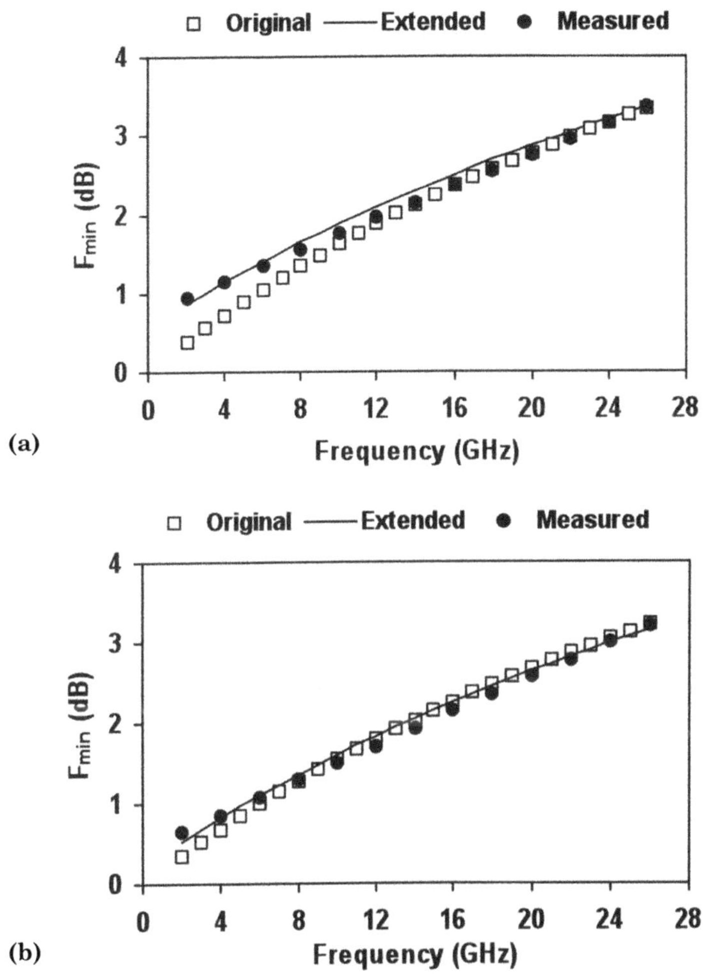

Figure 6.25 Optimum noise figure versus frequency. Comparison of original model, extended model, and measurements for (a) 2 × 20 μm and (b) 2 × 40 μm devices (bias condition: V_{gs} = 0 V, V_{ds} = 2 V; I_{ds} = 6 mA, 12 mA)

between modeled and measured optimum noise figures in the entire frequency range for 2 × 20 μm and 2 × 40 μm devices, too.

6.4 Relationships among CS, CG, and CD FETs

When used in amplifiers, MES, FETs and HEMTs are almost exclusively operated in a common-source (CS) configuration. However, a configuration in the CS mode usually produces more noise than its common-gate (CG) counterpart. This is primarily due to the pure matching ([34, 35]) between optimum and real source impedance with respect to noise. The situation is much more relaxed in CG configuration because the input impedance is $1/g_m$ and not nearly pure capacitive as in CS. Therefore, a CG configuration has the advantage of being able to match the input to 50 Ω and to the optimum (noise) source impedance at the same time, also especially in broadband applications. As a consequence it is much easier to achieve noise values close to F_{min} in a CG than in a CS configuration. In general this leads to lower noise values in CG than in CS configuration although F_{min} values of both configurations are close together, which we show in this chapter. Hence, a CG configuration is generally more suitable for optical and microwave broadband communication applications. The common-drain (CD) configuration has been widely used in the design of low thermal resistance oscillator and low distortion variable-gain amplifier [36, 37].

The complete characterization and modeling of the transistor (CS, CG and CD) in terms of noise and scattering parameters is necessary for CAD of microwave circuits. The S- and noise parameters for each configuration can be obtained by measuring test patterns of CS, CG and CD configurations. However, this method requires two special test structures (CG and CD) for each device size on the wafer, and the non-uniformity across the wafer has to be ignored. A complex mathematical method for determining the noise parameters of active devices based on two- and three-port relationships is proposed by [38, 39]. This method is complicated and time-consuming due to the full characterization of the three-port network, which has to be obtained first from the two-port network.

A simple but efficient transformation technique for microwave FET devices have developed [40]. This technique is based on the combination of equivalent-circuit model, and the conventional two-port network signal and noise correlation matrix techniques. The signal and noise parameters of the CG and CD configuration can be obtained directly by using a simple set of formulas from CS signal and noise parameters. All the relationships provide a bi-directional bridge for the transformation between CS, CG, and CD FETs, respectively.

6.4.1 Signal Parameter Relationships

The CS, CG, and CD configurations for the PHEMT devices considered in our studies are shown in Figure 6.26(a), (b), and (c), respectively,

(a)
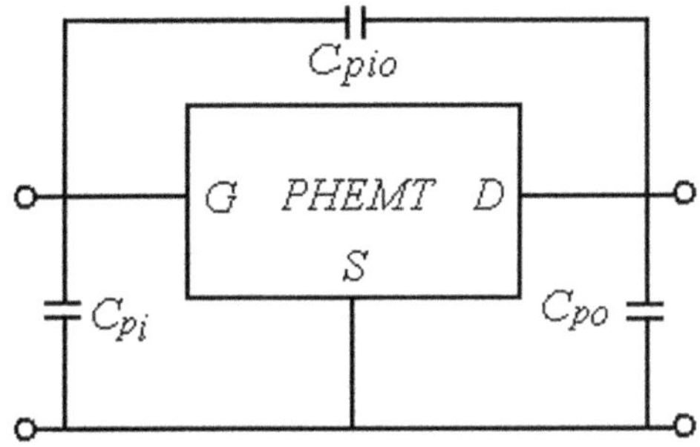

(b)
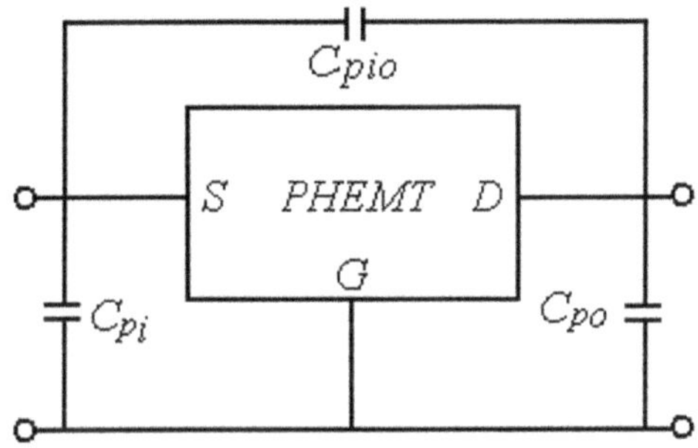

(c)
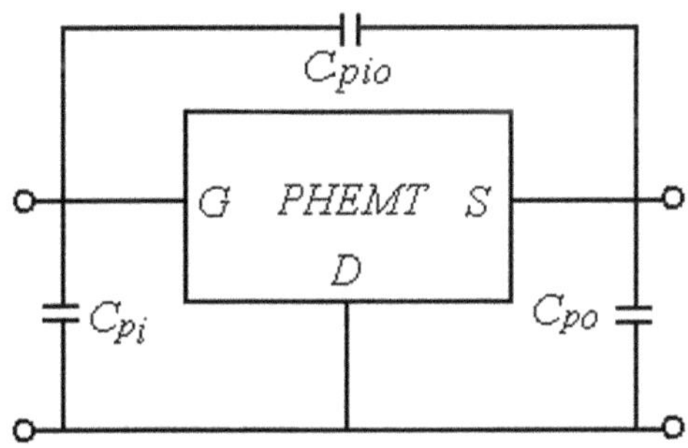

Figure 6.26 CS, CG, and CD configurations of PHEMT

where C_{pi}, C_{po}, and C_{pio} represent the input, output, and feedback pad capacitances. The Z-parameter expressions for the CS configuration can be expressed as follows:

$$Z_{11}^{CS} = Z_G + Z_S + \frac{Y_{22}}{\Delta Y} \tag{6.125}$$

$$Z_{12}^{CS} = Z_S - \frac{Y_{12}}{\Delta Y} \tag{6.126}$$

$$Z_{21}^{CS} = Z_S - \frac{Y_{21}}{\Delta Y} \tag{6.127}$$

$$Z_{22}^{CS} = Z_D + Z_S + \frac{Y_{11}}{\Delta Y} \tag{6.128}$$

where Y_{11}, Y_{12}, Y_{21}, and Y_{22} are the Y-parameters of the intrinsic part of the CS device:

$$Y_{11} = \frac{j\omega C_{gs}}{1 + j\omega R_i C_{gs}} + j\omega C_{gd} \tag{6.129}$$

$$Y_{12} = -j\omega C_{gd} \tag{6.130}$$

$$Y_{21} = \frac{g_m e^{-j\omega\tau}}{1 + j\omega R_i C_{gs}} - j\omega C_{gd} \tag{6.131}$$

$$Y_{22} = g_{ds} + j\omega(C_{gd} + C_{ds}) \tag{6.132}$$

with $\Delta Y = Y_{11}Y_{22} - Y_{12}Y_{21}$.

Similar to (6.125)–(6.128), the corresponding Z parameters for CG and CD configurations are as follows:

$$Z_{11}^{CG} = Z_G + Z_S + \frac{Y_{22}}{\Delta Y} \tag{6.133}$$

$$Z_{12}^{CG} = Z_G + \frac{Y_{12} + Y_{22}}{\Delta Y} \tag{6.134}$$

$$Z_{21}^{CG} = Z_G + \frac{Y_{21} + Y_{22}}{\Delta Y} \tag{6.135}$$

$$Z_{22}^{CG} = Z_D + Z_G + \frac{Y_{11} + Y_{12} + Y_{21} + Y_{22}}{\Delta Y} \tag{6.136}$$

$$Z_{11}^{CD} = Z_G + Z_D + \frac{Y_{11} + Y_{12} + Y_{21} + Y_{22}}{\Delta Y} \tag{6.137}$$

$$Z_{12}^{CD} = Z_D + \frac{Y_{12} + Y_{11}}{\Delta Y} \tag{6.138}$$

$$Z_{21}^{CD} = Z_D + \frac{Y_{21} + Y_{11}}{\Delta Y} \tag{6.139}$$

$$Z_{22}^{CG} = Z_D + Z_S + \frac{Y_{11}}{\Delta Y} \tag{6.140}$$

Comparing the Z-parameters of the CS, CG, and CD configurations, we can get the Z-parameter relationships as shown in Tables 6.6 and 6.7. The *ABCD*- and S-parameter relationships between the CS, CG, and CD configurations are also given in Tables 6.6 and 6.7 by using matrix conversion technique.

The extracted values of the bias-independent extrinsic small-signal elements are summarized in Table 6.8. Once the values of the parasitic elements are known, all bias-dependent elements can be easily determined by using the direct extraction technique. The corresponding intrinsic parameters g_m, τ, C_{gs}, C_{gd}, C_{ds}, R_{bi}, and g_{ds} are also summarized in Table 6.8 for a constant drain-source voltage V_{ds} = 2.0 V and gate-drain voltage V_{gs} = 0.0 V, respectively.

The measured S- and noise parameters without pad capacitances are obtained after de-embedding the pad capacitances.

Figure 6.27 compares the measured and modeled S-parameters for a 2 × 40 μm AlGaAs/InGaAs/GaAs PHEMT in the frequency range 1 to 26 GHz. An excellent agreement over the whole frequency range is obtained for CS configuration. To illustrate the efficiency of the S-parameter transformation formulas we compare the measured and predicted results for CG and CD configurations in Figures 6.28 and 6.29. Good agreement can be observed between calculated data from equivalent-circuit model and data predicted by the transformation formulas.

Table 6.6 *Z*-, *ABCD*- and *S*-parameter relationships between CS and CG configurations

	Common-source	Common-gate
Z	$Z_{11}^{CS} = Z_{11}^{CG}$ $Z_{11}^{CS} = Z_{11}^{CG}$ $Z_{21}^{CS} = Z_{11}^{CG} - Z_{21}^{CG}$ $Z_{22}^{CS} = Z_{11}^{CG} + Z_{22}^{CG} - Z_{21}^{CG} - Z_{12}^{CG}$	$Z_{11}^{CG} = Z_{11}^{CS}$ $Z_{12}^{CG} = Z_{11}^{CS} - Z_{12}^{CS}$ $Z_{21}^{CG} = Z_{11}^{CS} - Z_{21}^{CS}$ $Z_{22}^{CG} = Z_{11}^{CS} + Z_{22}^{CS} - Z_{21}^{CS} - Z_{12}^{CS}$
ABCD	$A^{CS} = A^{CG} / (A^{CG} - 1)$ $B^{CS} = B^{CG} / (A^{CG} - 1)$ $C^{CS} = C^{CG} / (A^{CG} - 1)$ $D^{CS} = 1 - D^{CG} + B^{CG}C^{CG} / (A^{CG} - 1)$	$A^{CG} = A^{CS} / (A^{CS} - 1)$ $B^{CG} = B^{CS} / (A^{CS} - 1)$ $C^{CG} = C^{CS} / (A^{CS} - 1)$ $D^{CG} = 1 - D^{CS} + B^{CS}C^{CS} / (A^{CS} - 1)$
S	$S_{11}^{CS} = \frac{1}{K}[3S_{11}^{CG} + 2(S_{12}^{CG} + S_{12}^{CG})$ $+S_{22}^{CG} + \Delta S^{CG} - 1]$ $S_{12}^{CS} = \frac{2}{K}[1 + S_{11}^{CG} - S_{22}^{CG} - 2S_{12}^{CG} - \Delta S^{CG}]$ $S_{21}^{CS} = \frac{2}{K}[1 + S_{11}^{CG} - S_{22}^{CG} - 2S_{21}^{CG} - \Delta S^{CG}]$ $S_{11}^{CS} = \frac{1}{K}[3S_{11}^{CG} + 2(S_{12}^{CG} + S_{12}^{CG})$ $+S_{22}^{CG} + \Delta S^{CG} - 1]$	$S_{11}^{CG} = \frac{1}{k}[3S_{11}^{CS} + 2(S_{12}^{CS} + S_{12}^{CS})$ $+S_{22}^{CS} + \Delta S^{CS} - 1]$ $S_{12}^{CG} = \frac{2}{k}[1 + S_{11}^{CS} - S_{22}^{CS} - 2S_{12}^{CS} - \Delta S^{CS}]$ $S_{21}^{CG} = \frac{2}{k}[1 + S_{11}^{CS} - S_{22}^{CS} - 2S_{21}^{CS} - \Delta S^{CS}]$ $S_{22}^{CG} = \frac{1}{k}[3S_{22}^{CS} - 2(S_{12}^{CS} + S_{12}^{CS})$ $+S_{11}^{CS} - \Delta S^{CS} + 1]$
	$\Delta S^{CG} = S_{11}^{CG}S_{22}^{CG} - S_{12}^{CG}S_{21}^{CG}$ $K = 5 - \Delta S^{CG} + S_{11}^{CG} - S_{22}^{CG}$ $-2(S_{12}^{CG} + S_{21}^{CG})$	$\Delta S^{CG} = S_{11}^{CS}S_{22}^{CS} - S_{12}^{CS}S_{21}^{CS}$ $k = 5 - \Delta S^{CS} + S_{11}^{CS} - S_{22}^{CS}$ $-2(S_{12}^{CS} + S_{21}^{CS})$

Table 6.7 Z-, A-, and S-parameter relationships between CS and CD configurations

	Common-source	Common-drain
Z	$Z_{11}^{CS} = Z_{11}^{CD} + Z_{22}^{CD} - Z_{21}^{CD} - Z_{12}^{CD}$ $Z_{12}^{CS} = Z_{22}^{CD} - Z_{12}^{CD}$ $Z_{21}^{CS} = Z_{22}^{CD} - Z_{21}^{CD}$ $Z_{22}^{CS} = Z_{22}^{CD}$	$Z_{11}^{CD} = Z_{11}^{CS} + Z_{22}^{CS} - Z_{21}^{CS} - Z_{12}^{CS}$ $Z_{12}^{CD} = Z_{22}^{CS} - Z_{12}^{CS}$ $Z_{21}^{CD} = Z_{22}^{CS} - Z_{21}^{CS}$ $Z_{22}^{CD} = Z_{22}^{CS}$
ABCD	$A^{CS} = 1 - A^{CD} + B^{CD}C^{CD} / (D^{CD} - 1)$ $B^{CS} = B^{CD} / (D^{CD} - 1)$ $C^{CS} = C^{CD} / (D^{CD} - 1)$ $D^{CS} = D^{CD} / (D^{CD} - 1)$	$A^{CD} = 1 - A^{CS} + B^{CS}C^{CS} / (D^{CS} - 1)$ $B^{CD} = B^{CS} / (D^{CS} - 1)$ $C^{CD} = C^{CS} / (D^{CS} - 1)$ $D^{CD} = D^{CS} / (D^{CS} - 1)$
S	$S_{11}^{CS} = \frac{1}{k}[3S_{11}^{CD} - 2(S_{12}^{CD} + S_{12}^{CD}) +$ $S_{22}^{CD} - \Delta S^{CD} + 1]$ $S_{12}^{CS} = \frac{2}{k}[1 - S_{11}^{CD} + S_{22}^{CD} - 2S_{12}^{CD} - \Delta S^{CD}]$ $S_{21}^{CS} = \frac{2}{k}[1 - S_{11}^{CD} + S_{22}^{CD} - 2S_{21}^{CD} - \Delta S^{CD}]$ $S_{22}^{CS} = \frac{1}{k}[3S_{22}^{CD} + 2(S_{12}^{CD} + S_{12}^{CD})$ $+S_{11}^{CD} + \Delta S^{CD} - 1]$	$S_{11}^{CD} = \frac{1}{k}[3S_{11}^{CS} - 2(S_{12}^{CS} + S_{12}^{CS})$ $+S_{22}^{CS} - \Delta S^{CS} + 1]$ $S_{12}^{CD} = \frac{2}{k}[1 - S_{11}^{CS} + S_{22}^{CS} - 2S_{12}^{CS} - \Delta S^{CS}]$ $S_{21}^{CD} = \frac{2}{k}[1 - S_{11}^{CS} + S_{22}^{CS} - 2S_{21}^{CS} - \Delta S^{CS}]$ $S_{22}^{CD} = \frac{1}{k}[3S_{22}^{CS} + 2(S_{12}^{CS} + S_{12}^{CS})$ $+S_{11}^{CS} + \Delta S^{CS} - 1]$
	$\Delta S^{CD} = S_{11}^{CD}S_{22}^{CD} - S_{12}^{CD}S_{21}^{CD}$ $K = 5 - \Delta S^{CD} - S_{11}^{CD} + S_{22}^{CD}$ $-2(S_{12}^{CD} + S_{21}^{CD})$	$\Delta S^{CD} = S_{11}^{CS}S_{22}^{CS} - S_{12}^{CS}S_{21}^{CS}$ $k = 5 - \Delta S^{CS} - S_{11}^{CS} + S_{22}^{CS}$ $-2(S_{12}^{CS} + S_{21}^{CS})$

Table 6.8 Extrinsic and Intrinsic PHEMT Parameters

Extrinsic Parameters	Values	Intrinsic Parameters	Values
C_{pi} (fF)	25.5	g_m (mS)	32.5
C_{po} (fF)	28	τ (pS)	0.7
C_{pio} (fF)	4.5	C_{gs} (fF)	63.5
L_g (pH)	60	C_{gd} (fF)	22.5
L_d (pH)	65	Cds (fF)	4
L_s (pH)	8	R_{bi} (Ω)	12
R_g (Ω)	5	g_{ds} (mS)	2
R_d (Ω)	5		
R_s (Ω)	2.5		

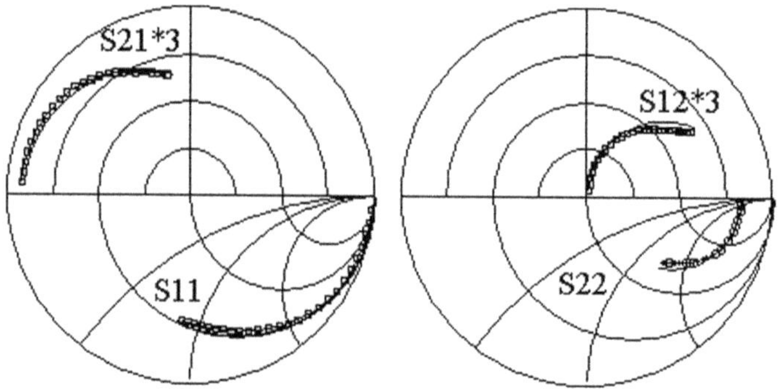

Figure 6.27 Comparison of modeled and measured S-parameters for the PHEMT in CS configuration (bias condition: $V_{gs} = 0.0$ V, $V_{ds} = 2.0$ V), — calculated data from equivalent circuit model, measured data.

6.4.2 Noise Parameter Relationships

Figure 6.30 shows the noise equivalent circuits of CS, CG, and CD configurations for FET devices. $\overline{\upsilon_{CS}^2}$ and $\overline{i_{CS}^2}$ are the two noise sources at the input of the noiseless FET device for CS configuration as shown in Figure 6.30(a), $\overline{\upsilon_{CG}^2}$ and $\overline{i_{CG}^2}$ are for CG configuration as shown in Figure 6.30(b), and $\overline{\upsilon_{CD}^2}$ and $\overline{i_{CD}^2}$ are for CD configuration as shown in Figure 6.30(c).

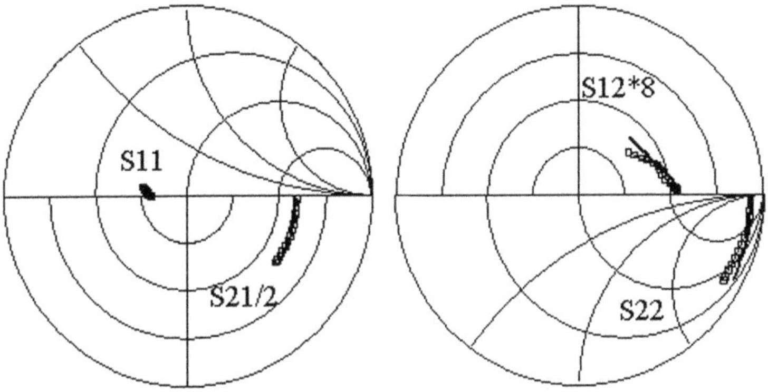

Figure 6.28 Comparison of modeled and predicted S-parameters for the PHEMT in CG configuration (bias condition: V_{gs} = 0.0 V, V_{ds} = 2.0 V; — calculated data from equivalent-circuit model, predicted data from CS configuration).

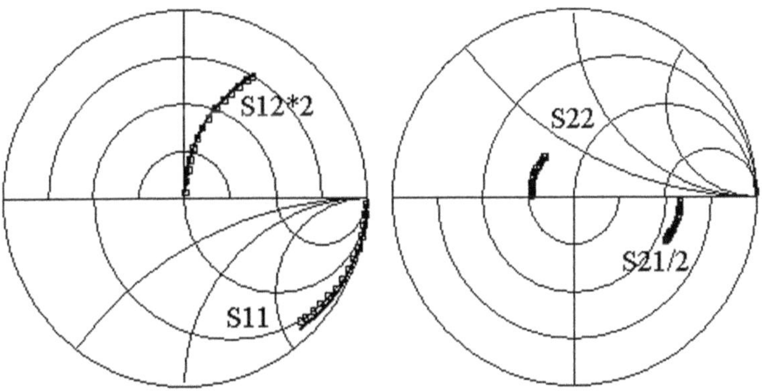

Figure 6.29 Comparison of modeled and predicted S parameter for the PHEMT in CD configuration (bias condition: V_{gs} = 0.0 V, V_{ds} = 2.0 V; — calculated data from equivalent circuit model, predicted data from CS configuration).

The self- and cross-power spectral densities of the two-port noise sources in ABCD-matrix form can be expressed as

$$C_A^{Ci} = \begin{bmatrix} \overline{v_{Ci}^2} & \overline{v_{Ci} i_{Ci}^*} \\ \overline{v_{Ci}^* i_{Ci}} & \overline{i_{Ci}^2} \end{bmatrix} \quad (i = S, G, D) \tag{6.141}$$

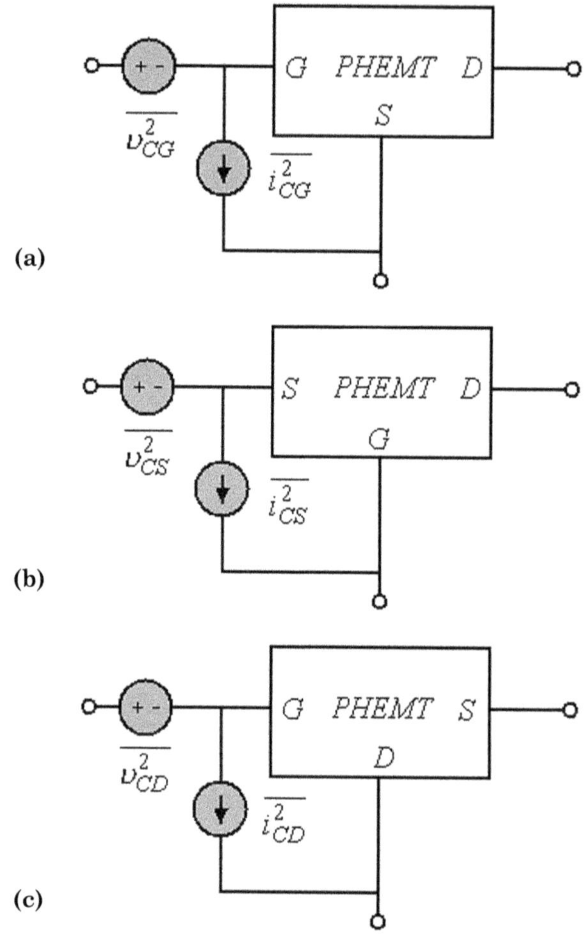

Figure 6.30 PHEMT noise equivalent-circuit models of CS, CG and CD configurations.

The chain noise correlation matrix is easier to obtain from the noise measurement because there is a direct relation between the measured noise parameters (F_{min} is minimum noise figure, R_n is noise resistance, G_{opt} is optimum source conductance and B_{opt} is optimum source susceptance, respectively. The chain noise correlation matrix can be expressed in terms of four noise parameters [23]:

$$C_A^{Ci} = 4kT \begin{bmatrix} R_n^{Ci} & \dfrac{F_{\min}^{Ci} - 1}{2} - R_n^{Ci} (Y_{opt}^{Ci})^* \\ \dfrac{F_{\min}^{Ci} - 1}{2} - R_n^{Ci} Y_{opt}^{Ci} & R_n^{Ci} \left| Y_{opt}^{Ci} \right|^2 \end{bmatrix} \quad (i = S, G, D) \tag{6.142}$$

Using transformation techniques for noise sources, the relationships between the noise sources in the noise equivalent-circuit models of the CS, CG, and CD configurations [39] can be expressed as follows:

$$v_{CG} = \frac{v_{CS}}{A^{CS} - 1} \tag{6.143}$$

$$i_{CG} = \frac{C^{CS}}{A^{CS} - 1} v_{CS} - i_{CS} \tag{6.144}$$

$$v_{CD} = v_{CS} + \frac{B^{CS}}{1 - D^{CS}} i_{CS} \tag{6.145}$$

$$i_{CD} = \frac{i_{CS}}{1 - D^{CS}} \tag{6.146}$$

The corresponding noise sources transformation between CG, CD, and CS configurations are shown in Figures 6.31 and 6.32. The relationship of the noise parameters between the CS, CG, and CD configurations are also given in Tables 6.9 and 6.10 by using noise source transformation matrix technique.

For the condition $\omega C_{gs} R_i \gg 1$ (valid for $f < 30$ GHz), the noise parameter relationships between CS, CG, and CD configuration for intrinsic FET device can be simplified by neglecting the influence of the parasitics:

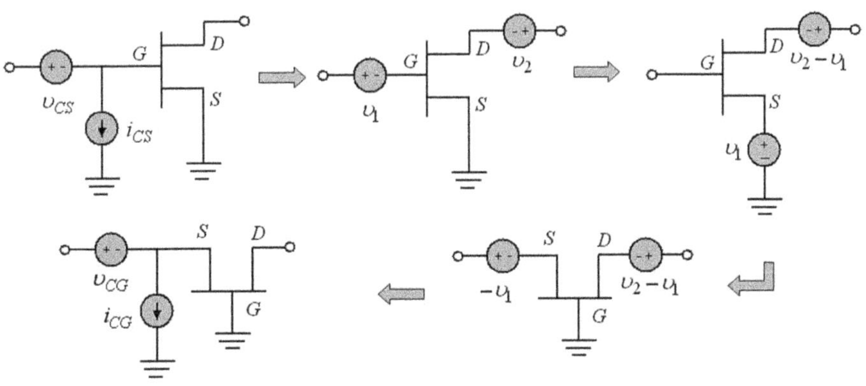

Figure 6.31 Noise sources transformation between CS and CG configurations

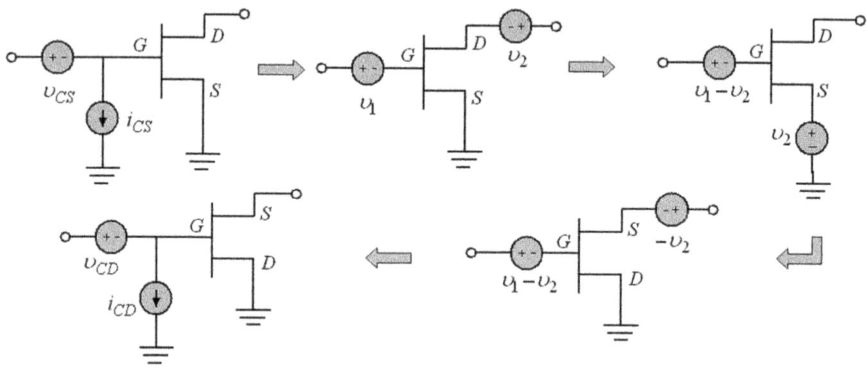

Figure 6.32 Noise sources transformation between CS and CD configurations

Table 6.9 Noise Parameter Relationships between CS and CG configuration

	Common Source	Common Gate
Noise Parameters	$R_n^{CS} = \frac{R_n^{CG}}{\left\|A^{CG} - 1\right\|^2}$	$R_n^{CG} = \frac{R_n^{CS}}{\left\|A^{CS} - 1\right\|^2}$
	$B_{opt}^{CS} = -\mathrm{Im}(C^{CG}) - B_{opt}^{CG}\mathrm{Re}(A^{CG} - 1) + \mathrm{Im}(A^{CG} - 1)[\frac{F_{min}^{CG} - 1}{2R_n^{CG}} - G_{opt}^{CG}]$	$B_{opt}^{CG} = -\mathrm{Im}(C^{CS}) - B_{opt}^{CS}\mathrm{Re}(A^{CS} - 1) + \mathrm{Im}(A^{CS} - 1)[\frac{F_{min}^{CS} - 1}{2R_n^{CS}} - G_{opt}^{CS}]$
	$G_{opt}^{CS} = \sqrt{\left\|C^{CG}\right\|^2 + \left\|Y_{opt}^{CG}\right\|^2 \left\|A^{CG} - 1\right\|^2 - 2K_1 - (B_{opt}^{CS})^2}$	$G_{opt}^{CG} = \sqrt{\left\|C^{CS}\right\|^2 + \left\|Y_{opt}^{CS}\right\|^2 \left\|A^{CS} - 1\right\|^2 - 2k_1 - (B_{opt}^{CG})^2}$
	$F_{min}^{CS} = 1 + 2(G_{opt}^{CS} R_n^{CS} + K_2)$	$F_{min}^{CG} = 1 + 2\mathrm{Re}(G_{opt}^{CG} R_n^{CG} + k_2)$
	$K_1 = \mathrm{Re}[C^{CG}(A^{CG} - 1)^*(\frac{F_{min}^{CG} - 1}{2R_n^{CG}} - Y_{opt}^{CG})^*]$	$k_1 = \mathrm{Re}[C^{CS}(A^{CS} - 1)^*(\frac{F_{min}^{CS} - 1}{2R_n^{CS}} - Y_{opt}^{CS})^*]$
	$K_2 = R_n^{CS}(C^{CG})^* - \frac{\frac{F_{min}^{CG} - 1}{2} - (Y_{opt}^{CG})^* R_n^{CG}}{A^{CG} - 1}$	$k_2 = R_n^{CG}(C^{CS})^* - \frac{\frac{F_{min}^{CS} - 1}{2} - (Y_{opt}^{CS})^* R_n^{CS}}{A^{CS} - 1}$

Table 6.10. Noise Parameter Relationships between CS and CD Configuration

	Common Source	Common Drain
Noise Parameters	$R_n^{CS} = R_n^{CD}[1+\left\|\frac{B^{CD}}{1-D^{CD}}\right\|^2 \left\|Y_{opt}^{CD}\right\|^2]+2K_2$	$R_n^{CD} = R_n^{CS}[1+\left\|\frac{B^{CS}}{1-D^{CS}}\right\|^2 \left\|Y_{opt}^{CS}\right\|^2]+2k_2$
	$B_{opt}^{CS} = \frac{\mathrm{Im}[\frac{k_1}{(1-D^{CS})^*}]}{(1-D^{CS})^*} + \frac{R_n^{CD}\left\|Y_{opt}^{CD}\right\|^2 \mathrm{Im}(B^{CD})}{R_n^{CS}\left\|1-D^{CD}\right\|^2}$	$B_{opt}^{CD} = \frac{\mathrm{Im}[\frac{k_1}{(1-D^{CS})^*}]}{R_n^{CD}} + \frac{R_n^{CS}\left\|Y_{opt}^{CS}\right\|^2 \mathrm{Im}(B^{CS})}{R_n^{CD}\left\|1-D^{CS}\right\|^2}$
	$G_{opt}^{CS} = \sqrt{\frac{R_n^{CD}\left\|Y_{opt}^{CD}\right\|^2}{\left\|1-D^{CD}\right\|^2} - (B_{opt}^{CS})^2}$	$G_{opt}^{CD} = \sqrt{\frac{R_n^{CS}\left\|Y_{opt}^{CS}\right\|^2}{R_n^{CD}\left\|1-D^{CS}\right\|^2} - (B_{opt}^{CD})^2}$
	$F_{\min}^{CS} = 1 + 2\mathrm{Re}(G_{opt}^{CS}R_n^{CS} + K_3)$	$F_{\min}^{CD} = 1 + 2\mathrm{Re}(G_{opt}^{CD}R_n^{CD} + k_3)$
	$K_1 = \frac{F_{\min}^{CD}-1}{2} - (Y_{opt}^{CD})^* R_n^{CD}$	$k_1 = \frac{F_{\min}^{CS}-1}{2} - (Y_{opt}^{CS})^* R_n^{CS}$
	$K_2 = \mathrm{Re}[(\frac{B^{CD}}{1-D^{CD}})^* K_1]$	$k_2 = \mathrm{Re}[(\frac{B^{CS}}{1-D^{CS}})^* k_1]$
	$K_3 = \frac{K_1}{(1-D^{CD})^*} + \frac{B^{CD}\left\|Y_{opt}^{CD}\right\|^2 R_n^{CD}}{\left\|1-D^{CD}\right\|^2}$	$k_3 = \frac{k_1}{(1-D^{CS})^*} + \frac{B^{CS}\left\|Y_{opt}^{CS}\right\|^2 R_n^{CS}}{\left\|1-D^{CS}\right\|^2}$

$$R_n^{CG} \approx R_n^{CS} / (1+\frac{g_{ds}}{g_m})^2 \tag{6.147}$$

$$R_n^{CD} \approx R_n^{CS}(1+\omega^2 C_{gs}^2 / g_m^2) \tag{6.148}$$

$$B_{opt}^{CG} \approx B_{opt}^{CS}(1-\frac{C_{gd}}{C_{gs}}) \tag{6.149}$$

$$B_{opt}^{CD} \approx B_{opt}^{CS} \tag{6.150}$$

$$G_{opt}^{CG} \approx G_{opt}^{CD} \approx G_{opt}^{CS} \tag{6.151}$$

$$F_{\min}^{CG} \approx F_{\min}^{CD} \approx F_{\min}^{CS} \tag{6.152}$$

From (6.147)–(6.152), it can be found that the noise parameters of CG and CD configuration for intrinsic FET device can be predicted directly by using CS configuration. The expressions show that up to moderate frequencies the noise parameters of the three different configurations become close to each other if the parasitic elements can be neglected.

Figure 6.33 shows the measured and computed noise parameters versus frequency for the PHEMT in the CS configuration at $V_{gs} = 0.0$ V and $V_{ds} = 2.0$ V. To illustrate the efficiency of the transformation formulas for noise parameters, we compare the predicted and modeled results for PHEMTs in CG and CD configurations in Figures 6.34 and 6.35. The predicted data for CG and CD configurations are obtained from the measured data of CS configuration by using the formulas in Tables 6.9 and 6.10. Good agreement is obtained between modeled and predicted data, which verifies the supposed approach for the noise parameters.

In Figure 6.36, a comparison among CS, CG, and CD configuration noise parameters for the intrinsic PHEMT device are given. At low frequencies all noise parameters are more or less identical, with one exception R_n. As can be taken from Eqns. (6.147) and (6.148), R_n in CS and CG configuration differ from each other by a frequency-independent value. With increasing frequency the noise parameter values of the different configurations disperse more and more. While the minimum noise figures F_{min} and optimum source conductances G_{opt} show a weak divergence, R_n and B_{opt} values disperse more strongly with increasing frequency.

6.5 Summary

In this chapter, we introduced the noise modeling and parameter extraction methods for FET device, as well as how to determine noise parameters, including the tuner-based method and the noise figure-based method.

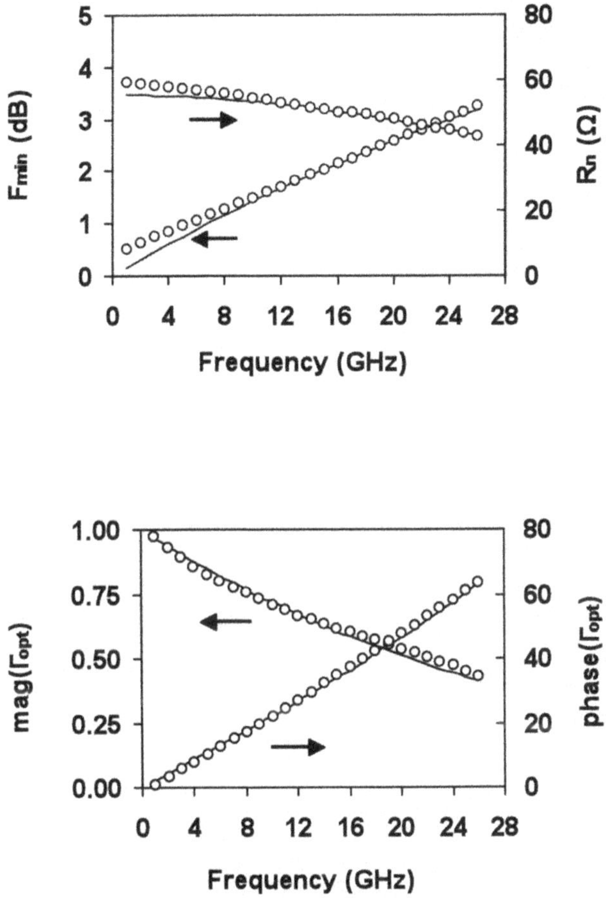

Figure 6.33 Comparison of modeled and measured noise parameter for the PHEMT in CS configuration (bias condition: V_{gs} = 0.0 V, V_{ds} = 2.0 V; — calculated data from equivalent circuit model, o o = measured data).

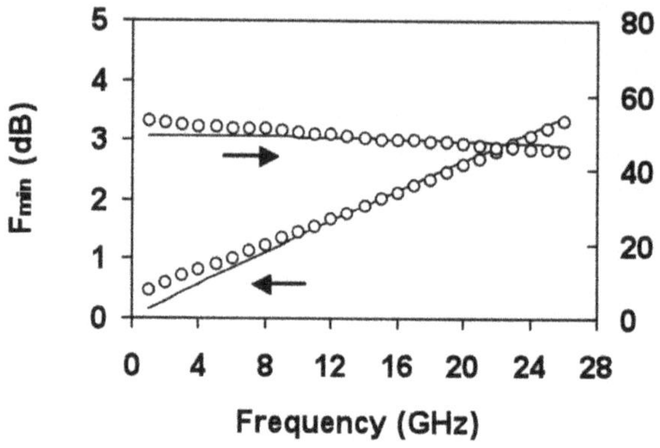

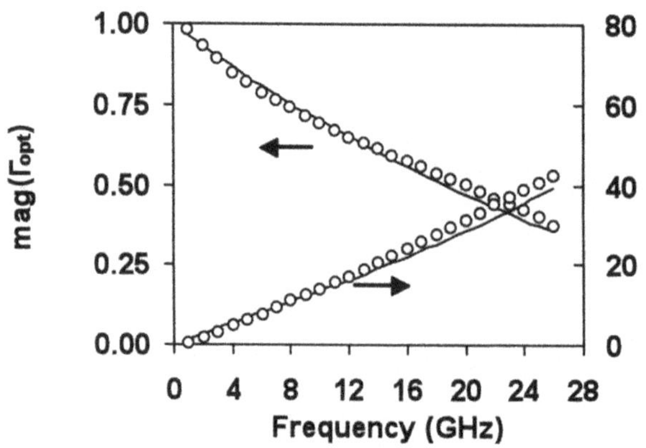

Figure 6.34 Comparison of modeled and predicted noise parameters for the PHEMT in CG configuration (bias condition: V_{gs} = 0.0 V, V_{ds} = 2.0 V; — calculated data from equivalent-circuit model, o o = predicted data from noise measurement of CS configuration).

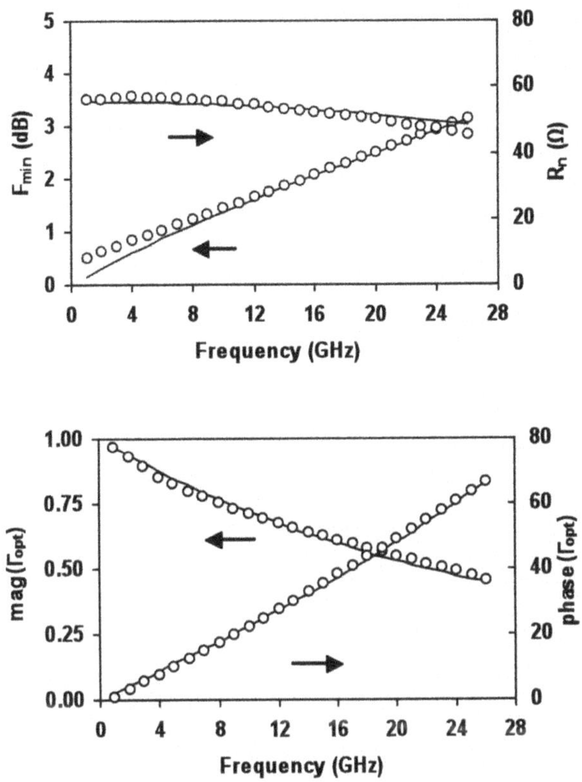

Figure 6.35 Comparison of modeled and predicted noise parameters for the PHEMT in CD configuration (bias condition: V_{gs} = 0.0 V, V_{ds} = 2.0 V, — calculated data from equivalent-circuit model, o o = predicted data from noise measurement of CS configuration).

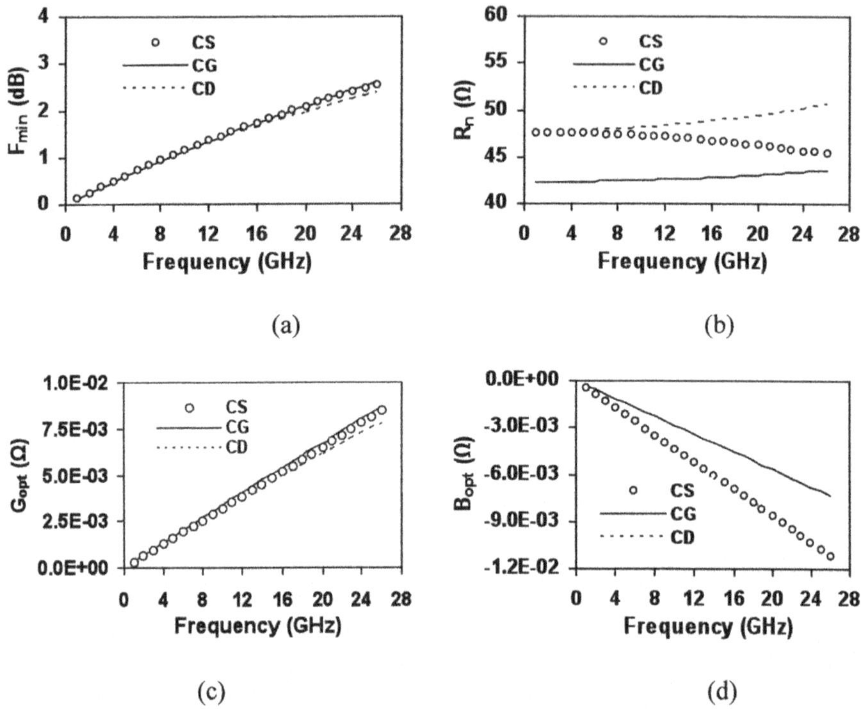

Figure 6.36 Comparison of noise parameters in CS, CG, and CD configuration for the intrinsic PHEMT device (bias condition: V_{gs} = 0.0 V, V_{ds} = 2.0 V).

Chapter 7

Artificial Neural Network Modeling Technique for FET

7.1 Overview of ANN Modeling Technique

Accurate small-signal and large-signal equivalent-circuit models of microwave active devices (e.g., diodes, FETs and HBTs) are very useful for device performance analysis (e.g., noise, gain) in designing microwave circuits and characterizing the device technological process. Nonlinear models of microwave devices are commonly in terms of state functions such as closed-form equations [1, 2], Volterra series [3], or look-up tables [4]. These quantities are classically determined via a small signal detour based on multi-bias S-parameters and DC measurements. The linear equivalent-circuit model is generally more accurate than a linearized nonlinear model for predicting the S-parameters.

Artificial neural networks (ANNs) have been used frequently in, for example, signal processing applications, speech and pattern recognition, and remote sensing for the last two decades. Recently, a computer-aided design (CAD) approach based on neural networks has been introduced in RF and microwave active linear/nonlinear modeling [5–10]. The ANN modeling techniques are efficient alternatives to conventional methods such as numerical modeling methods, which could be computationally expensive, or analytical methods, which could be difficult to obtain for new device or empirical models, whose ranges and accuracy could be limited. As in closed-form equation models [1, 2], ANNs have the adaptability necessary to represent a strong nonlinearity. However, closed-form equation models must be changed to fit the particular device type. ANNs can support any number and type of nonlinear features by simply changing their configurations. While the look-up table model requires a large database, which can sometimes reach several megabytes, the response of the ANN can be determined only by using a few hundred coefficients. The ANN model overcomes the drawbacks of the closed-form and look-up table models, while retaining their advantages.

However, it is difficult to determine the proper configuration with this attractive ANN. The neural network size required to develop an accurate model is not known a priori. Too many hidden neurons need more central processing units (CPUs) and too few neurons result in under-learning of the neural network. Moreover, the accuracy of the numerical optimization methods that minimize the difference between measured and modeled DC, S-parameters and harmonic performance versus frequency can vary depending upon the starting values, and the device is modeled by a black box for which small- and large-signal parameters are evaluated through a neural networks, so it is difficult to understand the physical mechanisms and evaluate the influence of the different parameters of the equivalent circuit model. Table 7.1 gives a comparison of ANN and conventional modeling techniques.

Table 7.1 Comparison of ANN and Conventional Modeling Techniques

	Physical Model	Empirical Model	ANN Model
Speed	Slow	Fast	Fast
Accuracy	High	Medium	High
Training data	None	little	Large
Physical meaning	Yes	Medium	No

Multilayer Perceptrons (MLPs) are a popularly used neural network structure. In the MLP neural network, the neurons are grouped into layers. The first and the last layers are called input and output layers, respectively, and the remaining layers are called hidden layers. Typically, an MLP neural network consists of an input layer, one or more hidden layers, and an output layer, as shown in Figure 7.1. For example, an MLP neural network with an input layer, one hidden layer, and an output layer, is referred to as a three-layered MLP, or MLP3. Suppose the total number of layers is L. The first layer is the input layer, the Lth layer is the output layer, and layers two to L–1 are hidden layers. Let the number of neurons in l^{th} layer be N_l, $l = 1, 2, \ldots, L$. Let W_{ij}^l represent the weight of the link between j^{th} neuron of l–1th layer and i^{th} neuron of lth layer.

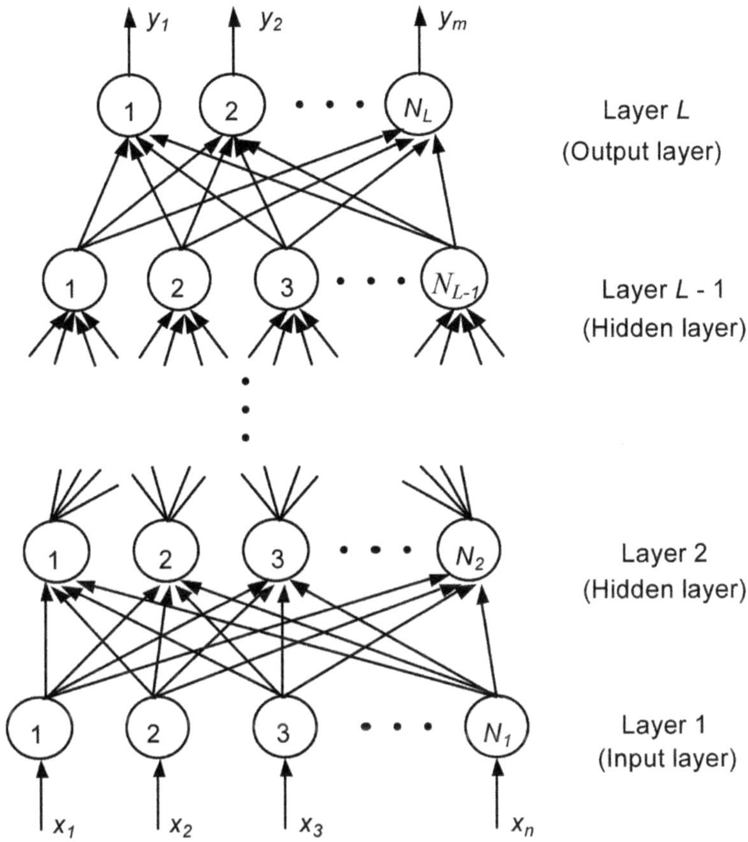

Figure 7.1 MLP neural network structure. Typically, an MLP network consists of an input layer, one or more hidden layers, and an output layer.

For given the input vector

$$x = \left[x_1, x_2, \cdots\cdots, x_n\right]^T$$

and the weight vector

$$w = \left[w_{10}^2, w_{11}^2, w_{12}^2 \cdots\cdots, w_{N_L N_{L-1}}^L\right]^T$$

The computation of the output vector $y = \left[y_1, y_2, \cdots\cdots, y_m\right]^T$ is given by [11]

$$z_i^1 = x_i \ (i = 1, 2, \ldots\ldots, N_1), \ n = N_1 \tag{7.1}$$

$$z_i^l = \sigma\left(\sum_{j=0}^{N_{l-1}} w_{ij}^l z_j^{l-1}\right) \quad (i = 1, 2, \ldots\ldots, N_1),\ l = 2, 3 \ldots\ldots, N_L \tag{7.2}$$

$$y_i = z_i^L \quad (i = 1, 2, \ldots\ldots, N_L),\ m = N_L \tag{7.3}$$

The most commonly used hidden neuron activation function is the sigmoid function given by

$$\sigma(\gamma) = \frac{1}{(1+e^{-\gamma})} \tag{7.4}$$

7.2 ANN-Based Linear Modeling

Figure 7.2 shows a standard MLP configuration for FET small-signal modeling [9]; to model the small-signal behavior of an active microwave

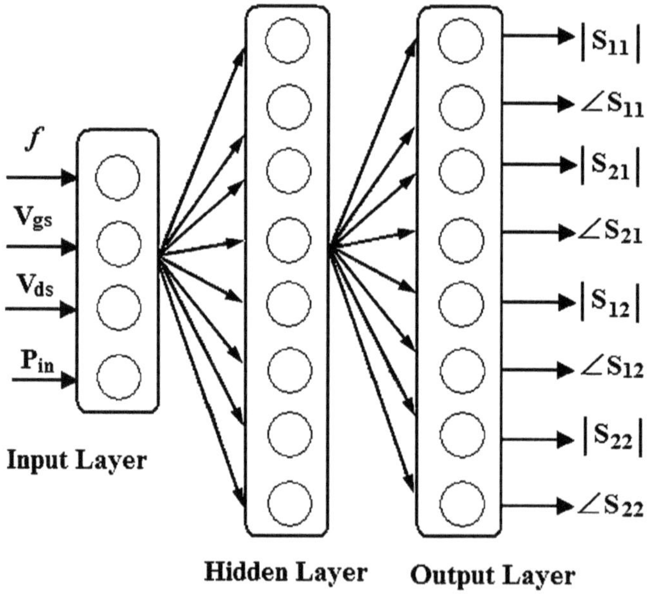

Figure 7.2 Neural network configuration for an S-parameter modeling.

device, an MLP with a single hidden layer having the same number of neurons as the output layer (eight in our case), has been found to be sufficient.

The input layer consists of four neurons, and can be expressed as follows:

$$x = \left[f, V_{ds}, V_{gs}, P_{in} \right]^T \tag{7.5}$$

where f is the frequency, V_{gs} and V_{ds} are the bias points, and P_{in} is the input power.

The output consists of S-parameter matrix in magnitude and phase formats.

$$y = \left[|S_{11}|, \angle S_{11}, |S_{21}|, \angle S_{21}, |S_{12}|, \angle S_{12}, |S_{22}|, \angle S_{22} \right]^T \tag{7.6}$$

where $|S_{ij}|$ and $\angle S_{ij}$ ($i, j = 1,2$) are the magnitude and phase of the S parameter.

7.3 ANN-Based Nonlinear Modeling

The ANN model overcomes the drawbacks of the closed-form and look-up table models, while retaining their advantages. However, it is difficult to determine the proper configuration if the multiple ANNs need to be involved in a device model. The neural network size required to develop an accurate model is not known a priori. If the device is modeled by a black box for which small- and large-signal parameters are evaluated through a neural network, it is difficult to understand the physical mechanisms because of the lack of analyzing the influence of the different parameters of the equivalent-circuit model. Therefore, it is proposed that the nonlinear modeling technique for FET devices should be based on the combination of the conventional equivalent-circuit modeling and ANN modeling technique. The advantages are as follows:

1. Information of equivalent circuit helps to preserve the physical interpretation of the model. In our model, the effects that become important at higher frequencies (e.g., the pad capacitances, series inductances) are taken into account.
2. Each nonlinear elements of FET is modeled by an ANN based on accurate small-signal parameter extraction. The multiple ANNs are trained separately, and it is convenient for determination of ANN structure.
3. The aforementioned ANNs are used as initial values for whole device training; therefore, efficient training of the combined ANN parts of equivalent-circuit model can be carried out.

To obtain an unified small-signal and large-signal model, a multi-goal training of the DC, S- parameters, and harmonics optimization are used simultaneously.

The new large-signal model is derived from the well-accepted small signal models. A 14-element equivalent circuit is used to represent the large-signal model [12]. The element list includes eight bias-independent extrinsic and six bias-dependent intrinsic elements, as shown in Figure 7.3, where L_g, L_d, and L_s represent the inductances of the gate, drain, and source feedline, respectively, C_{pg} and C_{pd} and represent the gate and drain pad capacitances, respectively, R_s and R_d are the source

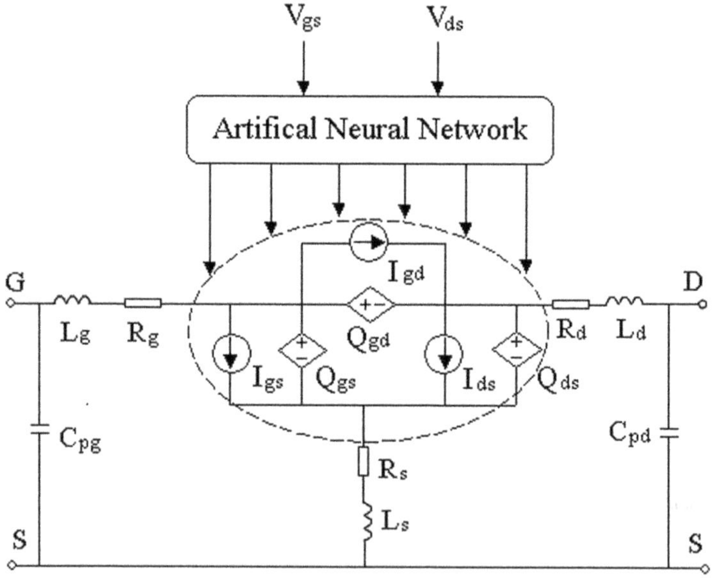

Figure 7.3 FET nonlinear model by using ANN

and drain resistances, respectively, and R_g is the distributed gate resistance. Three charge sources Q_{gs}, Q_{gd}, and Q_{ds} and three current sources I_{ds}, I_{gs}, and I_{gd} are used to represent the large signal performance.

The situation addressed here involves is when a new model needs to be created to fit device data. Because of the arbitrary variation of the nonlinear model parameters with respect to the biases, especially charge model parameters, it is human intensive to do a accurate modeling by creating the empirical closed-form equations. Therefore, the ANN approach is preferable for the nonlinear model parameters.

The multiplayer perceptron is a popularly used neural network structure, and is chosen for use in our model. The neurons are grouped into layers in the MLP neural network. The first and last layers are called input and output layers, respectively. Between input and output layers there exists a central part of the neural network, called a hidden layer. Depending on the complexity of the input response and desired output, the number of hidden layers and neurons at each layers can vary. To model the behavior of the model parameters of FETs, MLPs with a single hidden layer have been found to be sufficient.

As seen in Figure 7.3, we use six MLPs to model the nonlinear relationships between the nonlinear model parameters and the inputs. The outputs of the artificial neural network are used as the inputs of nonlinear model for FETs.

The current sources I_{ds}, I_{gs}, and I_{gd} can be expressed as follows by using artificial neural networks:

$$I_{gs} = f_{ANN}^{I_{gs}}(V_{gs}, w) \tag{7.7}$$

$$I_{gd} = f_{ANN}^{I_{gd}}(V_{gd}, w) \tag{7.8}$$

$$I_{ds} = f_{ANN}^{I_{ds}}(V_{ds}, V_{gs}, w) \tag{7.9}$$

The charge sources Q_{gs}, Q_{gd}, and Q_{ds} can be expressed as

$$Q_{gs} = f_{ANN}^{Q_{gs}}(V_{gs}, V_{gd}, w) \tag{7.10}$$

$$Q_{gd} = f_{ANN}^{Q_{gd}}(V_{gd}, V_{gs}, w) \tag{7.11}$$

$$Q_{ds} = f_{ANN}^{Q_{ds}}(V_{ds}, V_{gs}, w) \tag{7.12}$$

where f_{ANN} represents ANN of each element of FET device, and w represents the weighting factors between output layer and hidden layer; and the nonlinear elements are the values of the hidden neurons.

The overall nonlinear relationship between input and output is realized by various activation patterns of the neurons whose activation functions are typically a smooth switch function. Normally the neural model is then trained to learn the input-output relationship from the training data (sample of input-output data). Specifically, training is used to determine the neural model parameters, that is, neural network weights w, such that the ANN model predicted output best matches that of the training data. The testing data (new input-output samples) is used to test the accuracy of the ANN model. First, the extrinsic elements are extracted by using the Cold-FET approach [13], and the intrinsic elements are directly calculated from the measured S parameters after subtracting the extrinsic elements at various bias points. The adjoint ANN technology [8] has been used to obtain the electron charge of the intrinsic capacitance. The ANN models of each DC nonlinear elements of FET (e.g., I_{gs}, I_{gd} and I_{ds}) are obtained by using MLP structure. The solution of the individually trained MLPs are then used as effective initial values for next stage of training, where all the six MLPs are combined with the equivalent circuit to form an overall FET model. The overall combined model is then trained simultaneously with DC, small-signal S-parameters and large-signal harmonic balance (HB). The error function to be minimized is given by:

$$\begin{aligned} E = \min\{ & \sum_{bias}[I_{ds}^{ANN}(V_{gs}, V_{ds}, w) - I_{ds}^{data}(V_{gs}, V_{ds}, w)]^2 + \\ & \sum_{bais}\sum_{freq}[S^{ANN}(V_{gs}, V_{ds}, freq, w) - S^{data}(V_{gs}, V_{ds}, freq, w)]^2 + \\ & \sum_{Pin}\sum_{bias}\sum_{freq}[P_o^{ANN}(V_{gs}, V_{ds}, freq, P_{in}, w) - P_o^{data}(V_{gs}, V_{ds}, freq, P_{in} w)]^2 \} \end{aligned} \tag{7.13}$$

where P_{in} and P_o represent the input and output power level of the device, respectively. Figure 7.4 gives a flowchart for the proposed training process.

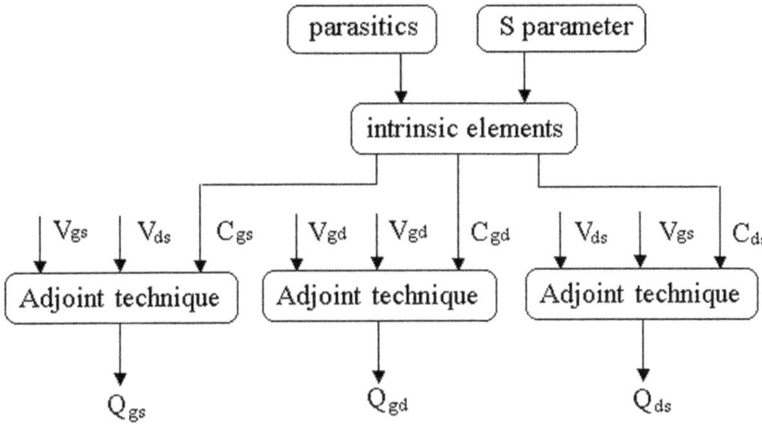

(a)

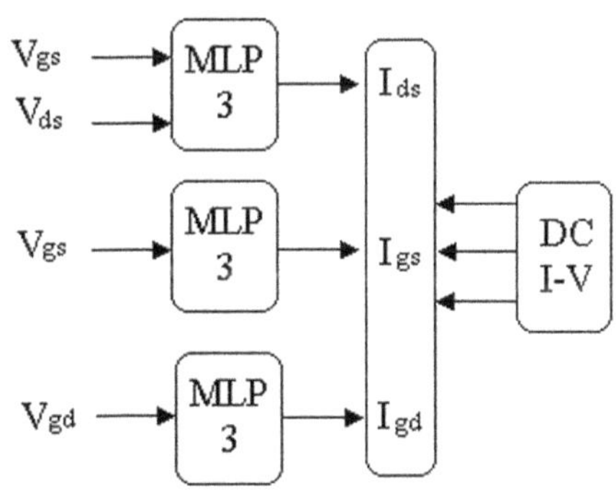

(b)

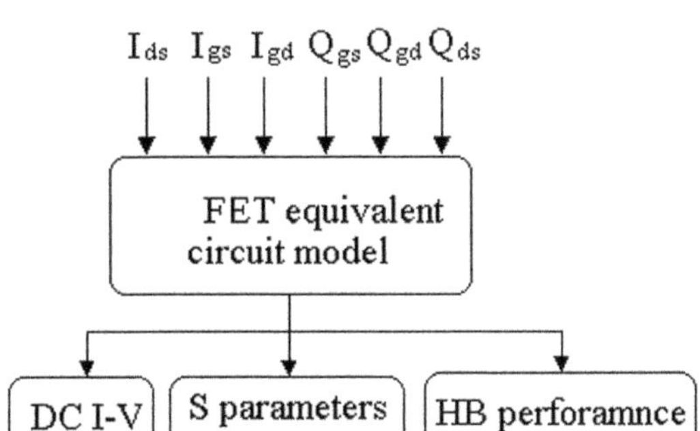

(c)

Figure 7.4 Flowchart of training process: (a) charge ANN model; (b) DC ANN model; (c) ANN circuit model

As an example to verify the previously proposed method, a built-in model for 200 μm gate width 0.25 μm HEMT in Agilent ADS has been used to generate small-signal and large-signal training data. Using the proposed method, the six MLPs are trained individually, and then the combined ANN is trained with DC, *S*-parameter, and HB data. A semi-analytical procedure is employed to determine the small-signal model parameters in this example. The initial values of most of the extrinsic and intrinsic model parameters are estimated from hot and cold S-parameter measurements. The final results are obtained from empirical optimization procedure.

The extracted values of the bias-independent small-signal elements for 200 μm gatewidth width 0.25 μm PHEMT are summarized in Table 7.2. Then the small-signal elements are extracted from the S-parameter measurements after de-embedding the parasitic elements. Figure 7.5 shows the extracted results of intrinsic elements versus gate-source and drain-source voltage. Figures 7.6 and 7.7 show the comparison between modeled and measured data. As shown in Figures 7.6 and 7.7, they agree very well.

Table 7.2 Parasitic Elements Used in Combined ANN Model

Elements	Values	Elements	Values
C_{pg} (fF)	20	R_g (Ω)	1
C_{pd} (fF)	20	L_d (pH)	60
C_{pgd} (fF)	5	L_s (pH)	5
R_d (Ω)	0.9	L_g (pH)	40
R_s (Ω)	0.9		

Figure 7.8 shows the comparison between the measured and simulated output power levels for the HEMT for the fundamental (2.0GHz), second, and third harmonics. The measured and simulated harmonics for the different bias points are shown in Figure 7.8. The simulations run fast, show good correspondence between measured and simulated data, and have no convergence problem. These indicate that our model simulates these figures of merit with high accuracy.

Table 7.3 shows the accuracy comparison of five different training processes; it can be found that the best way is #5, which is the trade-off way PHEMT large-signal modeling.

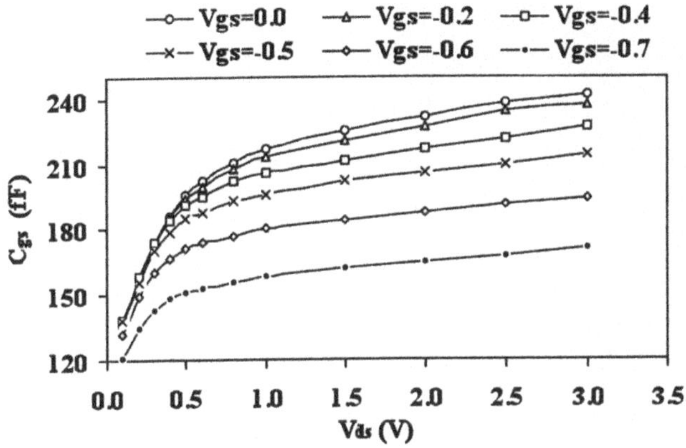

(a)

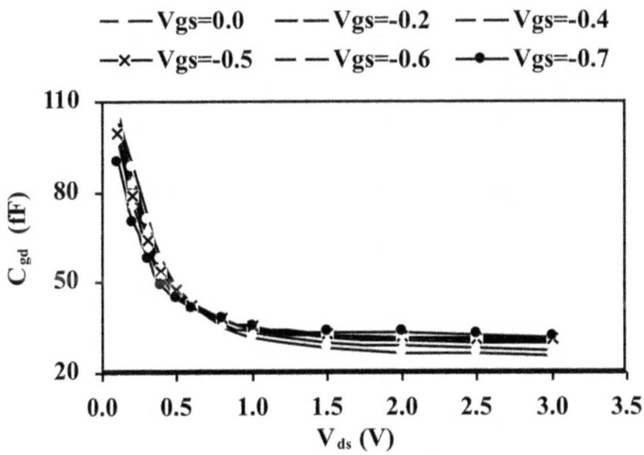

(b)

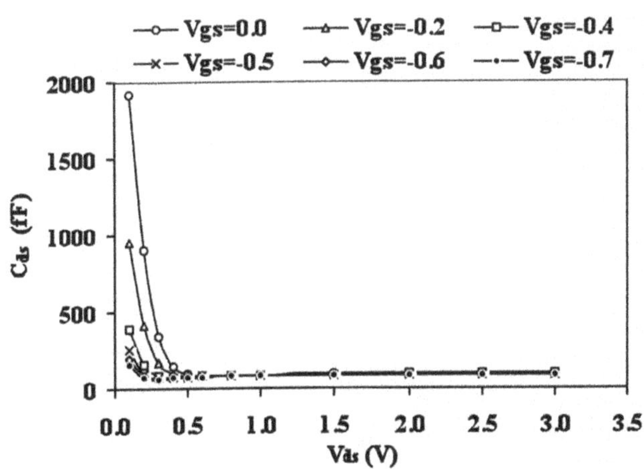

(c)

Figure 7.5 Extracted results of intrinsic elements *(continues)*

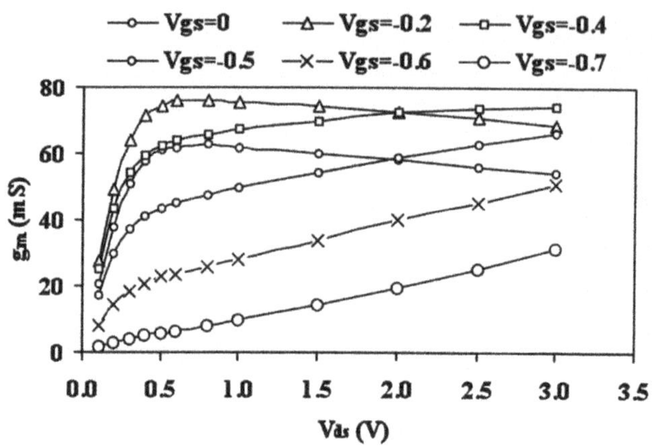

(d)

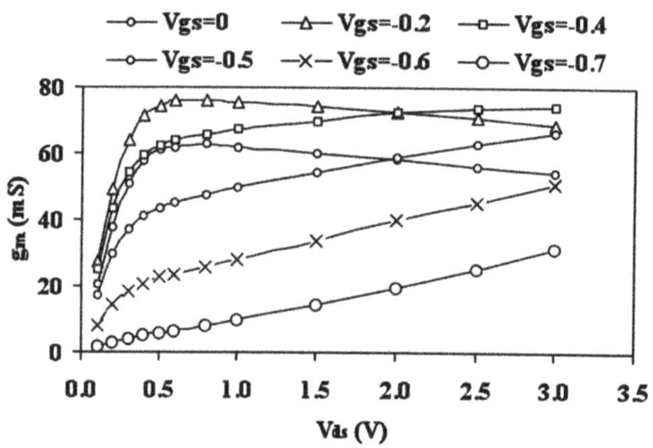

(e)

Figure 7.5 Extracted results of intrinsic elements *(continued)*

Table 7.3 Accuracy Comparison of Five Different Training Processes with Different Combinations of DC, *S*-parameters and HB Data

Training process	#1	#2	#3	#4	#5
Training objective	DC+S	DC+S	HB	HB	DC+S+HB
Testing information	DC+S	HB	HB	DC+S	DC+S+HB
Testing accuracy	3.2%	28%	3%	20%	4.6% HB 3.2% DC+S

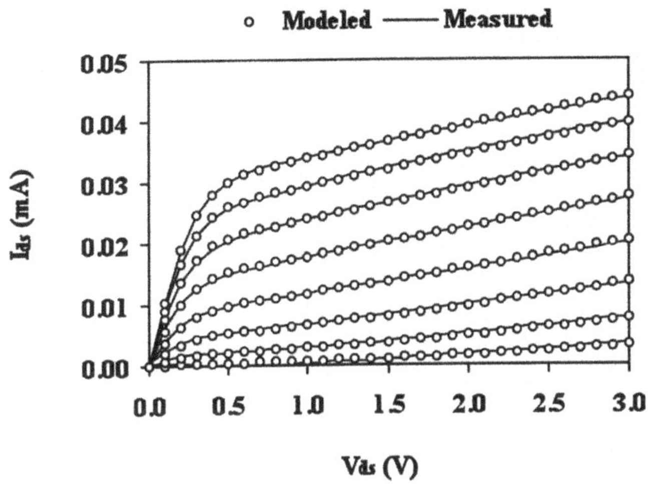

Figure 7.6 Comparison of modeled and measured DC performance

7.4 ANN-Based Noise Modeling

Based on the previous approach employed in [14–16], a noise model for FET that takes into account the behavior of the noise model parameters as a function of the frequency and output current is introduced. This model is a combination of the conventional equivalent-circuit and ANN modeling techniques. The proposed model overcomes the drawbacks of conventional equivalent circuit modeling and direct prediction ANN modeling techniques and retains their advantages.

Figure 7.9 shows the proposed FET noise modeling structure, which combines the conventional equivalent-circuit modeling and artificial neural network modeling technique. The complete noise modeling structure consists of two parts: (1) artificial neural network for noise model parameters; and (2) an improved noise equivalent circuit model for FET [17]. A three-layer MLP is used to model the nonlinear relationship between the noise model parameters and the inputs (including output drain-to-source current I_{ds} and frequency f). The outputs of the artificial neural network are regarded as the inputs of the improved noise model for an FET, which takes into account the real part of the noise correlation coefficient.

The proposed expressions for two correlated current noise source $\overline{i_g^2}$ and $\overline{i_d^2}$ can be written as:

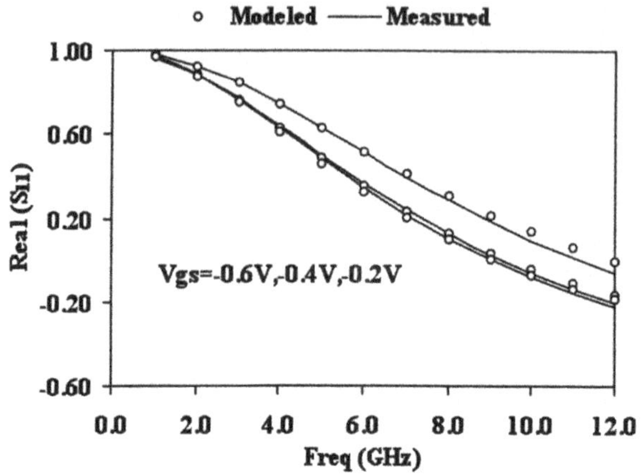

(a)

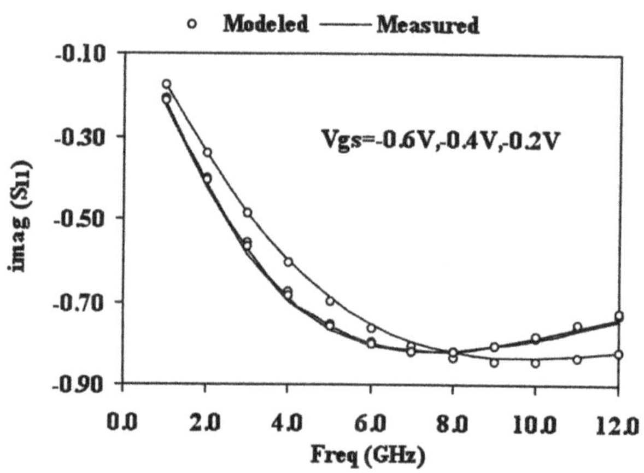

(b)

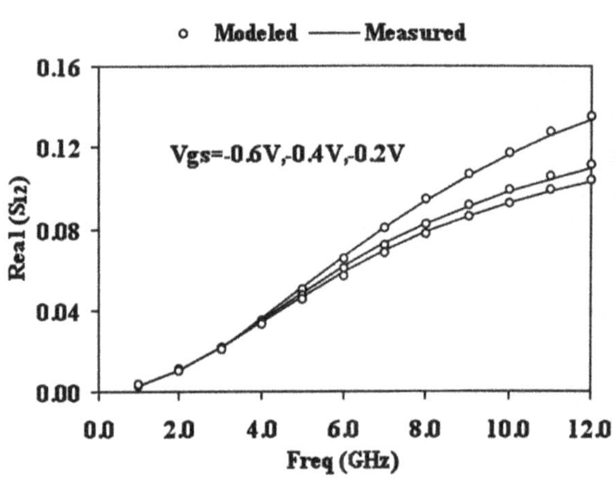

(c)

Figure 7.7 Comparison of measured and modeled S-parameters *(continues)*

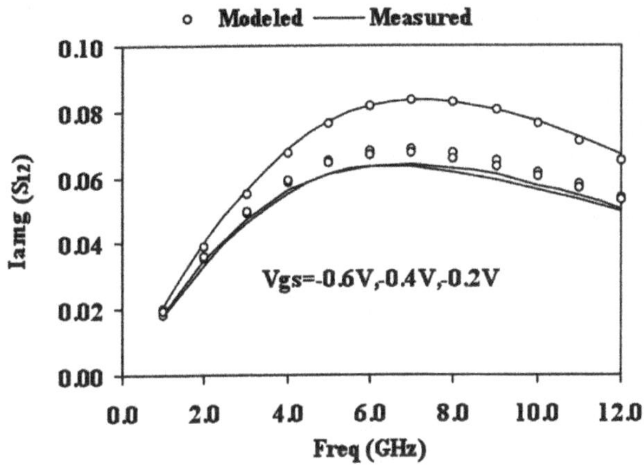

(d)

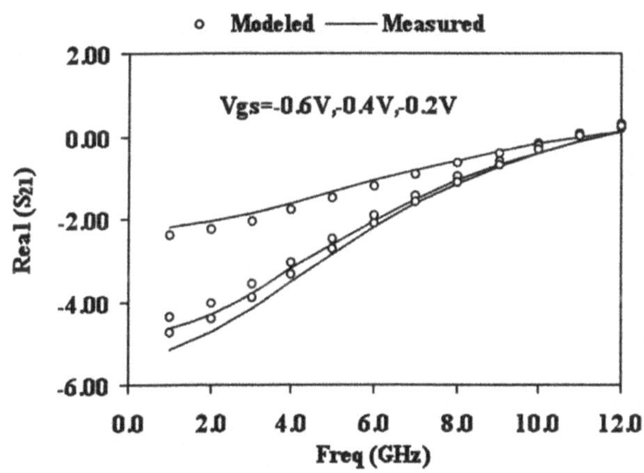

(e)

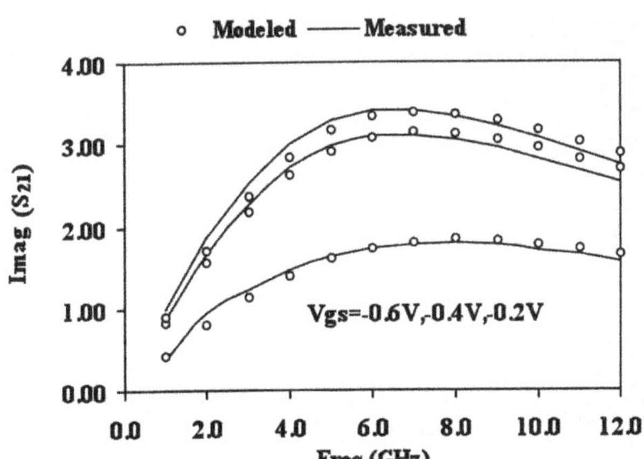

(f)

Figure 7.7 Comparison of measured and modeled S-parameters *(continued)*

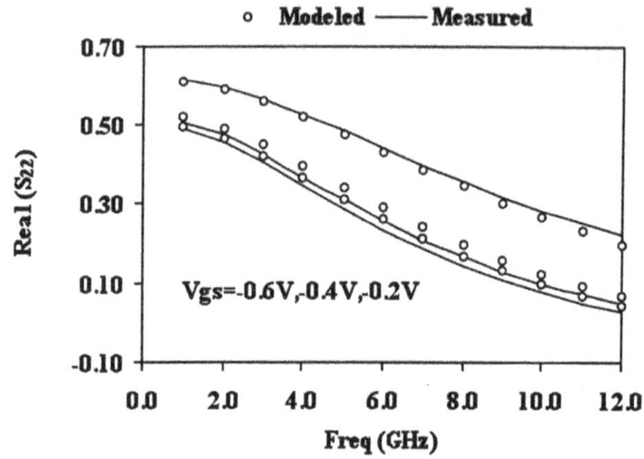

(g)

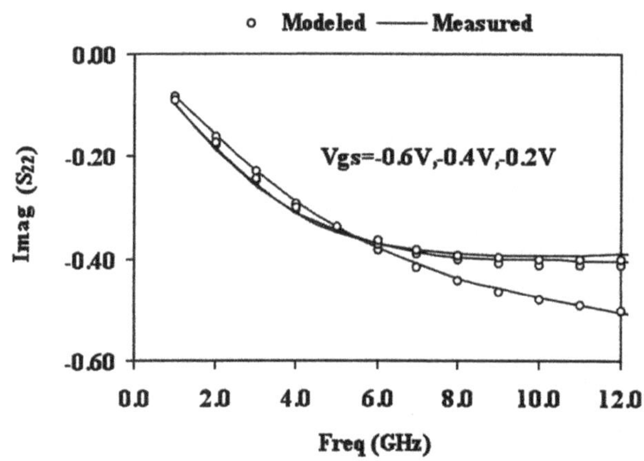

(h)

Figure 7.7 Comparison of measured and modeled S-parameters *(continued)*

$$\overline{i_g^2} = 4kT \frac{(\omega C_{gs})^2 R(I_{ds}, f)}{g_m} \Delta f \tag{7.14}$$

$$\overline{i_d^2} = 4kTg_m P(I_{ds}, f)\Delta f \tag{7.15}$$

The cross-correlation between $\overline{i_g^2}$ and $\overline{i_d^2}$ can be expressed as:

$$\overline{i_d^* i_g} = 4kT\omega C_{gs} C\sqrt{P(I_{ds}, f)R(I_{ds}, f)}\Delta f \tag{7.16}$$

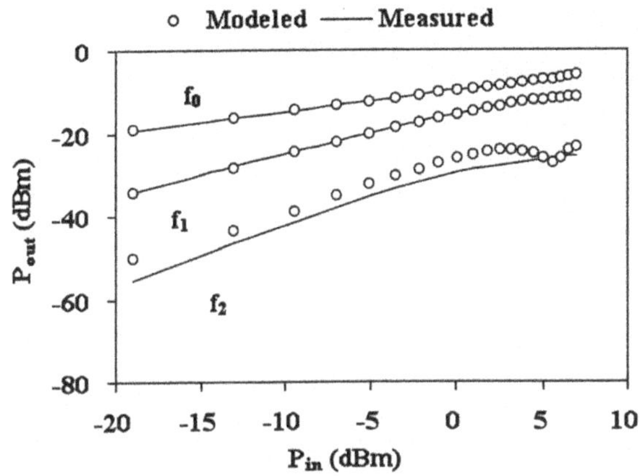

(a)

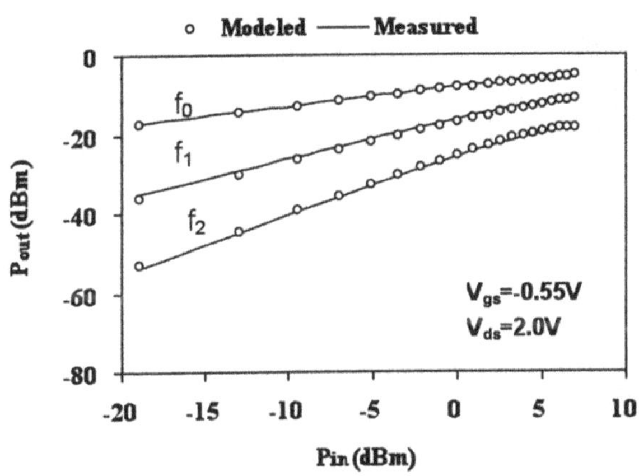

(b)

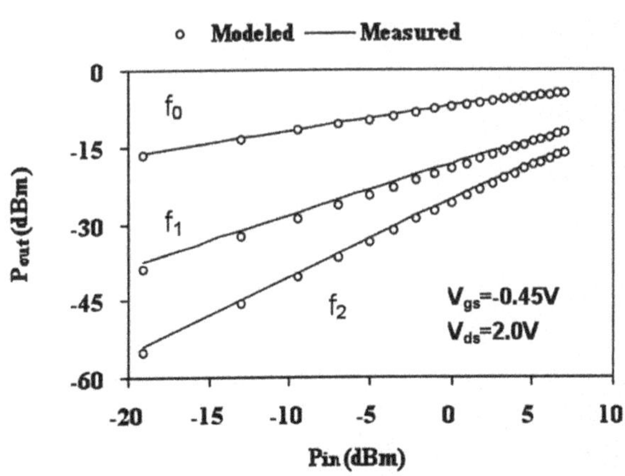

(c)

Figure 7.8 Comparison of measured and modeled output power levels (f_o = 2.0 GHz) *(continues)*

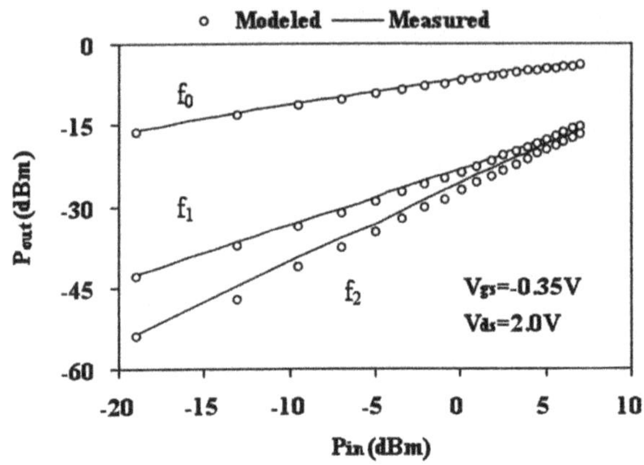

(d)

Figure 7.8 Comparison of measured and modeled output power levels (f_o = 2.0 GHz) *(continued)*

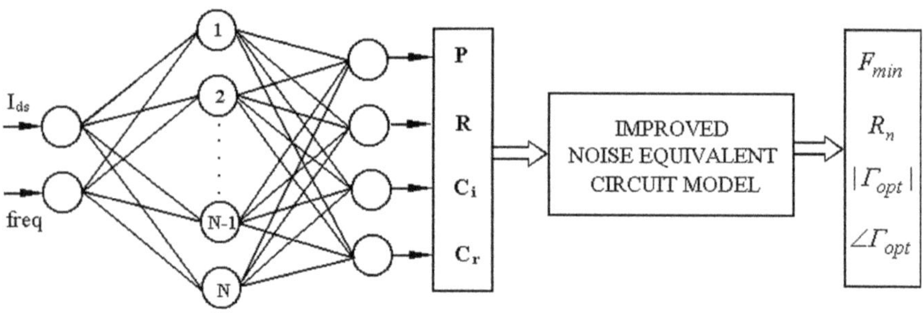

Figure 7.9 FET noise modeling structure based on conventional equivalent-circuit and ANN modeling techniques

where R and P are the gate and drain noise model parameters, and C is the correlation coefficient, defined by [11]:

$$C = C_r(I_{ds}, f) + jC_i(I_{ds}, f) \tag{7.17}$$

where C_r and C_i represent the real part and imaginary part, respectively.

It is noted that the proposed noise model is different from the conventional model [14–16]. All the noise model parameters are a function of the frequency f and drain-source current I_{ds}, and the real part of the correlation coefficient is also considered. Because of the arbitrary variation of the noise model parameters with respect to the frequency, it is not easy to do modeling by using the closed-form empirical equations.

Therefore, the ANN approach is proposed as a suitable method for the noise model parameters P, R, and C (including real part C_r and imaginary part C_i). FET noise modeling structure based on conventional equivalent-circuit and ANN modeling techniques

To model the behavior of the noise model parameters of PHEMT, an MLP with a single hidden layer is found to be sufficient. The input layer consists of two neurons, the inputs of which are frequency f and the drain-source current I_{ds}. The output consists of four noise model parameters P, R, C_r, and C_i.

For given input (frequency f and the drain-source current I_{ds}), the output of the MLP can be computed by

$$\begin{bmatrix} P \\ R \\ C_i \\ C_r \end{bmatrix} = \begin{bmatrix} w_{10} & w_{11} & \cdots\cdots & w_{1N} \\ w_{20} & w_{21} & \cdots\cdots & w_{2N} \\ w_{30} & w_{31} & \cdots\cdots & w_{3N} \\ w_{40} & w_{41} & \cdots\cdots & w_{4N} \end{bmatrix} \cdot \begin{bmatrix} Z_0 \\ Z_1 \\ \vdots \\ Z_N \end{bmatrix} \tag{7.18}$$

where $w_{ij}(i=1\sim 4, j=0\sim N)$ represents the weighting factors between output layer and hidden layer, and $Z_k(k=1\sim N)$ is the values of the hidden neurons and is computed as

$$Z_k = \sigma(\theta_k)$$

$$\theta_k = w_k^{I_{ds}} I_{ds} + w_k^f f \tag{7.19}$$

where $w_k^{I_{ds}}$ and $w_k^f(k=1\sim N)$ represent the weighting factors between input layer and hidden layer. $\sigma(\cdot)$ is an activation function. The overall nonlinear relationship between input and output is realized by various activation patterns of the neurons whose activation functions are typically a smooth switch function, for example, the sigmoid function can be expressed as:

$$\sigma(\theta_k) = \frac{1}{1+e^{-\theta_k}} = \frac{1}{1+e^{(-w_k^{I_{ds}} I_{ds} - w_k^f f)}} \tag{7.20}$$

The neural model is then trained to learn the input-output relationship from the training data (sample of input-output data). Specific training is needed to determine the neural model parameters (i.e., neural network weights w_{ij}), such that the ANN model predicted output best matches that of the training data. The testing data (new input-output samples) are used to test the accuracy of the ANN model. The training and testing data are obtained from the noise model parameters, which are determined directly from noise parameters on wafer measurement based on the noise correlation matrix technique.

The error function to be minimized during training is

$$E = \min\{\sum_{bias}\sum_{freq}[P_{ANN}(I_{ds}, freq, w) - P_{data}(I_{ds}, freq, w)]^2 + \sum_{bias}\sum_{freq}[R_{ANN}(I_{ds}, freq, w) - R_{data}(I_{ds}, freq, w)]^2 + \sum_{bias}\sum_{freq}[|C_{ANN}(I_{ds}, freq, w)| - |C_{data}(I_{ds}, freq, w)|]^2\} \tag{7.21}$$

where subscript *ANN* denotes the ANN noise model parameters, *data* is the extracted noise model parameters, *freq* is the frequency, and w is the neural network weights.

Figures 7.10 through 7.13 show the extracted results for noise model parameters P, R, C_r, and C_i versus frequency at different bias conditions, respectively. From Figure 7.10, it can be found that the drain noise factor P has a weak nonlinear relationship versus frequency, and can be regarded as frequency independent. The gate noise factor R exhibits a strong nonlinear relationship versus frequency as shown in Figure 7.11. R decreases with decreasing frequency and I_{ds}. From Figures 7.12 and 7.13, it can be found that the real part of the correlation coefficient C_r is in the same order with imaginary part C_i at low bias condition, and can be neglected only under high output current bias condition.

Up to now, a set of exact empirical closed-form expressions for noise model parameters P, R, C_r, and C_i are not available. Therefore, it is difficult to build an united empirical model for P, R, C_r and C_i by using the closed-form equations. ANNs are used to simulate the relationship between noise model parameters and frequency. Through experiments, we found that a three-layered configuration composed of five neurons adequately represents the four noise model parameters simultaneously. We adopted this neural network with the batch-mode back-propagation

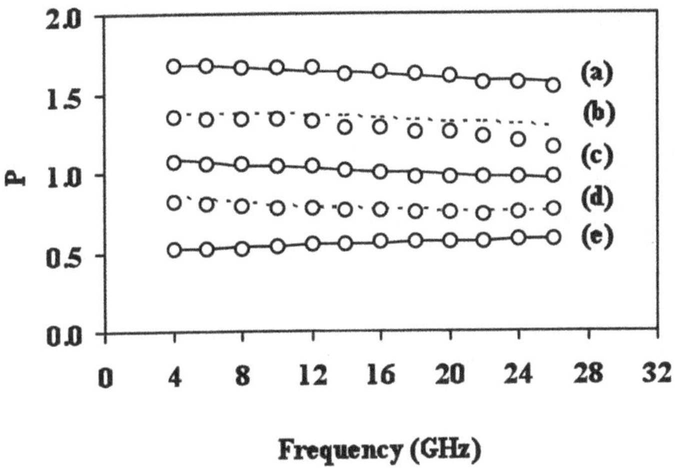

Figure 7.10 Comparison of measured and modeled drain noise model parameter P versus frequency (bias condition: $V_{ds} = 2\text{V}$): (a) $I_{ds} = 12.2$ mA; (b) $I_{ds} = 9.06$ mA; (c) $I_{ds} = 6.26$ mA; (d) $I_{ds} = 3.80$ mA; (e) $I_{ds} = 2.02$ mA.

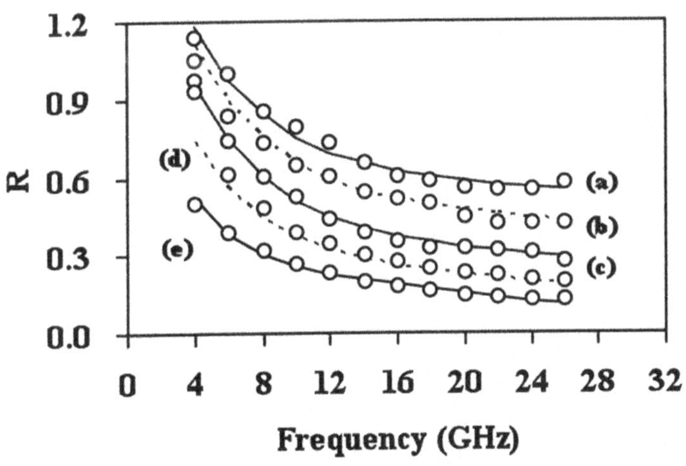

Figure 7.11 Comparison of measured and modeled gate noise model parameter R versus frequency (bias condition: $V_{ds} = 2\text{V}$): (a) $I_{ds} = 12.2$ mA; (b) $I_{ds} = 9.06$ mA; (c) $I_{ds} = 6.26$ mA; (d) $I_{ds} = 3.80$ mA; (e) $I_{ds} = 2.02$ mA.

algorithm and obtained agreement with the actual data. The three bias conditions are selected to train the neural network. Figure 7.10 to Figure 7.13 compares the measured and modeled noise parameters in the frequency range of 4 GHz to 26 GHz. The solid lines correspond to the training data by using training process, whereas dash lines denote val-

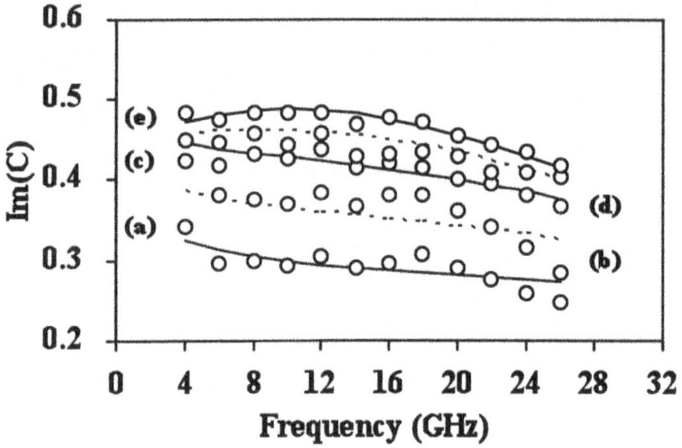

Figure 7.12 Comparison of measured and modeled imaginary part of correlation noise model parameter versus frequency (bias condition: $V_{ds} = 2\text{V}$): (a) $I_{ds} = 12.2$ mA; (b) $I_{ds} = 9.06$ mA; (c) $I_{ds} = 6.26$ mA; (d) $I_{ds} = 3.80$ mA; (e) $I_{ds} = 2.02$ mA.

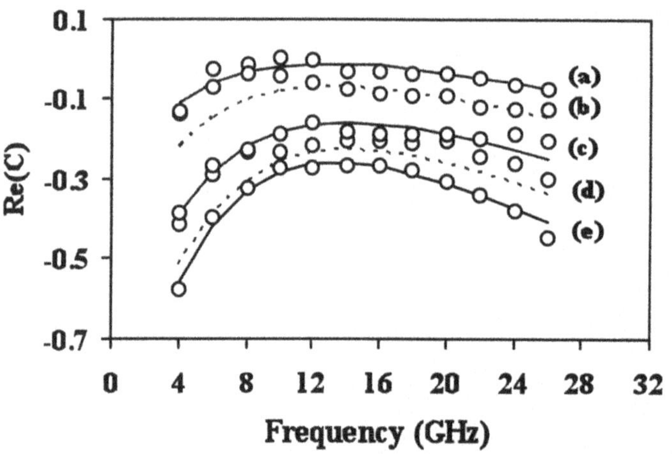

Figure 7.13 Comparison of measured and modeled real part of correlation noise parameter versus frequency (bias condition: $V_{ds} = 2\text{V}$): (a) $I_{ds} = 12.2$ mA; (b) $I_{ds} = 9.06$ mA; (c) $I_{ds} = 6.26$ mA; (d) $I_{ds} = 3.80$ mA; (e) $I_{ds} = 2.02$ mA.

ues predicted by the neural network for the data never used in training. A good agreement can be observed.

In Figures 7.14 and 7.15, we compare the measured and computed noise parameters versus frequency by using the new noise model for the tested PHEMT under two different bias conditions (I_{ds} = 9.06 mA, 3.80

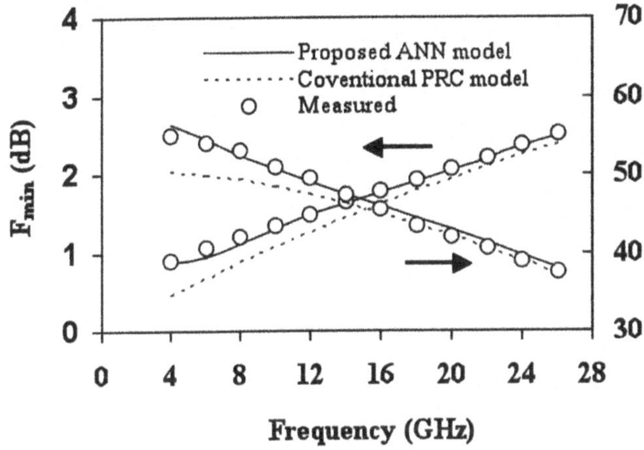

(a)

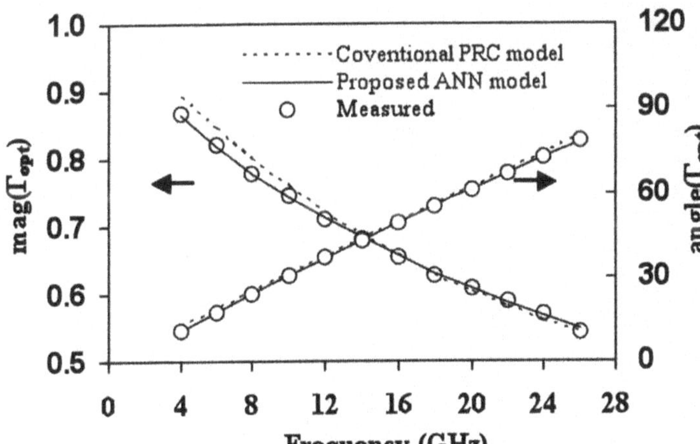

(b)

Figure 7.14 Comparison of measured and calculated noise parameters based on the new noise model. oo = Experimental data, — = calculated data from proposed model, --- = calculated data from PRC model (bias condition I_{ds} = 9.06 mA, V_{ds} = 2 V)

mA, V_{ds}=2V). A good agreement is observed. The new expressions are also compared with the conventional noise model ("PRC" model). It can be clearly found that the new model is more accurate than the PRC model, especially for minimum noise figure F_{min}, and noise resistance R_n. It is obvious that the influence of a gate-current can be taken into account by using the new model, whereas the PRC model cannot handle it directly. Table 7.4 shows comparison of the relative error percentage

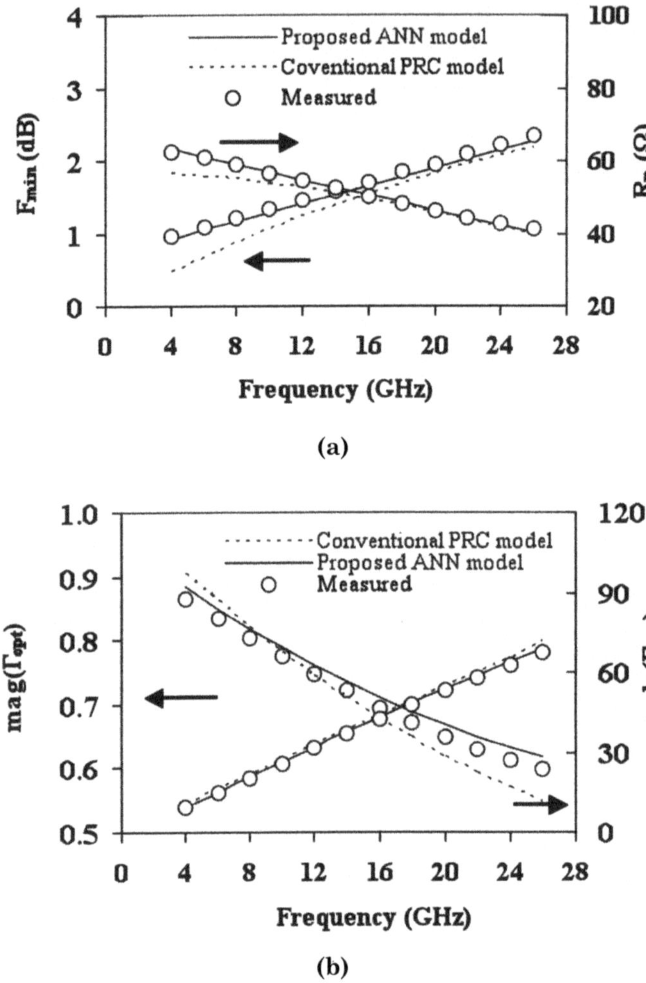

Figure 7.15 Comparison of measured and calculated noise parameters based on the proposed noise model. oo = Experimental data, — = calculated data from proposed model, --- = calculated data from PRC model (bias condition I_{ds} = 3.8 mA, V_{ds} = 2 V)

between the conventional model and the proposed model for PHEMT up to 26 GHz, significant improvements of the accuracy of noise parameters can be observed.

7.5 ANN Integration and Differential Technique

Neural-based microwave device modeling technique combines the conventional equivalent-circuit and artificial neural network modeling

Table 7.4 Model Accuracy Comparison between ANN Model and Conventional PRC Model (Unit %)

I_{ds} (mA)	F_{min}		R_n		Mag (Γ_{opt})		Phase (Γ_{opt})	
	PRC	ANN	PRC	ANN	PRC	ANN	PRC	ANN
2.02	6.90	1.41	3.46	1.60	4.08	0.68	10.1	2.26
3.80	5.35	1.32	2.68	0.74	3.84	2.36	6.69	1.77
6.26	5.08	1.17	3.20	1.11	2.06	0.54	5.98	0.87
9.06	3.28	0.38	2.23	1.16	1.54	0.63	4.25	1.67
12.2	2.41	0.95	1.35	0.75	1.71	0.28	5.09	0.96

technique. Each intrinsic nonlinear circuit elements can be modeled by using a sub-artificial neural network (SANN).

A typical neural-based Schottky diode nonlinear and linear equivalent-circuit model are shown in Figure 7.16(a) and (b), respectively, where I_d is the dc current, Q_d is the charge-storage, g_d is the output conductance, and C_d is the intrinsic capacitance.

The relationship between the linear and nonlinear model is as follows:

$$C_d = \frac{dQ_d(V_d)}{dV_d} \tag{7.22}$$

$$g_d = \frac{dI_d(V_d)}{dV_d} \tag{7.23}$$

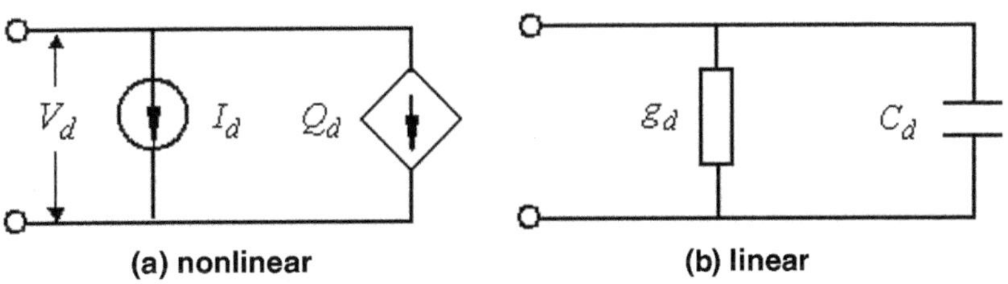

Figure 7.16 Neural-based Schottky diode nonlinear and linear equivalent-circuit model

where $Q_d(V_d)$ and $I_d(V_d)$ are the SANN and can be expressed as

$$I_d(V_d) = f_{ANN}^{I_d}(V_d) \tag{7.24}$$

$$Q_d(V_d) = f_{ANN}^{Q_d}(V_d) \tag{7.25}$$

where f_{ANN} represents the ANN of each element of Schottky diode.

The dc current I_d and capacitance C_d can be easily obtained from dc measurement and small-signal S-parameters measurements; the other intrinsic elements g_d and charge-storage Q_d can be calculated from the differential and integration of I_d and C_d, respectively. To build a united linear and nonlinear ANN model for diode, the differential and integration technique is necessary [18].

Another example is the neural-based FET nonlinear and linear equivalent-circuit model, as shown in Figures 7.17(a) and (b), respectively, where I_{gs}, I_{gd}, and I_{ds} are the dc currents, C_{gs} and C_{gd} are the intrinsic capacitances, Q_g is the gate charge-storage, g_m is the transconductance, and g_{ds} is the output conductance.

The dc current I_{gs}, I_{gd}, and I_{ds} can be obtained from dc measurement, and the capacitance C_{gs} and C_{gd} can be easily obtained from small-signal S-parameters measurement, the intrinsic elements g_m, g_{ds} can be calculated from the differential I_{ds}; and Q_g can be calculated from the integration of the intrinsic capacitances C_{gs} and C_{gd}. Please note that

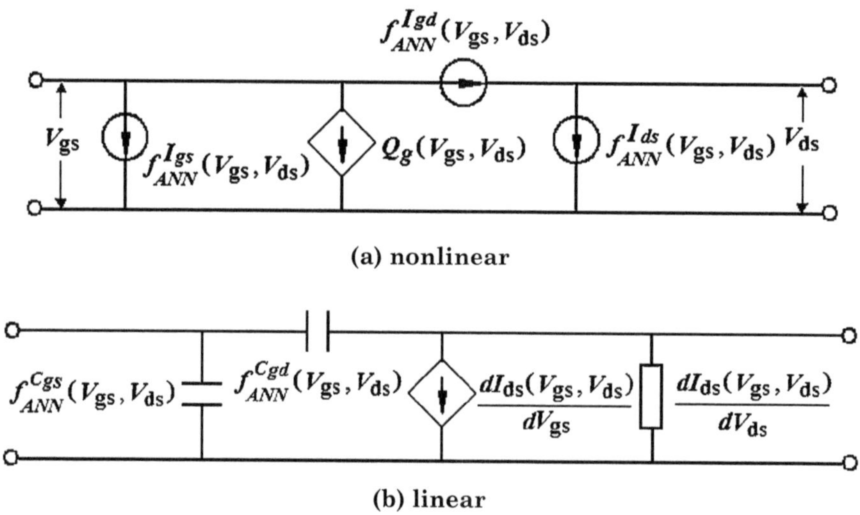

Figure 7.17 Neural-based FET nonlinear and linear equivalent-circuit model

Q_g is used to represent the total gate charge instead of the gate-to-source and gate-to-drain charge.

Where I_{ds}, I_{gs}, I_{gd}, C_{gs}, and C_{gd} are independent S-ANNs:

$$I_{ds}(V_{gs}, V_{ds}) = f_{ANN}^{I_{ds}}(V_{gs}, V_{ds}) \tag{7.26}$$

$$I_{gs}(V_{gs}, V_{ds}) = f_{ANN}^{I_{gs}}(V_{gs}, V_{ds}) \tag{7.27}$$

$$I_{gd}(V_{gs}, V_{ds}) = f_{ANN}^{I_{gd}}(V_{gs}, V_{ds}) \tag{7.28}$$

$$C_{gs}(V_{gs}, V_{ds}) = f_{ANN}^{C_{gs}}(V_{gs}, V_{ds}) \tag{7.29}$$

$$C_{gd}(V_{gs}, V_{ds}) = f_{ANN}^{C_{gd}}(V_{gs}, V_{ds}) \tag{7.30}$$

Therefore, to build a united linear and nonlinear ANN model for FET device, the differential and integration technique are necessary.

ANN models for RF/microwave components can be used in circuit design and optimization. To achieve this, the neural models are first incorporated into circuit simulators. Based on the aforementioned discussion, to build a unified microwave active device model that can predict the large-signal and small-signal performance, the integration and differential techniques for SANN is necessary. Therefore, our present task is to develop a technique for the integration and differential of y (ANN output) with respect to x (ANN input) for neural-based microwave active device modeling.

Depending on the complexity of the input response and desired output, the number of hidden layers and neurons at each layer can vary. Because there always exists a three-layer perceptron that can approximate an arbitrary nonlinear, continuous, multi-dimensional function f with any desired accuracy, a typical MLP neural network consists of an input layer, a hidden layer, and an output layer, as shown in Figure 7.18.

For given input x, the output of a three-layer MLP neural network can be computed by:

$$y = w_0^3 + \sum_{i=1}^{n} w_i^3 \sigma(w_{i0}^2 + \sum_{j=1}^{m} w_{ij}^2 x_j) \tag{7.31}$$

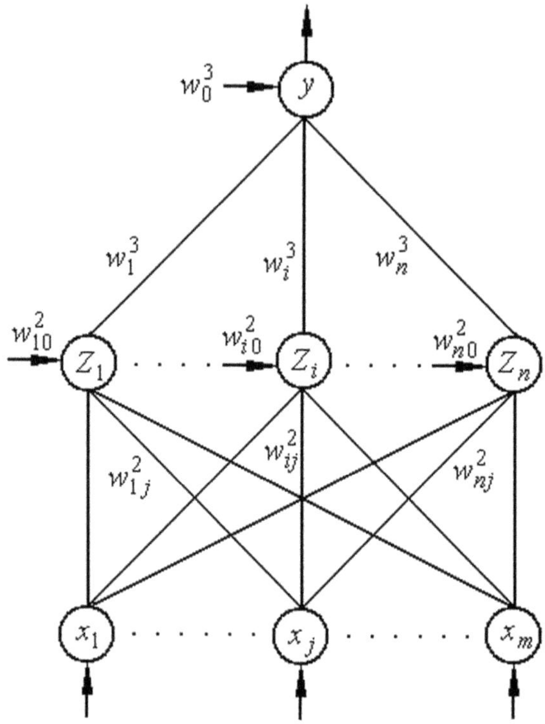

Figure 7.18 Three-layer MLP structure

that is,

$$y = [w_0^3, w_1^3 \dots, w_i^3, \dots, w_n^3][1, Z_1 \dots, Z_i \dots, Z_n]^{\mathrm{T}} \tag{7.32}$$

$$\begin{bmatrix} Z_1 \\ Z_2 \\ \dots \\ Z_i \\ \dots \\ Z_n \end{bmatrix} = \sigma \begin{bmatrix} w_{10}^2 & w_{11}^2 & \dots & w_{1j}^2 & \dots & w_{1m}^2 \\ w_{20}^2 & w_{21}^2 & \dots & w_{2j}^2 & \dots & w_{2m}^2 \\ \dots & \dots & \dots & \dots & \dots & \dots \\ w_{i0}^2 & w_{i1}^2 & \dots & w_{ij}^2 & \dots & w_{im}^2 \\ \dots & \dots & \dots & \dots & \dots & \dots \\ w_{n0}^2 & w_{n1}^2 & \dots & w_{nj} & \dots & w_{nm}^2 \end{bmatrix} \begin{bmatrix} 1 \\ x_1 \\ x_2 \\ \dots \\ x_j \\ \dots \\ x_m \end{bmatrix} \tag{7.33}$$

where $\sigma(\cdot)$ is an activation function. The overall nonlinear relationship between input and output is realized by various activation patterns of the neurons whose activation functions are typically a smooth switch function, for example the sigmoid function can be expressed as

$$\sigma(\gamma)=\frac{1}{1+e^{-\gamma}} \tag{7.34}$$

w_{ij}^{l} represents the weight of the link between jth neuron of $l-1$ th layer and ith neuron of lth layer, and c and w_{i0}^{2} represent the bias of each neurons of output and hidden layers, respectively.

7.5.1 Differential Artificial Neural Network (DANN)

The differential of output y with respect to input x_j for a three-layer MLP can be expressed as follows:

$$\frac{dy}{dx_j}=\sum_{i=1}^{n} w_i^3 w_{ij}^2 \Phi(w_{i0}^2+\sum_{j=1}^{m} w_{ij}^2 x_j) \tag{7.35}$$

that is,

$$dy/dx_j=[w_1^3 w_{1j}^2,\cdots,w_i^3 w_{ij}^2,\cdots,w_n^3 w_{nj}^2][Z_1^D,\cdots,Z_i^D,\cdots,Z_n^D]^{\mathrm{T}} \tag{7.36}$$

$$\begin{bmatrix} Z_1^D \\ Z_2^D \\ \cdots \\ Z_i^D \\ \cdots \\ Z_n^D \end{bmatrix}=\Phi\begin{bmatrix} w_{10}^2 & w_{11}^2 & \cdots & w_{1j}^2 & \cdots & w_{1m}^2 \\ w_{20}^2 & w_{21}^2 & \cdots & w_{2j}^2 & \cdots & w_{2m}^2 \\ \cdots & \cdots & \cdots & \cdots & \cdots & \cdots \\ w_{i0}^2 & w_{i1}^2 & \cdots & w_{ij}^2 & \cdots & w_{im}^2 \\ \cdots & \cdots & \cdots & \cdots & \cdots & \cdots \\ w_{n0}^2 & w_{n1}^2 & \cdots & w_{nj} & \cdots & w_{nm}^2 \end{bmatrix}\begin{bmatrix} 1 \\ x_1 \\ x_2 \\ \cdots \\ x_j \\ \cdots \\ x_m \end{bmatrix} \tag{7.37}$$

where $\Phi(\cdot)$ represents the activation function for a differential ANN (DANN) for three layer MLP and can be expressed as

$$\Phi(\lambda) = \sigma(\lambda)[1-\sigma(\lambda)] \tag{7.38}$$

The corresponding MLP structure of DANN is shown in Fig. 7.19, and a comparison of the MLP structure between DANN and original ANN is summarized in Table 7.5. It can be found that MLP structure and the weights of the hidden layer remain invariant, that the weights of the output layer are the product of the hidden layer weight w_{ij}^2 and the output layer weight w_i^3 of the original MLP neural network, and that the corresponding activation function is the product of $\sigma(\lambda)$ and $1-\sigma(\lambda)$.

Table 7.5 Comparison between DANN and Original ANN

	ANN	DANN
MLP	m-n-1	m-n-1
Weights of hidden layer	w_{ij}^2	w_{ij}^2
Weights of output layer	w_i^3 ($i = 0,1......n$)	$w_{ij}^2 w_i^3$ ($i = 1,2......n$)
Activation function	$\sigma(\lambda)$	$\sigma(\lambda)[1-\sigma(\lambda)]$

7.5.2 Integration of Artificial Neural Network (IANN)

Compared with DANN, the integration of ANN (IANN) is more complex; single input and multiple-input ANNs are discussed in this section separately.

7.5.2.1 Integration of Single Input ANN

Single-input ANN means the output y is dependent only on the single input x:

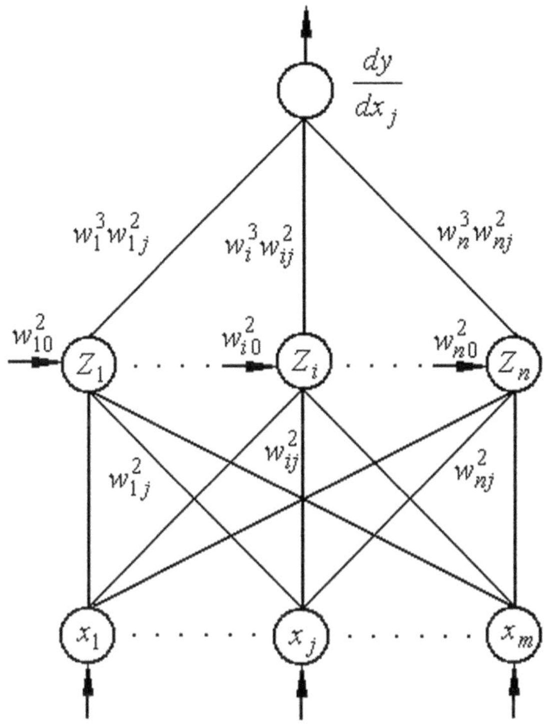

Figure 7.19 DANN of three-layer MLP structure

$$y = w_0^3 + \sum_{i=1}^{n} w_i^3 \sigma(w_{i0}^2 + w_i^2 x) \tag{7.39}$$

The integration of output y with respect to single-input x for a three-layer MLP can be expressed as follows:

$$\int y dx = w_0^3 x + \sum_{i=1}^{n} \frac{w_i^3}{w_{i1}^2} \Psi(w_{i0}^2 + w_{i1}^2 x) + C \tag{7.40}$$

that is,

$$\int y dx = [w_0^3, \frac{w_1^3}{w_1^2} ..., \frac{w_i^3}{w_i^2}, ..., \frac{w_n^3}{w_n^2}][x, Z_1^I ..., Z_i^I ..., Z_n^I]^{\mathrm{T}} \tag{7.41}$$

$$\begin{bmatrix} Z_1^I \\ Z_2^I \\ \cdots \\ Z_i^I \\ \cdots \\ Z_n^I \end{bmatrix} = \Psi \begin{bmatrix} w_{10}^2 & w_{11}^2 \\ w_{20}^2 & w_{21}^2 \\ \cdots\cdots & \cdots\cdots \\ w_{i0}^2 & w_{i1}^2 \\ \cdots\cdots & \cdots\cdots \\ w_{n0}^2 & w_{n1}^2 \end{bmatrix} \cdot \begin{bmatrix} 1 \\ x_1 \end{bmatrix} \tag{7.42}$$

where $\Psi(\cdot)$ represents the activation function for DANN for a three layer MLP, which can be expressed as

$$\Psi(\lambda) = \ln\left(\frac{1}{1-\sigma(\lambda)}\right) \tag{7.43}$$

The original and corresponding IANN are shown in Figures 7.20(a) and (b), respectively. A comparison between single-input IANN and original ANN is summarized in Table 7.6. It can be found that MLP structure and the weights of the hidden layer remain invariant, that the weights of the output layer are the ratio of output layer weight w_i^3

Table 7.6 Comparison between IANN and Original ANN for Single Input ANN

	ANN	IANN
MLP	1-n-1	1-n-1
Weights of hidden layer	w_{ij}^2	w_{ij}^2
Weights of output layer	w_i^3 $(i = 0,1\ldots\ldots n)$	w_i^3 / w_{i1}^2 $(i = 1,2\ldots\ldots n)$
Activation function	$\sigma(\lambda)$	$\ln\left(\frac{1}{1-\sigma(\lambda)}\right)$
Bias for output	w_0^3	$w_0^3 x + C$

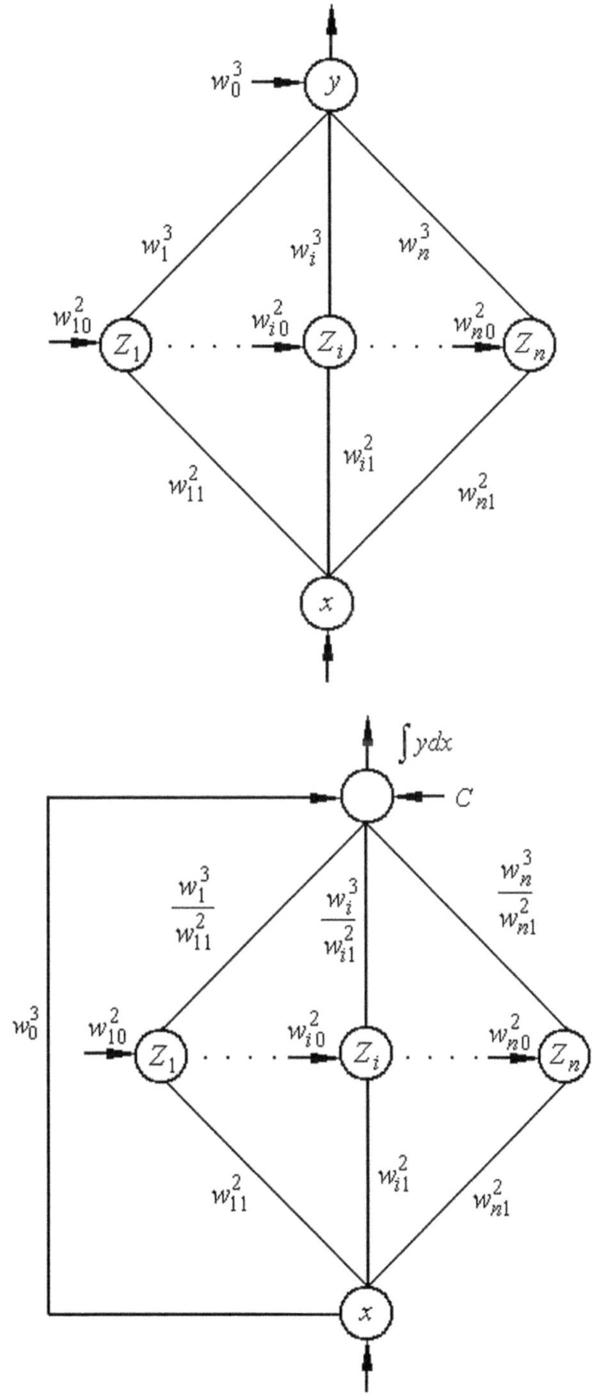

(a)

(b)

Figure 7.20 (a) Single input ANN and (b) integration ANN and the hidden layer weight w_i^2 and of the original MLP neural network, and that the activation function become

$$\ln\left(\frac{1}{1-\sigma(\lambda)}\right)$$

7.5.2.2 Integration of Multi-Input ANN

To obtain accurate integration of multi-input ANN, the differential of output y with respect to each input x_j (j = 1, 2,...m), that is dy/dx_j, must be known previously. For given input x, all the differentials can be regarded as the output of a three-layer MLP neural network and can be computed by:

$$\frac{dy}{dx_j} = \sum_{i=1}^{n} w_{ji}^3 z_i^M \quad j = 1,2,\cdots m \tag{7.44}$$

$$z_i^M = \sigma\left(\sum_{j=0}^{m} w_{ij}^2 x_j\right)$$

where w_{ji}^3 and w_{ij}^2 are the weights of output layer and the hidden layer, respectively.

To obtain the unique MLP of the output y, for given ith neuron of hidden layer, the weights w_{ji}^3 and w_{ij}^2 should be satisfied:

$$\frac{w_{1i}^3}{w_{i1}^2} = \frac{w_{2i}^3}{w_{i2}^2} = \cdots\cdots \frac{w_{ji}^3}{w_{ij}^2} = \cdots\cdots \frac{w_{mi}^3}{w_{im}^2} \tag{7.45}$$

Once such an MLP structure is determined, the integration of a multi-input ANN can be obtained directly as follows:

$$y = \sum_{i=1}^{n} \frac{w_{ji}^3}{w_{ij}^2} \Psi\left(w_{i0}^2 + \sum_{j=1}^{m} w_{ij}^2 x_j\right) + C \tag{7.46}$$

A comparison between multi-input IANN and the original ANN is summarized in Table 7.7. It can be found that the MLP structure has been changed, and that the weights of the output layer and hidden layer still remain invariant.

Table 7.7 Comparison between IANN and Original ANN for Multi-Input ANN

	ANN	IANN
MLP	m-n-m	m-n-1
Weights of hidden layer	w_{ij}^2	w_{ij}^2
Weights of output layer	w_i^3 $(i = 1,2......n)$	w_{ji}^3 / w_{ij}^2 $(i = 1,2......n)$
Activation function	$\sigma(\lambda)$	$\ln(\frac{1}{1-\sigma(\lambda)})$
Bias for output	0	C

7.5.3 Device Modeling Based on DANN and IANN Technique

Based on the approach for the DANN and IANN techniques, an improved ANN modeling concept for the Schottky diode and FET is introduced as shown in Figures 7.21(a) and (b), respectively. The modeling technique can be carried out using the following procedure:

1. Measurement of the DC I-V and S-parameter for microwave device.
2. Extraction of the intrinsic capacitances and DC current by using the de-embedding technique.
3. ANN (three-layers MLP) modeling of the intrinsic capacitances and DC current by using the back-propagation algorithm.
4. Calculation of the conductance (g_d for diode, and g_m and g_{ds} for FET) by using the DANN technique.
5. Calculation of the charge-storage (Q_d for diode, and Q_g for FET) by using the IANN technique.
6. Calculation of the S parameters by using the linear elements in the linear circuit simulator.

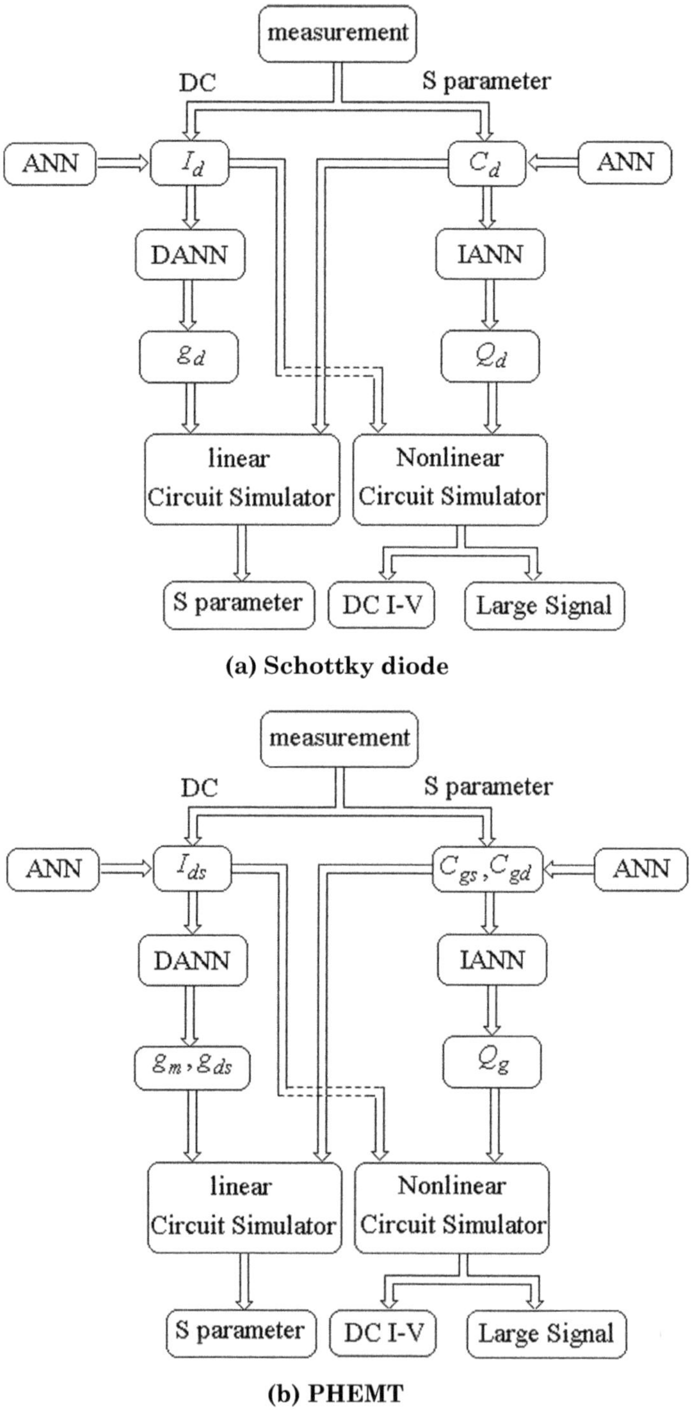

Figure 7.21 ANN modeling concept of (a) Schottky diode and (b) PHEMT

7. Calculation of the DC I-V and large-signal performance by using the nonlinear elements in the nonlinear circuit simulator.

Therefore, a unified neural-based small- and large-signal model for a microwave active device can be obtained easily by using the proposed DANN and IANN techniques.

To illustrate the method just proposed above for DANN and IANN, a Schottky diode and a 0.25 μm AlGaAs/InGaAs/GaAs PHEMT with 1 × 40 μm gatewidth are used in Agilent-ADS to generate small-signal and large-signal training data. The training data for the diode and PHEMT are created using an Agilent-ADS simulator.

After the Schottky diode DC current I_d and intrinsic capacitance C_d are obtained, the ANN models are trained using the NeuroModeler tool enclosed in [11]. During the training process using a combination of the conjugate-gradient and back propagation methods, the difference between training data and results obtained from ANN achieved was less than 1% error level for each component. Then the output conductance g_d and charge-storage Q_d can be determined by using the DANN and IANN techniques. Figure 7.22 shows the comparison of g_d and Q_d between the empirical model (ADS) and the ANN model for the Schottky diode; good agreement is obtained to validate the DANN and IANN techniques. The corresponding S parameter and harmonic performance are shown in Figures 7.23 and 7.24.

For PHEMT device, Figure 7.25 shows the comparison of g_m, g_{ds}, and Q_g between the empirical model (in ADS) and the ANN model for PHEMT, and the corresponding S parameter and harmonic perfor-

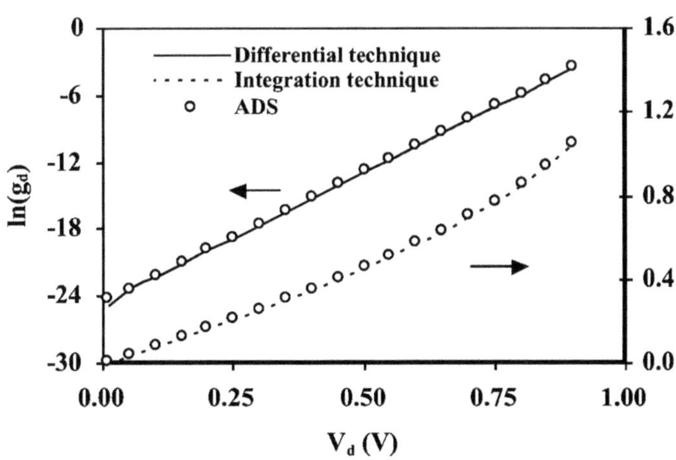

Figure 7.22 Comparison of g_d and Q_d between empirical model (ADS) and ANN model for Schottky diode

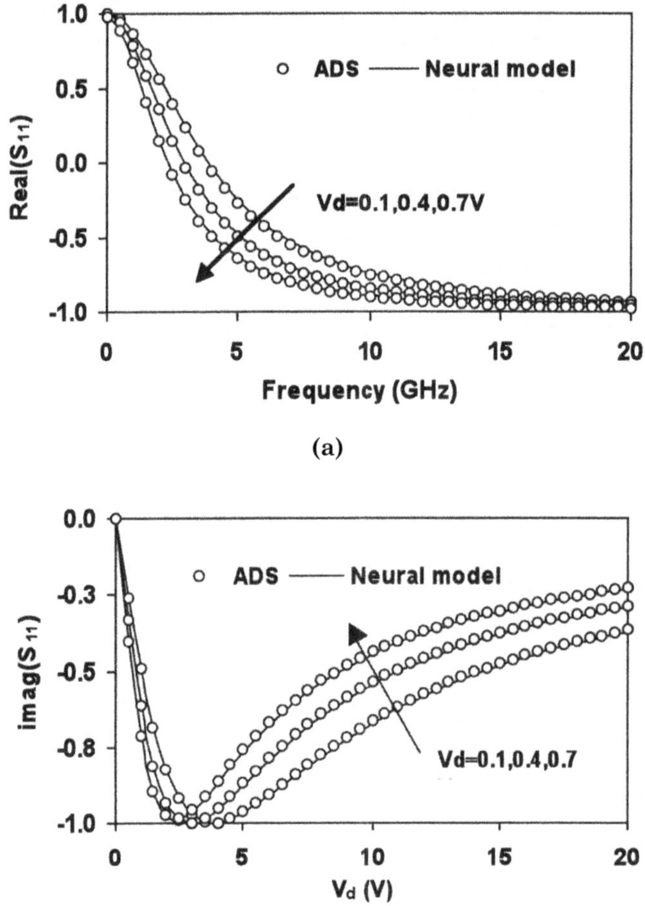

(a)

(b)

Figure 7.23 Comparison of S parameter between empirical model (ADS) and ANN model for Schottky diode

mance are shown in Figures 7.26 and 7.27. A good agreement between the empirical model and ANN results from using the proposed method can be indicated, and the validity of the method is confirmed.

7.6 Summary

In this chapter, the ANN-based linear, nonlinear, and noise models for FET were introduced. The corresponding integration artificial neural network and differential artificial neural network of the original ANN were developed.

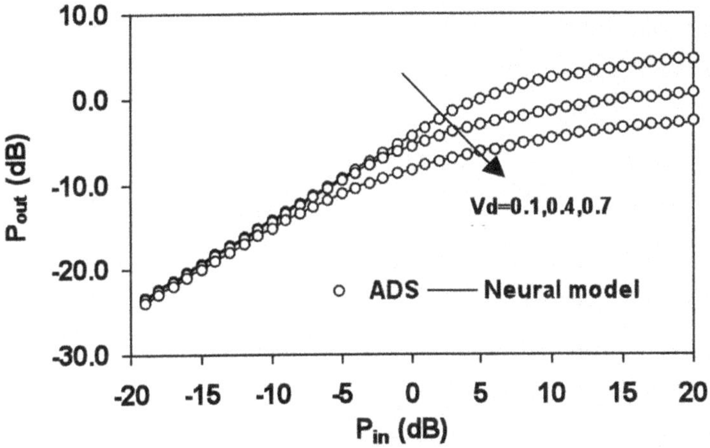

Figure 7.24 Comparison of S parameter between empirical model (ADS) and ANN model for Schottky diode

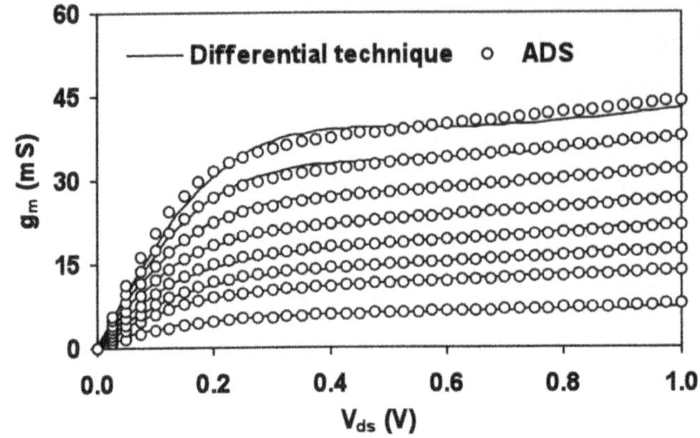

(a)

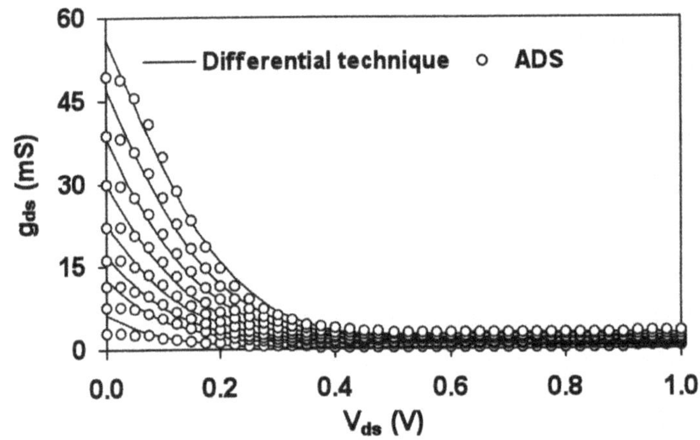

(b)

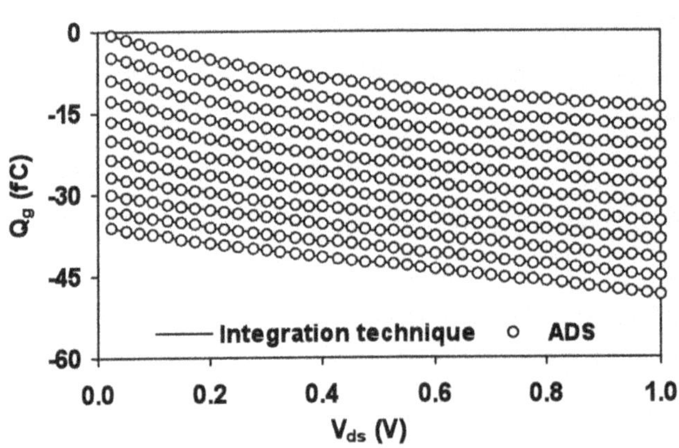

(c)

Figure 7.25 Comparison of harmonic performance between empirical model (ADS) and ANN model for Schottky diode (f = 1.0 GHz).

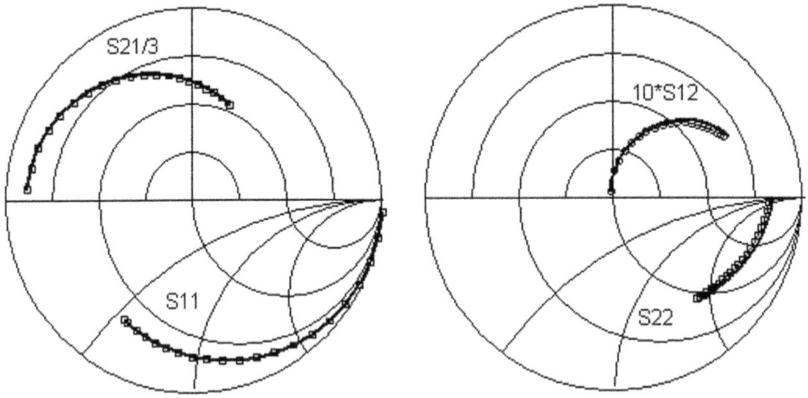

(a)

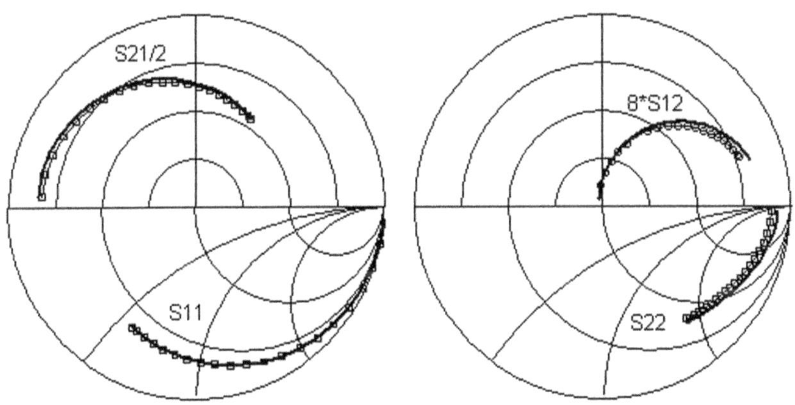

(b)

Figure 7.26 Comparison of S-parameter between empirical model (ADS) and ANN model for PHEMT. The squares indicate the data, and the lines the modeled ones (bias condition: (a) V_{gs} = 0 V, V_{ds}= 1 V (b) V_{gs}= –0.2 V, V_{ds} = 1 V)

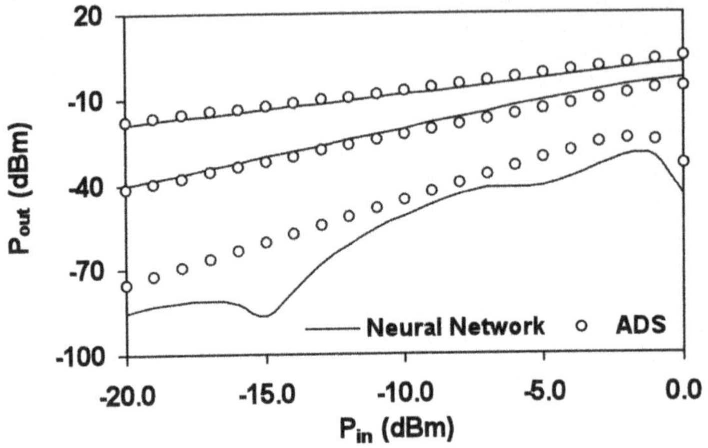

Figure 7.27 Comparison of harmonic performance between empirical model and ANN model for PHEMT ($f = 2.0$ GHz) (bias condition: $V_{gs} = -0.3$ V, $V_{ds} = 1$ V)

References

Chapter 1

[1] K. Chang, I. Bahl, and V. Nair, *RF and Microwave Circuit and Component Design for Wireless.* New York: John Wiley, 2002.

[2] T. Ytterdal, Y.Cheng, and T. A. Fjeldly, *Device Modeling for Analog and RF CMOS Circuit Design.* Chichester, UK: John Wiley & Sons Ltd, The Atrium, Southern Gate, 2003.

[3] W. Liu, *Fundamentals of III-V Devices.* New York: John Wiley & Sons Ltd, 1999.

[4] R. T. Thomas, *Behavioral Modeling of Nonlinear RF and Microwave Devices.* Norwood, MA: Artech House Inc., 2000.

[5] G. Massobrio and P. Antognetti, *Semiconductor Device Modeling with Spice.* New York: McGraw-Hill, 1989.

[6] J. J. Liou, *Advanced Semiconductor Device Physics and Modeling.* Norwood, MA: Artech House Inc., 1994.

[7] A. V. Ferris-Prabhu, *Introduction to Semiconductor Device Yield Modeling.* Norwood, MA: Artech House Inc., 1992.

[8] R. Ludwig and P. Bretchko, *RF Circuit Design: Theory and Applications.* NJ, USA: Pearson Education, 2000

[9] M. Feng, S.-C. Shen, D. C. Caruth, and J.-J. Huang, "Device Technologies for RF Front-End Circuits in Next-Generation Wireless Communications," in *Proceedings of the IEEE,* vol. 92, no. 2, pp. 354–375, 2004.

[10] L. M. Burns, "Application for GaAs and Silicon Integrated Circuit in Next Generation Wireless Communication Systems," *IEEE Journal of Solid-State Circuits,* vol. 30, no. 10, pp. 1088–1095, 1995.

[11] L. E. Larson, "Integrated Circuit Technology Options for RFICs-Present Status and Future Directions," *IEEE Journal of Solid-State Circuits,* vol. 33, no. 3, pp. 387–399, 1998.

Chapter 2

[1] D. A. Frickey, "Conversions between S, Z, Y, H, ABCD, and T parameters which Are Valid for Complex Source and Load Impedances," *IEEE Transactions on Microwave Theory and Techniques,* vol. 42, no. 2, pp. 205–211, 1994.

[2] H. Rothe and W. Dahlke, "Theory of Noisy Four Poles," *Proceedings of the IRE,* vol. 44, pp. 811–818, June 1956.

[3] H. A. Haus, et al., "Representation of Noise in Linear Two-Ports," *Proceedings of the IRE,* vol. 48, pp. 69–74, January 1960.

[4] R. P. Hecken, "Analysis of Linear Noisy Two-Ports Using Scattering Waves," *IEEE Transactions on Microwave Theory and Techniques,* vol. 27, no. 6, pp. 612–615, 1981.

[5] H. Bauer and H. Rothe, "Der Equivalene Rauschvierpol Als Wellenvierpol," *Archiv der Elektrischen Ueber Tragung,* vol. 10, pp. 241–252, 1956.

[6] H. Hillbrand and P. Russer, "An Efficient Method for Computer-Aided Noise Analysis of Linear Amplifier Networks," *IEEE Transactions on Circuits Systems*, vol. CAS-23, pp. 235–238, April 1976.

[7] J. A. Dobrowolski, *Introduction to Computer Methods for Microwave Circuit Analysis and Design.* Norwood, MA: Artech House Inc., 1991.

[8] J. B. Hagen, "Noise Parameter Transformations for Three-Terminal Amplifiers," *IEEE Transactions on Microwave Theory and Techniques,* vol. 38, no. 3, pp. 319–321, 1990.

[9] A. P. S. Khanna, "Three-Port S Parameter Ease GaAs FET Designing," *Microwave & RF* , vol. 24, no. 11, pp. 81–84, November 1985.

[10] T. O. Grosch and L. A. Carpenter, "Two-Port to Three-Port Noise-Wave Transformation for CAD Applications," *IEEE Transactions on Microwave Theory and Techniques,* vol. 41, no. 9, pp. 1543–1548, 1993.

[11] M. M. Radmanesh, *Radio Frequency Microwave Electronics Illustrated.* Upper Saddle River, NJ: Prentice Hall PTR, 2001.

[12] P. Ludwig and P. Bretchko, *RF Circuit Design: Theory and Applications..* Upper Saddle River, NJ: Prentice Hall PTR, 2000.

Chapter 3

[1] R .E. Ham, et al., "Microwave Measurements," in *The RF and Microwave Handbook,* ed. M. Golio. Boca Raton, FL: CRC Press, 2001.

[2] S. A. Wartenberg, *RF Measurements of Die and Packages.* Norwood, MA: Artech House, 2002.

[3] J. Williams, "Accuracy Enhancement Fundamentals for Vector Network Analyzer," *Microwave Journal,* vol. 32, no. 3, pp. 991–108, 1989.

[4] Hewlett-Packard, "Exploring the Architectures of Network Analyzers," Hewlett-Packard Application Note AN 1287-2 1997.

[5] P. C. L. Yip, *High-Frequency Circuit Design and Measurements.* Boca Raton, USA: Chapman and Hall, 1990.

[6] Microwave Measurements Division, "Application Note Three and Four Port S-Parameter Measurements." Atsugi, Japan: Anritsu Company, 2001

[7] P. Ludwig and P. Bretchko, *RF Circuit Design: Theory and Applications.* Upper Saddle River, NJ: Prentice Hall PTR, 2000.

[8] Hewlett-Packard, "In-Fixture Measurement Using Vector Network Analyzer," Hewlett-Packard Application Note AN 1287-9, pp. 1287–1289, 1997.

[9] Agilent Technologies,Inc.,"On-Wafer Multiport Calibration Using the ENA Series RF Network Analyzer with the Cascade Microtech Probing System." Agilent Product Note E5070/71-32003.

[10] H. Cho and D. E. Burk, "A Three-Step Method for the De-embedding of High-Frequency S-Parameter Measurements," *IEEE Transactions on Electron Devices*, vol. 38, no. 6, pp. 1371–1375, 1991.

[11] H. Friis, "Noise Figures of Radio Receivers," *Proceedings of the IRE,* vol. 32, no. 7, pp. 419–422, 1994.

[12] Agilent Technologies,Inc., "Fundamentals of RF and Microwave Noise Figure Measurements." Agilent Application Note 57-12003.

[13] Agilent Technologies,Inc., "Noise Figure Measurement Accuracy—The Y-Factor Method Application." Agilent Note 57-2, 2004.

[14] D. Boyd, "Calculated the Uncertainly of NF Measurements," *Microwave & RF* no. 10, pp. 93–102, 1999.

[15] M. M. Radmanesh, *Radio Frequency Microwave Electronics.* Upper Saddle River, NJ: Prentice Hall PTR, 2001.

[16] T. S. Laverghetta, *Handbook of Microwave Testing.* Norwood, MA: Artech, House, Inc., 1981.

[17] U. L. Rohde and D. P. Newkirk, *RF/Microwave Circuit Design for Wireless Applications.* New York: John Wiley& Sons, 2000.

[18] D. M. Pozar, *Microwave Engineering.* New York: John Wiley& Sons, 2005.

[19] J. Rogers and C. Plett, *Radio Frequency Integrated Circuit Design.* Norwood, MA: Artech House, 2003.

[20] J. F. Sevic, C. McGuire, G. M. Simpson, J. Pla, "Data-Based Load-Pull Simulation for Large-Signal Transistor Model Validation," *Microwave Journal,* vol. 40, no. 3, pp. 124–128, 1997.

Chapter 4

[1] K. Chang, *RF and Microwave Circuit and Component Design for Wireless System*. New York: Wiley Interscience, 2002.

[2] P.H.Ladbrooke, "MMIC Design GaAs FETs and HEMTs," Artech House, 1989

[3] D. Costa, W. U. Liu, and J. S. Harris, Jr., "Direct Extraction of the AlGaAs/GaAs Heterojunction Bipolar Transistors Small-Signal Equivalent Circuit," *IEEE Transactions on Electron Devices*, vol. 38, pp. 2018–2024, September 1991.

[4] J. Gao, C. L. Law, H. Wang, S. Aditya , and G. Boeck , "A New Method for PHEMT Noise Parameter Determination Based on 50-Ω Noise Measurement System" *IEEE Transactions on. Microwave Theory Technology,* vol. 55, no. 10, pp. 2079–2089, 2003.

[5] G. Dambrine, A. Cappy, F. Heilodore, and E. Playez, "A New Method for Determining the FET Small-Signal Equivalent Circuit," *IEEE Transactions on Microwave Theory Technology,* vol. 36, no. 7, pp. 1151–1159, 1988.

[6] P. M. White and R. M. Healy, "Improved Equivalent Circuit for Determination of MESFET and HEMT Parasitic Capacitances from 'Cold FET' Measurements," *IEEE Microwave Guided Wave Letters*, vol. 3, pp. 453–454, December 1993.

[7] Y. L. Lai and K. H. Hsu, "A New Pinch-Off Cold-FET Method to Determine Parasitic Capacitances of FET Equivalent Circuits," *IEEE Transactions on Microwave. Theory Technology,* vol. 49, no. 8, pp. 1410–1418, 2001.

[8] B. L. Ooi and J. Y. Ma, "Consistent and Reliable MESFET Parasitic Capacitance Extraction Method," *IEEE Proceedings—Microwaves, Antennas and Propagation,* vol.151, no. 1, pp. 81–84, 2004.

[9] S.-W. Chen, O. Aina, W. Li, L. Phelps, and T. Lee, "An Accuracy Scaled Small-Signal Model for Interdigitated Power P-HEMT up to 50GHz," *IEEE Transactions on Microwave Theory Technology,* vol. 45, no. 5, pp. 700–703, May 1997.

[10] J. Gao, X. Li, H. Wang, and G. Boeck, "A New Method for Determination of Parasitic Capacitances of PHEMTs," *IOP Semiconductor Science and Technology,* vol. 20, pp. 586–591, 2005.

[11] J. Gao, C. L. Law, H. Wang, and S. Aditya, " An Approach for Extracting Small Signal Equivalent Circuit of Double Heterojunction -Doped PHEMTs for Millimeter Wave Applications," *International Journal of Infrared and Millimeter Waves,* vol. 23, no. 3, pp. 345–364, March 2002.

[12] L. Yang and S. Long, "New Method to Measure the Source and Drain Resistance of the GaAs MESFET," *IEEE Electron Device Letters*, vol. EDL-7, pp. 75–77, February 1986.

[13] K. W. Lee, K. Lee, M. Shur, T. Vu, P. Roberts, et al., "Source, Drain and Gate Series Resistances and Electron Saturation Velocity in Ion-Implanted GaAs FETs," *IEEE Transactions on Electron Devices*, vol. ED-32, pp. 987–992, May 1985.

[14] R. J. Bennett, "Interpretation of Forward Bias Behavior of Schottky Barrier," *IEEE Transactions on Electron Devices*, vol. ED-34, no. 4, pp. 935–937, 1987.

[15] J. A. Reynoso-Hernandez, F. E. Rangel-Patino, and J. Perdomo, "Full RF Characterization for Extracting the Small-Signal Equivalent Circuit in Microwave FETs," *IEEE Transactions on Microwave Theory Technology*, vol. 44, no. 12, pp. 2625-2633, May 1996.

[16] P. Debie and L. Martens, "Fast and Accurate Extraction of Parasitic Resistances for Nonlinear GaAs MESFET Device Model," *IEEE Transactions on Electron Devices*, vol. ED-42, no. 12, pp. 2239–2242, December 1995.

[17] R. Anholt and S. Swirhun, "Equivalent Circuit Parameter Extraction for Cold GaAs MESFET's," *IEEE Transactions on Microwave Theory Technology*, vol. 39, no. 7, pp. 1243–1251, 1991.

[18] J. Gao, C. L. Law, H. Wang, and S. Aditya, "An Improved Pinchoff Equivalent Circuit Model for Determining Small-signal Model Paremeters of Double Heterojunction-Doped pHEMTs," *International Journal of Infrared and Millimeter Waves*, vol. 23, no. 11, pp. 1611–1626, November 2002.

[19] B.-S. Kim, "Existence of Optimum Cold FET Intrinsic Reference Plan for Active FET Small Signal Modeling," *IEEE Microwave and Wireless Component Letters*, vol. 11, no. 7, pp. 302–304, 2001.

[20] C. F. Campbell and S. A. Brown, "An Analytic Method to Determine GaAs FET Parasitic Inductances and Drain Resistance under Active Bias Condition," *IEEE Transactions on Microwave Theory Technology,* vol. 49, no. 7, pp. 1241–1247, 2001.

[21] J. Gao, C. L. Law, H. Wang, and S. Aditya, "An Approach to Linear Scalable DH-PHEMT Model," *International Journal of Infrared and Millimeter Waves,* vol. 23, no. 12, pp. 1787–1801, December 2002.

[22] K. Shirakawa, H. Oikawa, T. Shimura, T. Kawasaki, Y. Ohashi, T. Saito, et al., "An Approach to Determining an Equivalent Circuit for HEMTs," *IEEE Transactions on Microwave Theory Technology,* vol. 43, no. 3, pp. 499–503, March 1995.

[23] S. Yanagawa, H. Ishihara, and M. Ohtomo, "Analytical Method for Determining Equivalent Circuit Parameters of GaAs FETs," *IEEE Transactions on Microwave Theory Technology,* vol. 44, no. 10, pp. 1637–1641, July 1996.

[24] J. Gao, "An Approach for Determining PHEMT Small-Signal Circuit Model Parameters up to 110 GHz," *International Journal of Infrared and Millimeter Waves,* vol. 16, no. 7, pp. 1017–1029, 2005.

Chapter 5

[1] P. J. Rudge, R. E. Miles, M. B. Steer, and C. M. Snowden, "Investigation into Intermodulation Distortion in HEMTs Using a Quasi-2-D Physical Model," IEEE Transactions on Microwave Theory and Techniques, vol. 49, no. 12, pp. 2315–2321, 2001.

[2] M. H. Somerville, A. Ernst, and J. A. Alamo, "A Physical Model for the Kink Effect in InAlAs/InGaAs HEMTs," *IEEE Transactions on Electron Devices,* vol. 47, no. 5, pp. 922–930, 2000.

[3] P. R. Pantoja, M. J. Howes, J. R. Richardson, and C. M. Snowden, "A Large-Signal Physical MESFET Model for Computer-Aided Design and Its Applications," *IEEE Transactions on Microwave Theory and Techniques,* vol. 37, no. 12, pp. 2039–2045, 1989.

[4] G. Ghione, C. U. Naldi, and F. Filicori, "Physical Modeling of GaAs MESFETs in an Integrated CAD Environment: From Device Technology to Microwave Circuit Performance," *IEEE Transactions on Microwave Theory and Techniques,* vol. 37, no. 3, pp. 457-468, 1989.

[5] R. J. Trew, "MESFET Models for Microwave Computer-Aided Design," *Microwave Journal*, vol. 33, pp. 115–130, May 1990.

[6] D. E. Root and S. Fan, "Experimental Evaluation of Large-Signal Modeling Assumptions Based on Vector Analysis of Bias-Dependent S-Parameter Data from MESFETs and HEMTs," *IEEE MTT-S International Microwave Symposium Digest,* vol. 1, pp. 255–258, June 1–5, 1992.

[7] A. Orzati, D. M. M.-P. Schreurs, L. Pergola, and H. Benedickter, "A 110-GHz Large-Signal Lookup-Table Model for InP HEMTs Including Impact Ionization Effects," *IEEE Transactions on Microwave Theory and Techniques,* vol. 51, no. 2, pp. 468–474, 2003.

[8] H. Statz, P. Newman, I.W. Smith, R. A. Pucel, H. A. Haus, "GaAs FET Device and Circuit Simulation in SPICE," *Electron Devices, IEEE Transactions on,* vol. 34, no. 2, pp. 160–169, 1987.

[9] S. V. Bosch and L. Martens, "Fast Accurate Extraction of Capacitance Parameters for the Statz MESFET Model," *IEEE Transac-*

tions on Microwave Theory and Techniques, vol. 45, no. 8, pp. 1247–1249, 1997.

[10] A. J. McCamant, G. D. McCormack, and D. H. Smith, "An Improved GaAs MESFET Model for SPICE," *IEEE Transactions on Microwave Theory and Techniques,* vol. 38, no. 6, pp: 822–824, 1990.

[11] R. B. Hallgren and P. H. Litzenberg, "TOM3 Capacitance Model: Linking Large- and Small-Signal MESFET Models in SPICE," *IEEE Transactions on Microwave Theory and Techniques,* vol. 47, no. 5, pp. 556–561, 1999.

[12] W. R. Curtice, "A MESFET Model for Use in the Design of GaAs Integrated Circuits," *IEEE Transactions on Microwave Theory and Techniques,* vol. 28, no. 5, pp. 448–456, 1980.

[13] W. R. Curtice and M. Ettenberg, "A Nonlinear GaAs FET Model for Use in the Design of Output Circuits for Power Amplifiers," *IEEE Transactions on Microwave Theory and Techniques,* vol. 33, no. 12, pp. 1383–1394, 1985.

[14] W. R. Curtice, "GaAs MESFET Modeling and Nonlinear CAD," *IEEE Transactions on Microwave Theory and Techniques,* vol. 36, no. 2, pp. 220–230, 1988.

[15] Y. Tamijia, B. Wrona, and K. Mihima, "GaAs FET Large Signal Model and Its Application to Circuit Design," *Electron Devices, IEEE Transactions on,* vol. ED-28, pp. 171–175, 1981.

[16] T. Kacprzak and A. Materka, "Compact DC Model GaAs FET for Large Signal Computer Calculation," *IEEE Journal of Solid State Circuits,* vol. SC-18, no. 2, pp. 211–213, 1983.

[17] L. Angelov, N. Rorsman, J. Stenarson, M. Garcia, and H. Zirath, "An Empirical Table-Based FET Model," *IEEE Transactions on Microwave Theory and Techniques,* , vol. 47, no. 12, pp. 2350–2357, 1999.

[18] L. Angelov, L. Bengtsson, and M. Garcia, "Extensions of the Chalmers Nonlinear HEMT and MESFET Model," *IEEE Transactions on Microwave Theory and Techniques,* vol. 44, no. 10, pp. 1664–1674, 1996.

[19] L. Angelov, H. Zirath, and N. Rosman, "A New Empirical Nonlinear Model for HEMT and MESFET Devices," *IEEE Transactions on Microwave Theory and Techniques,* vol. 40, no. 12, pp. 2258–2266, 1992.

[20] J. Gao, X. Li, H. Wang, and G. Boeck, "Empirical All Region Current Based PHEMT DC Model," *SBMO/IEEE MTT-S International Microwave and Optoelectronics Conference, 2003. Proceedings of the 2003* ,vol. 1, no. 9, pp. 99–101.

[21] J. Gao, C. L. Law, H. Wang, and S. Aditya, "A Submicron PHEMT Nonlinear Model Suitable for Low Current Amplifier Design," *International Journal of Electronics,* vol. 90, no. 7, pp. 433–443, 2003.

Chapter 6

[1] A. van der Ziel, "Thermal Noise in Field-Effect Transistor," *Proceedings of the IRE,* .vol. 50, pp. 1808–1812, 1962.

[2] W. C. Bruncke and Z. van Ziel, "Thermal Noise in Junction Gate Field-Effect Transistor," *IEEE Transactions on Electron Device,* vol. 13, no.3, pp. 323–329, 1966.

[3] R. A. Pucel, H. A. Haus, and H. Statz, "Signal and Noise Properties of GaAs Microwave FET," in *Advances in Electronics and Electron Physics,* vol. 38, ed. L. Morton. New York: Academic Press, 1975.

[4] W. Baechtold, "Noise Behavior of GaAs Field-Effect Transistor with Short Gate Length," *IEEE Transactions on Electron Device,* vol. 19, no. 5, pp. 674–680, 1972.

[5] C. A. Liechti, "Microwave Field-Effect Transistor," *IEEE Transactions on Microwave Theory and Technique,* vol. 24, no. 6, pp. 279–300, 1976.

[6] T. M. Brookes, "The Noise Properties of High-Electron-Mobility Transistor," *IEEE Transactions on Electron Device,* vol. 33, no. 1, pp. 52–57, 1986.

[7] B. Carnez, A. Cappy, R. Fauquembergue, E. Constant, G. Salmer, "Noise Modeling in Submicrometer-Gate FETs," *IEEE Transactions on Electron Device,* vol. 28, no. 7, pp. 784–789, 1981.

[8] C. Moglestue, "A Monte-Carlo Particle Study of the Intrinsic Noise Figure in GaAs MESFET," *IEEE Transactions on Electron Device,* vol. 32, no. 10, pp. 2092–2096, 1985.

[9] H. Fukui, "Design of Microwave GaAs MESFETs for Broadband Low-Noise Amplifier," *IEEE Transactions on Microwave Theory and Techniques,* vol. 27, no. 7, pp. 643–650, 1979.

[10] H. Fukui, "Optimal Noise Figure of Microwave GaAs MESFETs," *IEEE Transactions on Electron Devices,* vol. 26, no. 7, pp. 1032–1037, 1979.

[11] M. W. Pospieszzlski and W. Wiatr, "Comment on Design of Microwave GaAs MESFETs for Broadband, Low-Noise Amplifiers," *IEEE Transactions on Microwave Theory and Techniques,* vol. 34, no. 1, pp. 194, 1986.

[12] D. Delagebeaudeuf, J. Chevrier, M. Lavron, and P. Delescluse, "A New Relationship between the Fukui Coefficient and Optimal Cur-

rent Value for Low-Noise Operation of Field-Effect Transistors," *IEEE Electron Device Letters,* vol. 6, no. 9, pp. 444–445, 1985.

[13] C. H. Oxley and A. J. Holden, "Simple Models for High Frequency MESFETs and Comparison with Experimental Results," *Microwaves, Antennas and Propagation, IEE Proceedings H,* vol. 133, no. 5, pp. 335–340, 1986.

[14] A. F. Podell, "A Functional GaAs FET Noise Model," *IEEE Transactions on Electron Devices,* vol. 28, no. 5, pp. 511–517, 1981.

[15] M. S. Gupta, O. Pitzalis, S. E. Rosenbaum, and P. T. Greiling, "Microwave Noise Characterization of GaAs MESFETs: Evaluation by On-Wafer, Low Frequency Output Noise Current Measurement," *IEEE Transactions on Microwave Theory and Techniques,* vol. 35, no. 12, pp. 1208–1217, 1987.

[16] M. S. Gupta and P. T. Greiling, "Microwave Noise Characterization of GaAs MESFETs: Determination of Extrinsic Noise Parameters," *IEEE Transactions on Microwave Theory and Techniques,* vol. 36, no. 4, pp. 745–751, 1988.

[17] A. Cappy, "Noise Modeling and Measurement Technique," *IEEE Transactions on Microwave Theory and Techniques,* vol. 36, no. 1, pp. 1–10, 1988.

[18] M. W. Pospieszalski, "Modeling of Noise Parameters of MESFETs and MODFETs and Their Frequency and Temperature Dependence," *IEEE Transactions on Microwave Theory and Techniques,* vol. 37, no. 9, pp. 1340–1350, 1989.

[19] M. Garcia, K. Yhland, H. Zirath, and I. Angelov, "A New Extraction Method for the Two-Parameter FET Temperature Noise Model," *IEEE Transactions on Microwave Theory and Techniques,* vol. 46, no. 11, pp. 1679–1685, 1998.

[20] P. Heymann and H. Prinzler, "Improved Noise Model for MESFETs and HEMTs in Lower Gigahertz Frequency Range," *Electronics Letters,* vol. 28, no. 7, pp. 611–612, 1992.

[21] R. Reuter, S. Van, and F. J. Tegude, "A New Noise Model of HFET with Special Emphasis on Gate-Leakage," *IEEE Electron Device Letters,* vol. 16, no. 2, pp. 74–76, February 1995.

[22] D. S. Shin, J. B. Lee, H. S. Min, J. E. Oh, Y. J. Park, W. Jung "Analytical Noise Model with the Influence of Shot Noise Induced by the Gate Leakage Current for Submicrometer Gate-Length High-Electron-Mobility Transistors," *IEEE Transactions on Electron Devices,* vol. 44, no. 11, pp. 1883–1887, 1997.

[23] H. Hillbrand and P. Russer, "An Efficient Method for Computer-Aided Noise Analysis of Linear Amplifier Networks," *IEEE Transactions on Circuits and Systems,* vol. 23, no. 4, pp. 235–238, 1976.

[24] S.-W. Chen, O. Aina,W. Li, L. Phelps, and T. Lee, "An Accuracy Scaled Small-Signal Model for Interdigitated Power P-HEMT up to 50GHz," *IEEE Transactions on Microwave Theory and Techniques,* vol. 45, no. 5, pp. 700–703, 1997.

[25] R. Q. Lane, "The Determination of Noise Parameters," *Proceedings of IEEE,* pp. 1461–1462, August 1969.

[26] L. Escotte, R. Plana, and J. Graffeuil, "Evaluation of Noise Parameter Extraction Methods," *IEEE Transactions on Microwave Theory and Techniques,* vol. 41, no. 3, pp. 382–387, 1993.

[27] G. Caruso and M. Sannino, "Computer-Aided Determination of Microwave Two-Port Noise Parameters," *IEEE Transactions on Microwave Theory and Techniques,* vol. 26, no. 9, pp. 639–642, 1978.

[28] A. C. Davidson, B. W. Leake, and E. Strid, "Accuracy Improvements in Microwave Noise-Parameter Measurements," *IEEE Transactions on Microwave Theory and Techniques,* vol. 37, no. 12, pp. 1973–1978, 1989.

[29] J. M. O'Callaghan and J. P. Mondal, "A Vector App. roach for Noise Parameter Fitting and Selection of Source Admittance," *IEEE Transactions on Microwave Theory and Techniques,* vol. 39, no. 8, pp. 1376–1382, 1991.

[30] P. J. Tasker, W. Reinert, B. Hughes, J. Braunstein, and M. Schlechtweg, "Direct Extraction of All Four Transistor Noise Parameters Using a 50 Measurement System," *IEEE MTT-S International Microwave Symposium*, pp. 1251–1254, 1993.

[31] G. Dambrine, H Happy, F. Danneville, and A. Cappy, " A New Method for On Wafer Noise Measurement," *IEEE Transactions on Microwave Theory and Techniques,* vol. 41, no. 3, pp. 375–381, 1993.

[32] A. Lazaro, L. Pradell, and J. M. O'Callaghan, "FET Noise-Parameter Determination Using a Novel Technique Based on a 50-Noise-Figure Measurements," *IEEE Transactions on Microwave Theory and Techniques,* vol. 47, no. 3, pp. 315–324, 1999.

[33] J. Gao, C. L. Law, H. Wang, S. Aditya, and G. Boeck, "A New Method for PHEMT Noise Parameter Determination Based on 50-Noise Measurement System," *IEEE Transactions on Microwave Theory and Techniques,* vol. 51, no. 10, pp. 2079–2089, 2003.

[34] K. W. Kobayaski, D. C. Streit, D. K. Umemoto, et al., "A Monolithic HEMT-HBT Direct-Coupled Amplifier with Active Input Matching," *IEEE Microwave and Guided Wave Letters,* vol. 6, no. 1, pp. 55–57, 1996.

[35] K. B. Niclas, "Active Matching with Common-Gate MESFETs," *IEEE Transactions on Microwave Theory and Techniques,* vol. 33, no. 6, pp. 492–499, 1985.

[36] R. L. Camisa and F. N. Sechi, "Common-Drain Flip-Chip GaAs FET Oscillators," *IEEE Transactions on Microwave Theory and Techniques,* vol. 27, no. 5, pp. 391–394, 1979.

[37] K. Nishikawa and T. Tokumitsu, "An MMIC Low-Distortion Variable-Gain Amplifier Using Active Feedback," *IEEE Transactions on Microwave Theory and Techniques,* vol. 43, no. 12, pp. 2812–2816, 1995.

[38] T. O. Grosch and L. A. Carpenter, "Two-Port to Three-Port Noise-Wave Transformation for CAD Applications," *IEEE Transactions on Microwave Theory and Techniques,* vol. 41, no. 9, pp. 1543–1548, 1993.

[39] J. B. Hagen, "Noise Parameter Transformations for Three-Terminal Amplifiers," *IEEE Transactions on Microwave Theory and Techniques,* vol. 38, no. 3, pp. 319–321, 1990.

[40] J. Gao and G. Boeck, "Relationships between Common Source, Common Gate and Common Drain FETs." *IEEE Transactions on Microwave Theory and Techniques,* vol. 53, no. 12, pp. 3825–3831, 2005.

[41] P. Heymann, "Experimental Evaluation of Microwave Field-Effect-Transistor Noise Models," *IEEE Transactions on Microwave Theory and Techniques,* vol. 47, no. 2, pp. 156–163, 1999 .

Chapter 7

[1] W. R. Curtice and M. Ettemberg, "A Nonlinear GaAs FET Model for Use in the Design of Output Circuits for Power Amplifiers," *IEEE Transactions on Microwave Theory Technology,* vol. 33, pp. 1383–1394, June 1985.

[2] L. Angelov, L. Bengtsson, and M. Garcia, "Extensions of the Chalmers Nonlinear HEMT and MESFET Model," *IEEE Transactions on Microwave Theory Technology,* vol. 44, pp. 1664–1674, October 1996.

[3] S. A. Maas and A. Crosmun, "Modeling the Gate I-V Characteristic of a GaAs MESFET for Volterra-Series Analysis," *IEEE Transactions on Microwave Theory Techniques*, vol. 37, pp. 1134–1136, July 1989.

[4] D. Root, S. Fan, and J. Meyer, "Technology Independent Large Signal Nonquasi-Static FET Models by Direct Construction from Automatically Characterized Device Data," in *Proceedings of the*

21st European Microwave Conference, pp. 923–927, Stuttgart, Germany, 1991.

[5] J. Rousset, Y. Harkouss, J. M. Collantes, M. Campovecchio, "An Accurate Neural Network Model of FET for Inter Modulation and Power Analysis," in *Proceedings of the 26th European Microwave Conference* ,Prague, Czechoslovakia, 1996.

[6] K. Shirakawa, M. Shimiz, N. Okubo, Y. Daido, "A Large Signal Characterization of an HEMT Using a Multilayered Neural Network," *IEEE Transactions on Microwave Theory Technology*, vol. 45, pp. 1630–1633, September 1997.

[7] . K. Shirakawa, M. Shimiz, N. Okubo, Y. Daido, "Structural Determination of Multilayered Large-Signal Neural-Network HEMT Model," *IEEE Transactions on Microwave Theory Technology*, vol. 46, pp. 1367–1375, October 1998.

[8] J. Xu, M. C. E. Yagoub, R. Ding, and Q.-J. Zhang, "Exact Adjoint Sensitivity Analysis for Neural-Based Microwave Modeling and Design," *IEEE Transactions on Microwave Theory Technology*, vol. 51, pp. 226–237, January 2003.

[9] M. Lazaro, L. Santamaria, and C. Pantaleon, "Neural Networks for Large- and Small-Signal Modeling of MESFET/HEMT Transistors," *IEEE Transactions on Instrumentation and Measurement*, vol. 50, pp. 1587–1593, December 2001.

[10] D. M. M.-P. Schreurs, J. Verspecht, S. Vandenberghe, and E. Vandamme, "Straightforward and Accurate Nonlinear Device Model Parameter-Estimation Method Based on Vectorial Large-Signal Measurements," *IEEE Transactions on Microwave Theory Technology*, vol. 50, pp. 2315—2319, October 2002.

[11] Q. J. Zhang and K. C. Gupta, *Neural Networks for RF and Microwave Design*. Norwood, MA: Artech House, 2000.

[12] J. Gao, L. Zhang, J. Xu, and Q.-J. Zhang, "Nonlinear HEMT Modeling Using Artificial Neural Network Technique," *IEEE International Microwave Symposium Digest,* vol. 1, pp. 469-472, 2005 .

[13] P. M. White and R. M. Healy, "Improved Equivalent Circuit for Determination of MESFET and HEMT Parasitic Capacitances from 'Cold-FET' Measurements," *IEEE Microwave Guided Wave Letters*, vol. 3, pp. 453–454, December 1993.

[14] R. A. Pucel, H. A. Haus, and H. Statz, "Signal and Noise Properties of Gallium Arsenide Microwave Field-Effect Transistors," in *Advances in Electronics and Electron Physics*, vol. 38. New York: Academic Press, 1975, pp. 195–265.

[15] A. Cappy, "Noise Modeling and Measurement Techniques," *IEEE Transactions on Microwave Theory Technology*, vol. MTT-36, pp. 1–10, January 1988.

[16] P. Heymann, M. Rudolph, H. Prinzler, R. Doerner, L. Klapproth, and G. Boeck, "Experimental Evaluation of Microwave Field Effect Transistor Noise Models," *IEEE Transactions on Microwave Theory Technology*, vol. 47, pp. 156–163, February 1999.

[17] X. Li, J. Gao, and Q.-J. Zhang, "Microwave Noise Modeling for PHEMT Using Artificial Neural Network Technique," *International Journal of RF and Microwave Computer-Aided Engineering*, vol.19, no.2, pp.187-196, 2009.

[18] X. Li, J. Gao, and G. Boeck, "PHEMT Modeling by Using Neural Network Technique," *IOP Semiconductor Science Technology,* vol. 21, no. 6, pp. 833–840, 2006.

Index

L

M

N

O

P

www.ingramcontent.com/pod-product-compliance
Ingram Content Group UK Ltd.
Pitfield, Milton Keynes, MK11 3LW, UK
UKHW022215051026
15624UKWH00001B/164

9 781891 121890